Handbook of Inductively Coupled
Plasma Mass Spectrometry

Handbook of Inductively Coupled Plasma Mass Spectrometry

K.E. JARVIS and A.L. GRAY
Geology Department
Royal Holloway and Bedford New College
University of London

R.S. HOUK
Department of Chemistry
Iowa State University

with additional invited chapters from
I. JARVIS, J.W. McLAREN and J.G. WILLIAMS

Springer Science+Business Media, LLC

© 1992 Springer Science+Business Media New York
Originally published by Blackie & Son Ltd in 1992
Softcover reprint of the hardcover 1st edition 1992
First published 1992

British Library Cataloguing in Publication Data

Handbook of ICP–MS.
  I. Jarvis, K.E.   II. Gray, A.L.
  543

  ISBN 978-94-010-5355-6

Library of Congress Cataloging-in-Publication Data

Jarvis, K.E.
  Handbook of inductively coupled plasma mass spectrometery / K.E.
Jarvis and A.L. Gray, R.S. Houk : with additional invited chapters
from I. Jarvis. . . [et al.].
    p.   cm.
  Includes bibliographical references and index.
  ISBN 978-94-010-5355-6   ISBN 978-94-011-3046-2 (eBook)
  DOI 10.1007/978-94-011-3046-2
  1. Plasma spectroscopy. 2. Mass spectrometry. I. Gray, A.L.
(Alan L.) II. Houk, R.S. III. Title. IV. Title: Inductively
coupled plasma mass spectrometry.
QD96.P62J37 1991
543'.0873—dc20   91-25745
                                              CIP

Typesetting by Thomson Press (India) Limited, New Delhi

Preface

Since the introduction of the first commercial inductively coupled plasma mass spectrometry (ICP–MS) instruments in 1983, the technique has gained rapid and wide acceptance in many analytical laboratories. There are now well over 400 instruments installed worldwide, which are being used in a range of disciplines for the analysis of geological, environmental, water, medical, biological, metallurgical, nuclear and industrial samples. Experience of ICP–MS in many laboratories is limited, and there is therefore a need for a handbook containing practical advice in addition to fundamental information. Such a handbook would be useful not only to users new to the technique, but also to users with some experience who wish to expand their knowledge of the subject. Therefore we have written this book for users in a variety of fields with differing levels of experience and expertise.

The first two chapters provide a brief history of ICP–MS and discussions of design concepts, ICP physical processes, and fundamental principles of instrument operation. Armed with this background knowledge, users will be better equipped to evaluate advantages and limitations of the technique. Detailed descriptions and information for instrumental components are provided in chapter 3. Subsequent chapters deal with the practical aspects of sample analysis by ICP–MS. Whether samples are to be analysed in liquid, solid or gaseous form is always an important consideration, and there is a wide choice of sample introduction techniques. Limitations placed on the accuracy and precision of analytical data by interference effects are discussed in detail. As sample introduction is frequently the main limiting factor of ICP–MS analysis, recommended methods are outlined in chapter 7. Although most analyses are currently performed in liquid samples, the direct analysis of solids (e.g. using laser ablation as a means of sample introduction) holds great potential for the future, and so a chapter has been dedicated to this particular subject.

We would like to acknowledge support in the UK by the Natural Environment Research Council (NERC) for the ICP–MS facility. One of us (RSH) acknowledges support from the Ames Laboratory which is operated by Iowa State University for the US Department of Energy under contract No. W-7405-ENG-82.

K.E.J.
A.L.G.
R.S.H.

Contributors

Dr A.L. Gray NERC ICP–MS Facility, Geology Department, Royal Holloway and Bedford New College, University of London, Egham, Surrey, TW20 0EX

Professor R.S. Houk Ames Laboratory, Department of Chemistry, Iowa State University, Ames, IA 50011

Dr I. Jarvis School of Geological Sciences, Kingston Polytechnic, Kingston upon Thames, Surrey, KT1 2EE

Dr K.E. Jarvis NERC ICP–MS Facility, Geology Department, Royal Holloway and Bedford New College, University of London, Egham, Surrey, TW20 0EX

Dr J.W. McLaren Institute for Environmental Chemistry, National Research Council Canada, Ottawa, K1A 0R6

Dr J.G. Williams NERC ICP–MS Facility, Geology Department, Royal Holloway and Bedford New College, University of London, Egham, Surrey, TW20 0EX

Contents

1 Origins and development   **1**

  1.1. Introduction   1
  1.2 The ICP–MS system   5

2 Instrumentation for ICP–MS   **10**

  2.1 The inductively coupled plasma   10
    2.1.1 Torch and plasma   10
    2.1.2 RF coupling   13
    2.1.3 Sample introduction   13
    2.1.4 Sample history   14
    2.1.5 Plasma populations   15
    2.1.6 Distribution of ions in the plasma   18
    2.1.7 Other plasmas   21
  2.2 Ion extraction   21
    2.2.1 Boundary layer and sheath   22
    2.2.2 Plasma potential and secondary discharge   23
    2.2.3 Supersonic jet   26
    2.2.4 Gas dynamics   28
    2.2.5 Ion kinetic energies   30
  2.3 Ion focusing   31
    2.3.1 Operation of ion lenses   31
    2.3.2 Ion lenses in ICP–MS   33
    2.3.3 Space charge effects   35
  2.4 Quadrupole mass spectrometers   37
    2.4.1 Quadrupole configuration   37
    2.4.2 Ion trajectories and stability diagrams   38
    2.4.3 Characteristics of mass spectra from quadrupoles 41
    2.4.4 RF-only quadrupoles   42
    2.4.5 Scanning and data acquisition   44
  2.5 Other mass spectrometers   45
  2.6 Ion detection   48
    2.6.1 Channeltron electron multipliers   48
    2.6.2 Signal measurement by pulse counting   50
    2.6.3 Other detectors   51
  2.7 Vacuum considerations   51
    2.7.1 Properties and flow of gases   51
    2.7.2 A vacuum system for ICP–MS   53
    2.7.3 Pumps used in ICP–MS   54

3 Instrument options   **58**

  3.1 Introduction   58
  3.2 Nebulisers   58
    3.2.1 Introduction   58
    3.2.2 Concentric nebulisers   59
    3.2.3 Cross-flow nebulisers   64
    3.2.4 Babington type nebuliser   65
    3.2.5 Frit type nebuliser   66
    3.2.6 Ultrasonic nebuliser   67

3.3 Spray chambers 68
  3.3.1 Principles 68
  3.3.2 Operation 69
  3.3.3 Thermally stabilised spray chambers for ICP–MS 71
3.4 Torches 75
  3.4.1 Construction 75
  3.4.2 Demountable torches 75
  3.4.3 Alignment 76
  3.4.4 Specialised torches 76
3.5 Interface 78
  3.5.1 Introduction 78
  3.5.2 Sampling cones 78
  3.5.3 Skimmer cones 80

4 Sample introduction for liquids and gases 81
4.1 Introduction 81
4.2 Electrothermal vaporisation 82
  4.2.1 Principles 82
  4.2.2 Instrumentation 83
  4.2.3 Operating parameters 89
  4.2.4 Applications and analytical performance of ETV–ICP–MS 93
4.3 Vapour generation and gas phase sample introduction 98
  4.3.1 Introduction 98
  4.3.2 Hydride generation 98
  4.3.3 Osmium tetroxide vapour generation 105
  4.3.4 Reactive gases 110
4.4 Liquid chromatography 112
  4.4.1 Introduction 112
  4.4.2 Principles 113
  4.4.3 Instrumentation, reagents and operating parameters 118
  4.4.4 Applications 118
4.5 Flow injection 119
  4.5.1 Introduction 119
  4.5.2 Apparatus 119
  4.5.3 Sample introduction 120
  4.5.4 Operating parameters 120
  4.5.5 Applications 122
4.6 Direct sample insertion 124
  4.6.1 Principles 124
  4.6.2 Applications 124

5 Interferences 125
5.1 Introduction 125
5.2 Spectroscopic interferences 125
  5.2.1 Isobaric overlap 125
  5.2.2 Polyatomic ions 129
  5.2.3 Refractory oxides 134
  5.2.4 Doubly charged ions 143
  5.2.5 Alleviation of spectroscopic interferences 145
5.3 Non-spectroscopic interferences 148
  5.3.1 High dissolved solids 148
  5.3.2 Suppression and enhancement effects 150

6 Calibration and data handling 153
6.1 Introduction 153

6.2   General concepts   153
      6.2.1   Mass scale calibration   153
      6.2.2   Accuracy, precision and reproducibility   154
6.3   Instrumental modes of data collection   154
      6.3.1   Peak hopping   154
      6.3.2   Scanning   155
6.4   Linearity of response   157
6.5   Blanks   158
6.6   Factors affecting signal stability   158
6.7   Qualitative analysis   160
6.8   Semi-quantitative calibration   160
6.9   Quantitative analysis   162
      6.9.1   External calibration techniques   162
      6.9.2   Raw data correction procedures   162
      6.9.3   Standard additions   167
      6.9.4   Isotope dilution   168

7   **Sample preparation for ICP–MS**   **172**
7.1   Introduction   173
7.2   General considerations   173
      7.2.1   Laboratory equipment and practices   173
      7.2.2   Choice of mineral acids   174
      7.2.3   Limits of quantitative analysis   180
      7.2.4   Precision and accuracy: assessing a digestion procedure   181
7.3   Digestion procedures   181
      7.3.1   Open vessel digestions   182
      7.3.2   Closed vessel digestions   192
      7.3.3   Alkali fusions   196
      7.3.4   Microwave digestion   202
7.4   Separation and pre-concentration methods   209
      7.4.1   Rare earth elements   210
      7.4.2   Precious metals   216
      7.4.3   Petrogenic disciminators: Hf, Nb, Ta, Zr   221
7.5   Conclusions and overview   224

8   **Elemental analysis of solutions and applications**   **225**
8.1   Introduction   225
8.2   Multi-element determinations   225
8.3   Geological applications   228
      8.3.1   Rare earth elements   229
      8.3.2   Platinum group metals   235
      8.3.3   Zirconium, niobium, hafnium, tantalum, thorium and uranium 240
      8.3.4   Molybdenum, tungsten and thallium   242
      8.3.5   Analysis of specific sample types   244
8.4   Environmental applications   247
      8.4.1   Multi-element applications   247
      8.4.2   Single-element applications   249
8.5   Nuclear applications   251
      8.5.1   Uranium matrices   251
      8.5.2   Lithium and boron matrices   252
      8.5.3   Zirconium and hafnium alloys   253
8.6   Industrial applications   253
      8.6.1   Metals   254
      8.6.2   Hydrocarbons   256
      8.6.3   Other sample types   259

8.7   Biological applications   260
      8.7.1   Foods   260
      8.7.2   Animal tissue   261
      8.7.3   Medical applications   263
8.8   Summary   264

9   The analysis of natural waters by ICP–MS   265

9.1   Introduction   265
9.2   Water sampling procedures for ICP–MS   267
      9.2.1   Filtration, acidification and storage   267
9.3   Direct water analysis by ICP–MS   269
      9.3.1   Pneumatic nebulisation   269
      9.3.2   Electrothermal vaporisation and direct sample insertion 269
      9.3.3   Gas phase injection   270
9.4   Water analysis with chemical separation and/or pre-concentration 271
      9.4.1   Seawater   271
      9.4.2   Freshwater   273
      9.4.3   On-line separation and pre-concentration   274
9.5   Calibration strategies   276
      9.5.1   External calibration   276
      9.5.2   Standard additions   277
      9.5.3   Isotope dilution   277

10   Analysis of solid samples   279

10.1   Introduction   279
       10.1.1   Calibration   280
10.2   Slurry nebulisation   281
       10.2.1   Grinding techniques   281
       10.2.2   Dispersing agents   283
       10.2.3   Particle size distributions   283
       10.2.4   Applications of slurry nebulisation   284
10.3   Laser ablation   290
       10.3.1   What is a laser?   291
       10.3.2   Modes of operation   291
       10.3.3   System configuration   292
       10.3.4   Laser operation   294
       10.3.5   Sample preparation   295
       10.3.6   Calibration   296
       10.3.7   Interferences   299
       10.3.8   Detection limits   301
       10.3.9   Practical considerations   302
       10.3.10 Applications   305
10.4   Direct sample insertion   308
10.5   Powdered solids   308
10.6   Arc nebulisation   309

11   Isotope ratio measurement   310

11.1   Introduction   310
       11.1.1   Traditional methods of isotope ratio determination   311
11.2   Instrument performance   312
       11.2.1   Sensitivity and counting statistics   312
       11.2.2   Dead time   312
       11.2.3   Resolution and abundance sensitivity   313
       11.2.4   Mass bias   315

11.3 Applications and methods of isotope analysis   315
      11.3.1 Lithium   315
      11.3.2 Boron   316
      11.3.3 Iron   320
      11.3.4 Copper   324
      11.3.5 Zinc   324
      11.3.6 Rhenium and osmium   327
      11.3.7 Lead   331
      11.3.8 Uranium   334
      11.3.9 Other isotopic ratios determined by ICP–MS   336

Appendices   **338**
Appendix 1 Originators of reference material cited in the text   338
Appendix 2 Naturally-occurring isotopes—useful data   341
Appendix 3 Glossary   348

References   **355**

Index   **377**

1 Origins and development

1.1 Introduction

The motivation for the development of a new technique for instrumental analysis may originate from a variety of causes, extending from the perception and analysis of a market requirement, to the adaptation of an existing scientific method to a new application. The former approach is more likely to lead to scientific and commercial exploitation within a reasonable time scale, and such an approach was adopted in the case of inductively coupled plasma source mass spectrometry, as the technique was initially called. The use of the word 'source' in the title follows general practice in descriptions of mass spectrometry ion sources. However, the technique was taken up by analytical chemists who were used to omitting the term 'source' in optical spectrometry and replacing it with a dash, hence ICP–MS became the acronym and is used from here on.

At the end of the 1960s, atomic emission spectrometry, using the inductively coupled plasma source became a very important 'future' technique for multi-element analysis at trace levels (Greenfield *et al.*, 1964; Wendt and Fassel, 1965). Although it was 1974 before commercial instruments were launched, first by Applied Research Laboratories in the United States, a number of companies were deeply involved by 1970 and it was clear that the potential market was considerable. At that time, as it has largely remained, the technique was directed to the analysis of solutions and important applications in environmental monitoring, mineral prospecting and medical research were envisaged.

It was already clear by 1970, however, that the analysis of trace elements in rock samples, using the currently available spectrometers, presented unique problems because of the relatively high matrix concentrations encountered (up to about 30% w/w in the solid). Thus even weak lines from matrix elements, which are found right across the spectrum, could, because of the high concentration present, interfere significantly with the wanted trace element responses. This was particularly serious with the very line rich spectrum of calcium, a common matrix element. The appreciation of this potential limitation, in the context of mineral exploration, stimulated a search for alternative methods of multi-element trace analysis which could still offer the ease of sample introduction and speed of analysis of ICP–AES.

Of the alternative spectrometric methods suitable for multi-element analysis, it appeared to the author (Gray, 1989a) that only mass spectrometry could offer the combination of simple spectra, adequate resolution and low detection limits that was desirable for trace determination in complex matrices. Spark source mass spectrometry (SSMS), using a radio frequency (RF) spark in vacuum, was already extensively used for trace element analysis. It provided detection limits in the solid of 1 in 10^6 or below and gave spectra containing only a few singly or multiply ionised peaks for each elemental isotope, and simple polyatomic ion peaks such as M^{2+}, MO^+, MOH^+, etc. There were thus far fewer ion peaks from matrix elements to interfere with trace elements and the resolution available from the large Mattauch–Herzog mass analysers used was usually quite adequate.

A number of problems limited its wider adoption, among which were cost, the mainly photographic readout needed, relatively poor precision and, most importantly, low sample throughput rates. Samples, in the form of solid conducting compacts (using graphite or aluminium for non-conducting materials), required skilled sample preparation and, for the lowest detection limits, could only be analysed at the rate of a few per day at best.

It was clear that these limitations were mainly a consequence of the type of ion source used, but a study of sources described in the literature revealed no potential alternatives. The major requirement for elemental analysis was that the source used could dissociate the sample as completely as possible, produce a high yield of singly charged ions but the minimum output of polyatomic fragments and multiply charged ions. These requirements are to some extent mutually exclusive. It is difficult to control the energy transfer to the sample in a vacuum, where normal ion sources operate, to provide adequate dissociation for refractory matrices, while at the same time avoiding excessive degrees of multiple ionisation for elements with low second and higher ionisation energies.

Attention therefore turned to the possibilities of higher ion source pressures at which the problems of energy transfer to the analyte molecules necessary for adequate dissociation might be expected to be less intractable. It was becoming clear around this time that atmospheric pressure direct current (DC) and RF plasmas already performed this function for emission spectrometry and that the ICP in particular owed much of its promise as an emission source to efficient excitation of analyte ionic species in the plasma and that these must be present at high concentrations. The problems of extracting these from the hostile atmosphere of the ICP and transferring them to the vacuum environment necessary for mass analysis and ion detection appeared formidable because of the combination of a temperature of about 8000 K and the high RF fields in the plasma (Fassel, 1977). However, the methods developed by Sugden's group at Cambridge for extracting ions from atmospheric pressure chemical flames at temperatures up to 3000 K (Knewstubb, 1963; Hayhurst and Sugden, 1966) appeared to offer a possible solution and the advice of Knewstubb and Hayhurst was sought. From this and some simple calculations it appeared that an inert gas, atmospheric pressure, electrical plasma might prove to be a very good ion source and a feasibility study was initiated using a small quadrupole mass analyser and channeltron ion detector.

Because of its ready availability in the ARL laboratory where the author was working, and its electrical simplicity, the feasibility study, led by J. Moruzzi at the University of Liverpool, was based on a small capillary DC arc source. This provided a small tail flame from which it was hoped to extract ions. The temperature in this source was lower than that in the ICP but was expected to be considerably higher in the core of the discharge than that of a chemical flame, possibly as much as 5000 K. This should produce a high degree of ionisation in elements with lower first ionisation energies, such as the transition elements. For the initial study a fine aerosol was produced from solid metal samples by means of an auxiliary arc and this was introduced to the capillary arc part way along the discharge. This study demonstrated that representative spectra could be obtained from metal samples and that the concept showed promise.

While the feasibility study was proceeding, consultation with potential users experienced in rock analysis by SSMS led to the production of a target for instrument performance which is shown in outline in Table 1.1. It was considered that if such a target could be achieved it would represent a marked advance on anything available at that time.

Although encouraging, the initial study yielded no information on potential sensitivity. The work, continued at ARL using solution samples, introduced by both pneumatic and ultrasonic nebulisers, showed that very low detection limits should be possible, partly because of the very low background levels obtained. Pulse counting ion detectors were used and the count rate from a blank solution at elemental masses absent from the plasma gas or solvent was effectively zero. The response obtained from a calibration solution at

Table 1.1 Initial design requirements for a new analytical mass spectrometer—March 1971

Initially the design programme should aim at producing an instrument with the following performance:

  a. Speed of analysis: 4–6 samples per hour
  b. Recorded m/z range: 6–238
  c. Sensitivity for monoisotopic elements: $0.1 \ \mu g \, g^{-1}$
  d. Precision: $\sim 25\%$

In addition it should possess the following features:

  e. Minimal operator control of the excitation source
  f. Automatic translation (scanning) through the mass range
  g. Print out of m/z intensity values.

It would require the development of an excitation source producing predominately singly charged, monatomic ions and a multiple sample loading device

---

Table 1.2 Achieved and calculated background equivalent concentration levels for plasma source mass spectrometer in $ng\,ml^{-1}$ $(1975)^a$

Element	First ionisation energy (eV)	Capillary arc plasma experimental	ICP calculated
Ag	< 8.5	0.03	0.02
Al		0.006	0.003
Bi		0.03	0.001
Co		0.008	0.006
Cr		0.008	0.006
Fe		0.008	0.006
Mg		0.003	0.003
Ni		0.01	0.008
Pb		0.04	0.04
Cd	> 8.5	0.44	0.04
Hg		90	0.1
Se		9	0.03
Zn		1	0.02

^{a}Based on background count rate of $1\,s^{-1}$.

the $\mu g\,ml^{-1}$ level was such that if a background count rate of $1\,s^{-1}$ was assumed, then the blank equivalent concentration for a range of elements was as shown in the third column of Table 1.2. For elements of ionisation energy below 8.5 eV very low values were obtained. To refer these back to the concentration in the solid, the values should be corrected by the solution dilution factor, typically 10^3. Even then the sensitivity requirements of Table 1.1 could be met comfortably for those favourable elements. However, it was clear that at the temperatures achieved in this DC plasma, the sensitivity for elements with ionisation energies above 8.5 eV was much lower. In addition because the discharge through the plasma gas consisted of a single high temperature central channel only a small part of the introduced sample actually reached the core against the steep temperature gradient and most of it skirted the discharge through the cooler outer regions. the effective temperature that the sample experienced was thus probably closer to 3000 K than 5000 K. Thus not only was a poorer degree of ionisation achieved than expected but dissociation of the sample molecules was poor and matrix effects severe. Nevertheless this stage of the work did demonstrate that an atmospheric pressure ion source could be used to give high sensitivity for those elements which were adequately ionised (Gray, 1974, 1975). It also showed clearly that a hotter, more suitable, plasma such as the ICP was necessary (Gray, 1978). In the ICP, the sample is introduced to a region where the gas temperature approaches 7000 K and the degree of ionisation for the elements poorly ionised in the DC plasma is correspondingly greater. Thus, assuming similar ion extraction parameters to those from the DC plasma, the corresponding blank equivalent concentrations for the ICP were predicted to be as shown in the fourth column of Table 1.2. Although still better for elements of lower ionisation energy, the overall range of values was expected to be greatly reduced and the use of the ICP appeared to be well worthwhile.

Plans were therefore made to extend the work but in the meantime the publication of the DC plasma work (Gray, 1974, 1975) attracted attention in both the United States and Canada. Projects were started in both those countries, using an ICP in the United States and initially a microwave induced plasma (MIP) in Canada (Douglas and French, 1981) which was later changed to an ICP (Douglas, 1983). Close collaboration was set up between the UK and US projects and the detailed history of these developments and the solutions found to the problems posed by the high temperatures and RF fields of the ICP are discussed in the literature (Houk et al., 1980; Date and Gray, 1981; Douglas et al., 1983b; Gray, 1985a, 1986a, 1989a) and will not be repeated here in detail. In summary, however, the problems centred around the need to provide an aperture robust enough to form the boundary between the centre channel of the plasma and the vacuum system at the temperatures involved which was also large enough to avoid the formation of a cool boundary layer in front of the aperture. This needed diameters of 0.2 mm or more and the volume of entering gas required improved pumping ability. This· was provided by the provision of an intermediate expansion stage pumped by a rotary pump and operating at a pressure of about 2 mbar and this step at the University of Surrey enabled the first spectra to be obtained in 1981 (Gray, 1982) which truly showed the performance to be expected from the ICP source (Date and Gray, 1983a; Gray and Date, 1983; Gray, 1986a). At this point aperture diameters of 0.4 mm and over were used and these permitted sufficient thickness of metal to give good heat conduction, and good aperture life was obtained. Initially this system resulted in plasma potentials above ground of about 30 V but this did not prevent excellent resolution being obtained from the large quadrupole mass analyser used at this stage. Subsequent developments in grounding methods and coil design greatly reduced this potential (Gray et al., 1987). During this period the Canadian group developed an ingenious alternative method of coil grounding which resulted in plasma potentials close to zero (Douglas and French, 1986; Houk et al., 1987).

Both the Surrey system, initially marketed by VG Isotopes Ltd., and the Canadian Sciex system reached the market place during the first half of 1983, although it was not until 1984 that instruments were first installed in customers' laboratories. Since then the technique has become increasingly accepted by analytical chemists and at the time of writing the number of instruments worldwide exceeds 450.

1.2 The ICP–MS system

Apart from their orientation and coil grounding arrangements the ICPs used in MS instruments are essentially the same as the AES versions. Similarly

the mass analysers, ion detectors and data collection systems used are similar
to those developed for quadrupole GC–MS instruments. Even the heart of
the ICP–MS instrument, the interface which transfers ions from atmospheric
pressure to the mass analyser in vacuum, bears a family resemblance to
systems used in molecular beam studies. However, there were many novel
problems to be overcome in its development and it represents the one system
component which was not already available from existing practice.

Although differing in detail, the two commercial systems, developed from
the parallel work in Canada and the United Kingdom, are essentially similar
and these, and later versions of them, make up the vast majority of the
systems in use today. A number of 'home-made' instruments have been built,
notably at the Ames Laboratory (USA) by Houk, which are closest in design
to those developed in Surrey, and there have been a number of other
instruments developed, particularly in Japan, which are also similar to the
Surrey instruments or their commercial versions.

The essential features of these instruments are shown in Figure 1.1. The
plasma system has already been discussed but it must be mentioned here
that an essential difference between the Sciex Elan instrument and the VG
PlasmaQuads (based on the Surrey system) lies in the load coil grounding.
The Elan is equipped with a centre tapped load coil or, in its latest version,
a virtual centre tap in the matching box, so that plasma potential arising
from capacitative coupling between the coil and the plasma is balanced out.
This produces a plasma jet entering the interface at close to ground potential
(Douglas and French, 1986; Fulford and Douglas, 1986; Houk et al., 1987).
The other instrument types all use a simple grounding of the load coil at
one end which, in the usual two turn coils, gives a plasma potential of about
5 V. The analytical consequences of this difference have been a cause of much
discussion at meetings and in the literature but in practice it proves to be of
little significance. Both systems use similar interface stages which extract gas

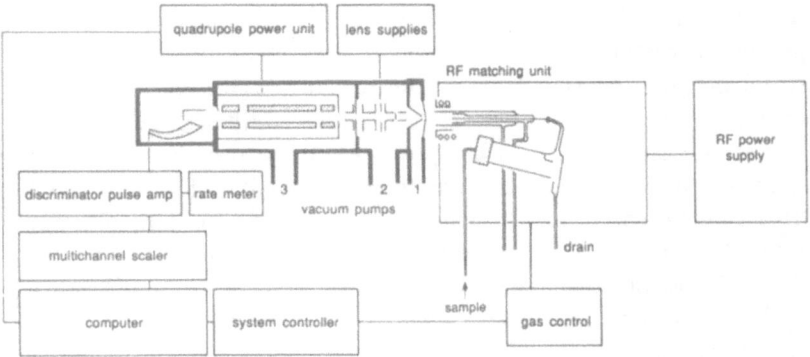

Figure 1.1 Schematic diagram of typical ICP–MS instrument showing main components.
Redrawn and adapted from Gray (1989a).

from the plasma. This is operated at a pressure of about 2.5 mbar and in both systems an extraction aperture of 1 or 1.2 mm diameter mounted on the water-cooled front plate of the stage is used. Behind this aperture a second skimmer cone is located, on the axis and between 6 and 7 mm from it. This sharp edged cone selects the centre of the expanding jet from the first aperture and allows it to pass into the next vacuum stage in which an electrostatic lens forms the expanding ion cloud into a focused beam for the mass analyser. This second vacuum stage is treated differently in the two systems. In the original Elan system, which includes all such systems installed up to the end of 1989, it is the final stage and contains the lens, the mass analyser and the ion detector, a fast pulse counting channel multiplier. This stage must be pumped to a pressure at which both quadrupole analyser and multiplier can operate without risk of electrical flashover at the high potentials employed and at which scattering of the beam is not significant. A pressure of just above 10^{-5} mbar is maintained by a large helium cryopump in this stage. The ion lens which precedes the quadrupole contains a photon stop disc on the axis which casts a shadow on the entrance aperture of the quadrupole to prevent direct photons from the plasma reaching the multiplier and creating a large background count. The ions are deflected around this disc and back again on to the axis but this results in some mass discrimination in the lens. In the VG PlasmaQuad system the stage immediately following the skimmer contains only an ion lens which focuses the ion beam through a differential aperture of 2 mm diameter into the final stage containing the quadrupole and multiplier. The intermediate stage is pumped by a large vapour pump to below 10^{-4} mbar and the lens also contains a photon stop which, because the differential aperture is relatively small, is itself smaller and obstructs less of the ion beam. No attempt is made to focus ions round it and less mass discrimination results. The final stage is also pumped by a vapour pump, to a pressure of about 2×10^{-6} mbar. These two stages are shown in Figure 1.1 which also shows the general arrangement of the control and data handling electronics of an instrument based on the Surrey system.

The differences in plasma grounding and vacuum system between the two types of instrument are less significant than they appear at first. Both instruments give signal count rates of more than 1 MHz from a mono-isotopic fully ionised element in solution at $1 \mu g ml^{-1}$, although there may be a difference in sensitivity when compromise settings are used to give a level response across as much of the mass range as possible. The most significant difference between the two systems lies in the method of handling the data produced by the multiplier response at each element mass.

Typical pulse rates from the multiplier range from background levels, which may be between 5 and 50 Hz, to well above 1 MHz, although above this detector and electronic dead times introduce non-linearity as concentration rises. The quadrupole is a sequential analyser and thus must be set to each mass of interest in turn and ions collected for a measured period which may be longer for weak responses than strong ones. The simplest method of achieving this is to program the control computer to drive the analyser to each mass in turn and store the response at the appropriate memory address. Usually several points are taken across the mass of each element. This peak hopping method of operation, used in the Elan, is convenient for multi-element analysis and isotope ratio determination but cannot scan the whole mass range very quickly because of the time constants involved in data storage in successive computer memory addresses. Thus transient responses from electrothermal evaporation or laser ablation, where the analyte is only present in the plasma for a few seconds, are difficult to measure for more than a limited number of elements. An alternative approach using a multi-channel scaler was adopted in the Surrey work where a fast scan rate was chosen to avoid plasma fluctuations. The maximum rate of scan possible is set by the time constants of the control circuits of the quadrupole and it was found that the practical upper limit permitted a full 250 m/z scan, accommodating the full mass range of the periodic table, in as little as 0.1 s. This enabled many scans to be accumulated within the period of slow plasma fluctuations or analyte transients. Even if only one integral were to be accumulated per mass, however, the rate at which the mass changed is too fast for a normal data handling microprocessor. A multi-channel analyser (MCA) in the multi-channel scaling (MCS) mode was therefore used as a buffer store for the counts collected. These are fast enough for the whole spectrum to be spread across a 2 K or 4 K memory so that 10 or more points could be accumulated across each peak every 0.1 s and a well defined spectrum could be built up, and displayed if desired, to reveal elements not previously expected in the sample. This system was adopted in the first VG PlasmaQuad instruments but in later versions a peak hopping facility was also provided. The pros and cons of these alternatives are discussed in more detail later but for most users the consequences of these different philosophies of data handling probably constitute the most significant distinction between the machines.

In order to provide the highest sensitivity, ion counting was universally adopted in the first instruments, but the upper limit of the dynamic range, determined by detector dead time, and thus limited to about 6 decades, occurred at increasingly lower analyte concentration as sensitivities improved. Again this is handled differently in the two instruments. Extended range is provided in the Elan by reducing ion lens transmission by a pre-set ratio when necessary but in the PlasmaQuad by using the detector in a mean current mode. In either case the overall dynamic range is extended to about 8 decades.

Both the mainstream instruments have evolved continuously since they were introduced. The most recent versions, introduced by both companies at the beginning of 1990, use turbopumps in the high vacuum stages. This required no change to the PlasmaQuad vacuum system but in the Elan it required a new three stage differentially pumped vacuum system similar to that of the Surrey based instruments because of the economic limitations on turbopump size. Changes of company organisation have also occurred in recent years and the Elan, although still made by Sciex, is now marketed by the Perkin Elmer Corporation, while the VG ICP–MS instruments are now manufactured by VG Elemental Ltd. This latter manufacturer has also extended the range of instruments beyond the basic PlasmaQuad. Two versions of this are made, the more sophisticated aimed at applications in chromatography and in laser ablation for surface analysis, and equipped with a 80386 microprocessor to provide the processing speed needed to give multi-element scans on successive eluent peaks or surface spots. In addition to these, a much larger instrument has been introduced using a magnetic sector mass analyser with a resolution approaching 10 000. This enables interfering polyatomic peaks to be resolved from the elements with which they interfere, and has solved one of the crucial limitations of ICP–MS with a quadrupole analyser. Sensitivity in the high resolution mode is similar to a conventional instrument but in a low resolution mode, corresponding to the quadrupole instruments, considerably lower detection limits are reported because of the low background count of a few counts min^{-1}. This is obtained in spite of the omission of a photon stop because of the long curved ion path. Such instruments have very specialised applications but a much wider market in pollution control probably exists for the latest instrument to be launched by VG, the PQe. This is a quadrupole instrument with a free running plasma using only one turbo pumped high vacuum stage, and equipped with a Faraday cup ion detector. Although of lower sensitivity than conventional systems this instrument still provides detection limits below $1\,ng\,ml^{-1}$. Because of the long time constants of the DC amplifier following the detector this instrument does not provide a scanned spectrum but rather a peak selection menu for the operator to choose from.

Although a number of other instruments have been announced by manufacturers, none has yet come into service, except in Japan, and little or no independent data are available on their performance. Clearly the situation is still an expanding one and over the next few years the number of users will probably continue to grow at an increasing rate.

2 Instrumentation for inductively coupled plasma mass spectrometry

2.1 The inductively coupled plasma

The characteristics of the ICP which make it so suitable as an ion source, with only minor changes from the form used for ICP–AES, have been extensively described in the literature and a bibliography of this would be too long for this volume. However, the nature of the ICP is important to an understanding of the basis and characteristics of ICP–MS and therefore an outline is given below. Brief reviews have been published by Fassel (1977, 1978) and fuller treatments given by Montaser and Golightly (1987), Boumans (1987) and Moore (1989) which may be followed up by those interested.

2.1.1 *Torch and plasma*

The inductively coupled plasma is an electrodeless discharge in a gas at atmospheric pressure, maintained by energy coupled to it from a radio frequency generator. This is done by a suitable coupling coil, which functions as the primary of a radio frequency transformer, the secondary of which is created by the discharge itself. The gas used is commonly argon although other gases are occasionally used, sometimes as additions to the main supply. The plasma is generated inside and at the open end of an assembly of quartz tubes known as the torch. A typical arrangement of a torch used for emission spectrometry is shown in Figure 2.1. Only minor changes have been made in the systems used for mass spectrometry, which are usually limited to mounting the torch with the axis horizontal for convenience and making some changes to the grounding point of the coupling (load) coil circuit, to control the plasma electrical potential with respect to the grounded mass spectrometer system. The torch commonly used, based on the 'Scott Fassel' design (Scott *et al.*, 1974), has an outer tube of inner diameter 18 mm and is about 100 mm long (Figure 2.2). Within this are two concentric tubes of 13 and 1.5 mm inner diameter which terminate short of the torch mouth. Each annular region formed by the tubes is supplied with gas by a side tube entering tangentially so that it creates a vorticular flow. The centre tube, through which the sample is introduced to the plasma, is brought out along the axis. The outer gas flow, termed the coolant flow, protects the tube walls and acts

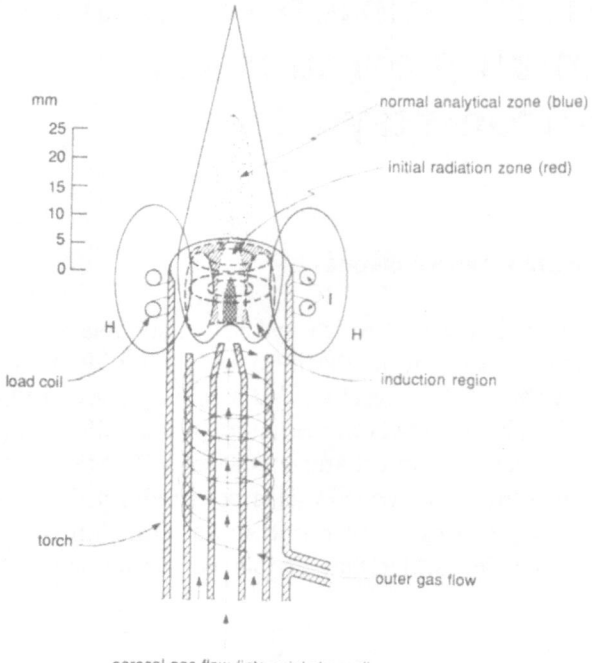

mm
25
20
15
10
5
0

normal analytical zone (blue)

initial radiation zone (red)

H   H

load coil   induction region

torch   outer gas flow

aerosol gas flow (into axial channel)

Figure 2.1 Schematic of ICP torch, gas flows and the induced magnetic field. The shaded zones are observed when a nebulised sample containing Y is introduced along the central channel in the injector gas flow. Reproduced with permission from Houk (1990).

as the main plasma support gas and is usually of $10-15\,l\,min^{-1}$. The second gas flow which is introduced to the inner annular space, termed the auxiliary flow, is mainly used to ensure that the hot plasma is kept clear of the tip of the central capillary injector tube, to prevent its being melted. The flow used depends on the precise torch and load coil geometry and values between 0 and $1.5\,l\,min^{-1}$ are common. The central gas flow, often called the injector, nebuliser or carrier flow, conveys the aerosol from the sample introduction system, and is usually about $1\,l\,min^{-1}$. This is sufficient, in the small diameter injector tube, to produce a high velocity jet of gas which punches a cooler hole through the centre of the plasma, termed the central or axial channel.

The coupling or load coil of 2–4 turns of fine copper tube, cooled by a water or gas flow, is located with its outer turn a few millimetres below the mouth of the torch. The RF current supplied from the generator produces a magnetic field which varies in time at the generator frequency, usually 27 or 40 MHz in the systems used for ICP–MS, so that within the torch, the field lies along the axis. The discharge is usually initiated in a cold torch by a spark from a Tesla coil, which provides free electrons to couple with the magnetic field. Electrons in the plasma precess around the magnetic field

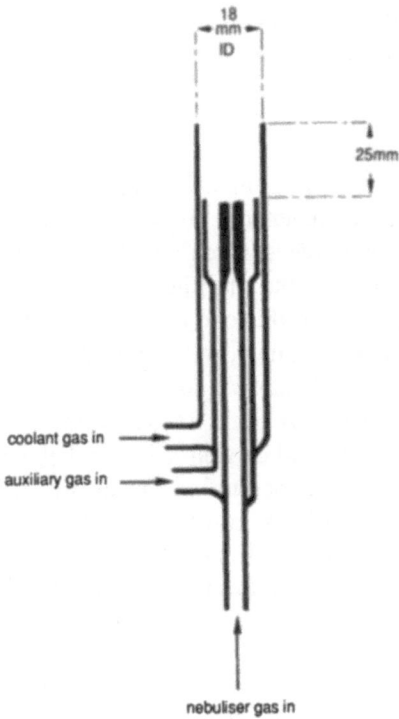

Figure 2.2 Schematic of a Fassel type ICP torch.

lines in circular orbits and the electrical energy supplied to the coil is converted into kinetic energy of electrons. At atmospheric pressure the free electron path before collision with an argon atom, to which its energy is transferred, is only about 10^{-3} mm, and thus the plasma is heated, forming a bright discharge or fireball. At the frequencies used, the skin effect occurring in RF induction heating ensures that most of the energy is coupled into the outer or induction region of the plasma. The cool injector gas flow punches a channel through the centre of the plasma, carrying most of the sample aerosol, so that little appears in the outer annular part of the plasma shown in Figure 2.1. Gas in the centre channel is heated mainly by radiation and conduction from the annulus and, while the temperature in the induction region of the plasma may be as high as 10 000 K, in the central channel the gas kinetic temperature is probably between 5000 K and 7000 K at the mouth of the torch. Note that the power is coupled mainly into the outer region, which is physically distinct from the central channel through which the sample aerosol travels. Thus the chemical composition of the sample solution can vary substantially without greatly affecting the electrical processes that sustain the plasma. This physical separation between the region where the electrical energy is added and the region containing the sample is one reason for the mildness of physical and chemical interferences in the ICP compared to those seen in most other spectrochemical sources.

2.1.2 *RF coupling*

The load coil and plasma present a low electrical impedance to the RF generator which feeds them energy. In order to provide efficient energy transfer, and avoid mis-matches which could produce high potentials from the reflected power, it is essential that the generator 'sees' a matched load at the end of the coupling line to the load coil, which is essentially resistive and absorbs the power delivered. The power required to maintain such a plasma usually lies between 0.75 and 2.0 kW and it is desirable to reduce the reflected power to a few watts or less. A basic RF generator consists of an oscillator 'tank' circuit, usually a coil and capacitor in parallel, which resonates at the desired frequency. Resistive losses and power drawn by the plasma load coil are made up by energy fed back into the circuit from an amplifier, so that instead of dying away, the oscillation continues as long as power is supplied. The amplifier is usually a large thermionic tube but for moderate powers (< 1.5 kW) some recent systems use solid state components. Two types of generator are used, free running systems, where the frequency is controlled by the oscillating circuit and load coil parameters (which include the composition of the plasma, which affects its impedance), and crystal controlled systems where the operating frequency is determined by an oscillating quartz crystal and a servo controlled matching circuit is used to ensure correct matching to the load. Both systems are used in ICP–MS. Because there are strict controls on allowed frequencies for RF systems which radiate any portion of the power fed to the load, free running systems need to be very well screened, usually with a double metal case, whereas crystal controlled systems using the permitted frequencies have less stringent screening requirements. Sufficient screening is always needed however to avoid interference with other electronic circuits used for system control, ion detection and data handling, etc. The interface system in ICP–MS provides an additional RF leakage path not present in ICP–AES and free running generators, although simpler and cheaper than crystal controlled ones, have only been introduced into the more recently developed ICP–MS systems. Most ICP–MS RF systems operate at power levels between 1.0 and 1.5 kW and the operating power level is usually stabilised to about 1%. A fuller discussion of ICP generator and torch systems is given by Greenfield (1987).

2.1.3 *Sample introduction*

The ICP requires any sample to be introduced into the central channel gas flow as a gas, vapour or aerosol of fine droplets or solid particles. A wide variety of methods may be used to produce these such as pneumatic or ultrasonic nebulisation of a solution, electrothermal volatilisation of micro-samples from a hot surface, laser or spark ablation from a solid, and generation of volatile hydrides or oxides from a reaction vessel among others. However, most systems are equipped with a pneumatic nebuliser as standard, in which a high velocity gas stream produces a fine droplet dispersion of the analyte solution. The larger droplets are removed by a spray chamber which allows only those below about 8 μm diameter to pass on to the plasma. These small droplets carry only about 1% of the solution which is usually metered to the nebuliser by a peristaltic pump. Although universally recognised as a very inefficient system, the pneumatic nebuliser, of which there are many forms, retains its popularity because of its convenience, reasonable stability, if correctly operated, and ease of use with multiple sample changers. Other introduction methods are only used to meet more specialised requirements. A detailed discussion of nebuliser operation is given by Gustavsson (1987) and Sharp (1988a) and more general discussions of other methods of sample introduction extend over several chapters in each of Montaser and Golightly (1987), Boumans (1987) and Moore (1989).

2.1.4 Sample history

Whichever method is used for sample introduction, the ultimate aim is to produce sample ions at the entrance to the mass spectrometer, usually by volatilisating, atomising and ionising a dispersion of fine solid particles in a carrier gas stream. In the most common case of pneumatic nebulisation the aerosol leaving the injector tube in the torch may still contain small liquid droplets, but these are quickly dried to solid microparticulates and as an increasingly higher temperature is experienced these are vaporised and the resulting vapour phase compounds dissociated. The transit through the centre of the plasma takes several milliseconds and once atomised the sample is substantially ionised at the high temperature experienced. This process may be seen in a plasma when a solution of yttrium is introduced. The small dark cone seen in the centre of the plasma immediately beyond the tip of the injector tube glows red with emission from YO bands and YI lines, in what is termed the initial radiation zone. Beyond this the yttrium ions are seen to emit in the blue in the normal analytical zone for emission spectrometry. These regions are shown shaded in Figure 2.1.

The sample atoms represent only 10^{-6} or less of the total atom population of the plasma. Their degree of ionisation is dependent on the ionisation conditions in the plasma, which are dominated by the major constituents, usually Ar, H, O and electrons, and the ionisation constant and partition functions for the atom concerned. The temperature in the central channel is high enough to produce almost complete ionisation of many elements and a significant level for those of higher ionisation energy. The Saha equation may be used to give an insight into the equilibrium in the plasma and the numbers of ions produced. A simple outline of this is given by Boumans (1966) and a more detailed discussion of the recent developments in the still incomplete understanding of this complex system is contained in Boumans (1987).

At the plasma powers usually used, thermal equilibrium is not strictly achieved. However, if it is assumed, a reasonable estimate of experimental values for degree of ionisation may be obtained from the Saha equation using the generally accepted values from the literature for ionisation temperature T_i and electron population n_e. Values for most of the elements of the periodic table shown by Houk and Thompson (1988) are given in Figure 2.3. The general form of the dependence of degree of ionisation on ionisation energy for singly charged ions is shown in Figure 2.4, from which it may be seen that the response falls away rapidly above 9 eV. As will be seen from Table 2.1 most elements have first ionisation energies below 10 eV, corresponding to $> 50\%$ ionisation while there are none whose second ionisation energies fall below 10 eV. Thus although there are a number of elements such as the alkaline and rare earths, thorium and uranium which undergo some double ionisation, the majority do not and doubly charged ions do not present serious problems.

As well as ionisation, high levels of excitation of both atoms and ions occur, and for optical emission analysis the point of observation in the plasma is suitably optimised. For mass spectrometry excitation is not necessary and the optimum point for ion extraction may be different from that for emission, in fact it is usually closer to the tip of the initial radiation zone.

2.1.5 Plasma populations

It is useful to consider the relative number densities of atoms and ions of the important species in the plasma which is to operate as an ion reservoir. The gas pressure is 1 bar and at a gas kinetic temperature of 5000 K the total particle density is given by the gas laws as 1.5×18^{18} cm^{-3}. The majority of this is argon. At an ionisation temperature of 7500 K the degree of ionisation of argon is about 0.1%. The population of Ar^{2+} ions is negligible as the second ionisation energy of argon is very high (27 eV). In a 'dry' plasma $n_{Ar^+} = n_e = 1 \times 10^{15}$ cm^{-3} and such a value may be typical for sample introduction by 'dry' methods, but if a nebulised solution is introduced additional electrons are contributed by the ionisation of H and O from the solvent, as well as H^+ and O^+ ions. At a nebuliser uptake of 1 ml min^{-1} and an efficiency of 1% the populations of H^+ and O^+ are respectively about 2×10^{14} cm^{-3} and 1×10^{14} cm^{-3}. In addition if the solution had been acidified with 1% HNO_3, as is commonly done, there would be a population of N^+ of about 1×10^{12} cm^{-3}. These all contribute to the electron population so the value of n_e rises to about 1.3×10^{15} cm^{-3}. The presence of water vapour in the aerosol thus contributes significantly to the ion and electron population of the axial channel.

1	2	3	4	5	6	7	8	9	10	11	12	13	14	15	16	17	18
H 0.1																	He
Li 100	Be 75											B 58	C 5	N 0.1	O 0.1	F 9×10⁻⁴	Ne 6×10⁻⁶
Na 100	Mg 98											Al 98	Si 85	P 33	S 14	Cl 0.9	Ar 0.04
K 100	Ca 99(1)	Sc 100	Ti 99	V 99	Cr 98	Mn 95	Fe 96	Co 93	Ni 91	Cu 90	Zn 75	Ga 98	Ge 90	As 52	Se 33	Br 5	Kr 0.6
Rb 100	Sr 96(4)	Y 98	Zr 99	Nb 98	Mo 98	Tc	Ru 96	Rh 94	Pd 93	Ag 93	Cd 65	In 99	Sn 96	Sb 78	Te 66	I 29	Xe 8.5
Cs 100	Ba 91(9)	La 90(10)	Hf 96	Ta 95	W 94	Re 93	Os 78	Ir	Pt 62	Au 51	Hg 38	Tl 100	Pb 97(.01)	Bi 92	Po	At	Rn
Fr	Ra	Ac	Unq	Unp	Unh	Uns	Uno										

Ce 96(2)	Pr 90(10)	Nd 99*	Pm	Sm 97(3)	Eu 100*	Gd 93(7)	Tb 99*	Dy 100*	Ho	Er 99*	Tm 91(9)	Yb 92(8)	Lu
Th 100*	Pa	U 100*	Np	Pu	Am	Cm	Bk	Cf	Es	Fm	Md	No	Lr

Figure 2.3 Calculated values for degree of ionisation (%) of M^+ and M^{2+} at $T_i = 7500\,K$, $n_e = 1 \times 10^{15}\,cm^3$. Elements marked by an asterisk yield significant amounts of M^{2+} but partition functions are not available (after Houk, 1986).

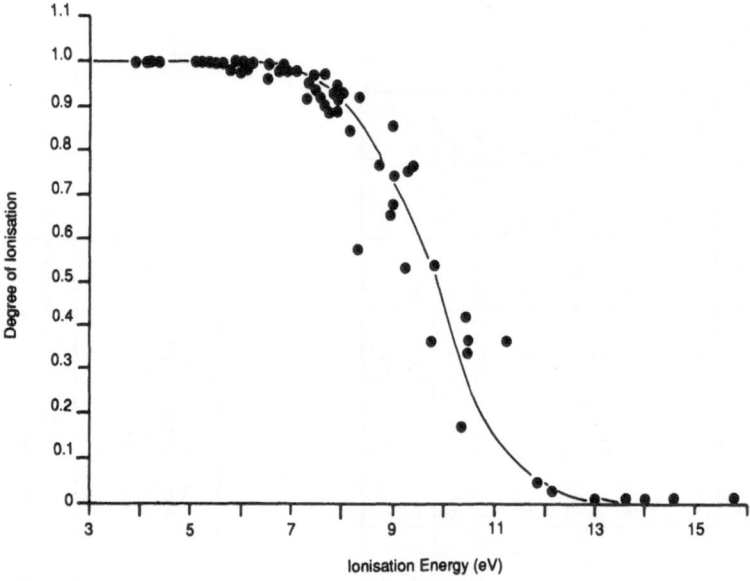

Figure 2.4 Degree of ionisation versus ionisation energy for singly charged ions in the ICP, calculated for a representative selection of elements. Redrawn and adapted from Gray (1989a).

Table 2.1 Distribution of ionisation energies among the elements for singly and doubly charged ions at 1 eV intervals (from Gray, 1989a)

Ionisation energy (eV)	Elements	
< 7	Li, Na, Al, K, Ca, Sc, Ti, V, Cr, Ga, Rb, Sr, Y, Zr, Nb, In, Cs, Ba, La, Ce, Pr, Nd, Pm, Sm, Eu, Gd, Tb, Dy, Ho, Er, Tm, Yb, Lu, Hf, Tl, Ra, Ac, Th, U	
7–8	Mg, Mn, Fe, Co, Ni, Cu, Ge, Mo, Tc, Ru, Rh, Ag, Sn, Sb, Ta, W, Re, Pb, Bi	
8–9	B, Si, Pd, Cd, Os, Ir, Pt, Po	
9–10	Be, Zn, As, Se, Te, Au	2^+ *ions*
10–11	P, S, I, Hg, Rn	Ba, Ce, Pr, Nd, Ra
11–12	C, Br	Ca, Sr, La, Sm, Eu, Tb, Dy, Ho, Er
12–13	Xe	Sc, Y, Gd, Tm, Yb, Th, U, Ac
13–14	H, O, Cl, Kr	Ti, Zr, Lu
14–15	N	V, Nb, Hf
15–16	Ar	Mg, Mn, Ge, Pb
> 16	He, F, Ne	All other elements

The addition of trace elements to the nebulised solution produces far lower populations of the elements to be determined against this background of the 'permanent' ions. An element at a concentration of $1 \, \mu g \, ml^{-1}$ in the sample solution, which is fully ionised in the plasma, contributes about 1×10^{10} ions cm^{-3} and the number is correspondingly lower for elements of higher ionisation energy. Thus a fully ionised matrix element at $5000 \, \mu g \, ml^{-1}$ in the solution only contributes about $5 \times 10^{13} \, cm^{-3}$ to the total level of n_e of $1.3 \times 10^{15} \, cm^{-3}$ and produces a barely significant shift in the equilibrium.

Thus ionisation suppression in the plasma is generally not a major cause of matrix interference in the ICP (Olivares and Houk, 1986), unless the concentration of the matrix element is extremely high.

As an ion source the ICP may thus be seen to have several valuable properties. The samples are introduced at atmospheric pressure, and may be readily interchanged. The degree of ionisation across the periodic table is relatively uniform and is mainly to singly charged ions. Sample dissociation is very efficient at the gas temperatures experienced and few sample molecular fragments remain. High ion populations of trace concentrations are produced and therefore potential sensitivity is high. The main disadvantages are the high gas temperature and pressure at which the ions are produced which require an appropriate interface design to transfer the ions without significant distortion of their relative populations to a mass analyser.

2.1.6 Distribution of ions in the plasma

Once the plasma leaves the mouth of the torch it becomes accessible for ion extraction into the mass spectrometer. The distribution of ions within it has been studied by a variety of methods but a simple visualisation may be obtained from spacially resolved profiles made by moving the ion extraction interface of an ICP–MS system across the plasma. When no sample is introduced, and the central channel contains only dry argon, the transverse profile of Ar^+ ions across the mouth of the torch is shown by the Ar^+ response in Figure 2.5. On the torch axis the cooler central gas stream shows a relatively low Ar^+ population but each side of the centre the higher degree of ionisation of Ar in the hotter induction region, the plasma annulus, is

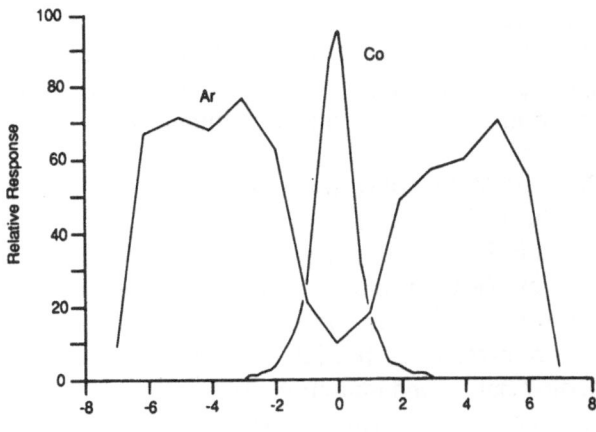

Transverse Displacement from Axis (mm)

Figure 2.5 Transverse profiles of ions across the mouth of the plasma torch. Profile for Ar^+ shown for dry argon only. Profile for Co^+ from a nebulised solution at $100\,ng\,ml^{-1}$.

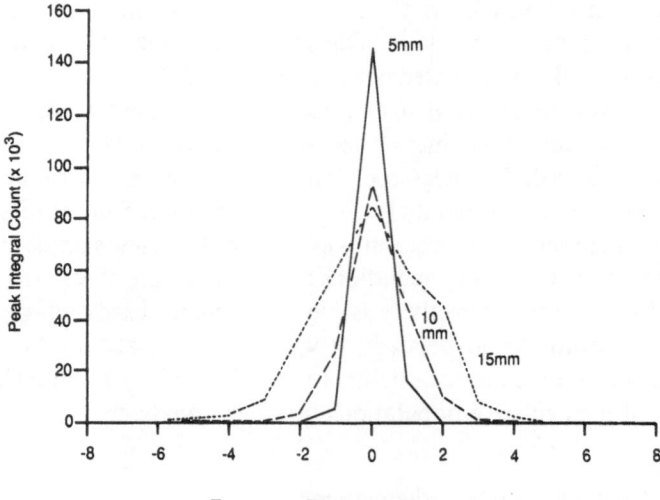

Figure 2.6 Transverse profile across plasma flame at 5, 10 and 15 mm from load coil in steps of 1 mm. Nebulised solution containing Co at 100 ng ml^{-1}.

clearly shown. The central channel may be seen to be about 3 mm wide. Beyond the overall diameter of about 12 mm the ion population drops sharply again at the edge of the plasma. This particular plasma appears to be slightly asymmetrical due to poor concentricity of the torch. If a similar plot is performed across the narrow central channel when a sample is being introduced, a profile such as that shown for Co$^+$ is obtained. Here Co solution at a concentration of 100 ng ml^{-1} is being nebulised and the Co$^+$ ions are concentrated mainly within 1 mm of the axis. Further along the axis away from the torch, the central channel diffuses into the annulus and similar profiles for Co at 1 μg ml^{-1} at 5, 10 and 15 mm from the load coil are shown in Figure 2.6. This method may be extended to cover most of the species occurring in the plasma and the optimum ion extraction position determined.

The ideal location of the orifice along the axis is less immediately obvious but to ensure the highest possible ion density in the extracted gas, the distribution plots in Figure 2.6 suggest that the position should be as close to the load coil as the torch mouth permits. However, during its passage along the central channel of the plasma the sample must be converted to atomic ions as completely as possible and the processes of desolvation, volatilisation, dissociation and ionisation take several milliseconds. The time required depends particularly on the size of the initially desolvated microparticulates in the aerosol, which in turn depends on the level of dissolved solids in a nebulised sample solution, and on the bond strengths of the molecular species (which may be refractory) in the sample. The time available, while

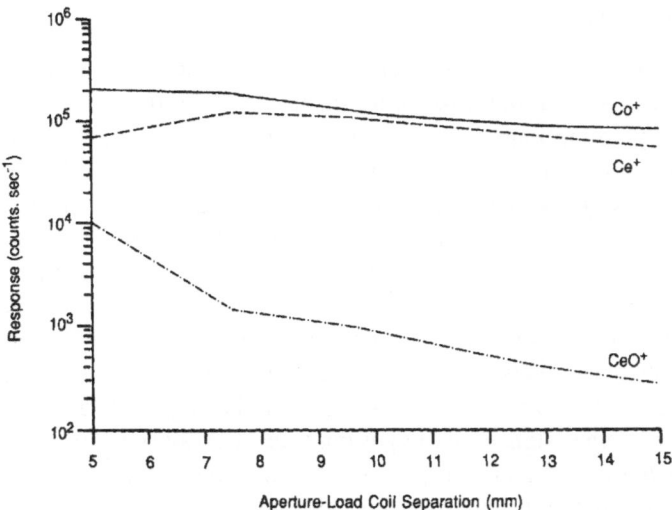

Figure 2.7 Longitudinal profile along axis of Co$^+$, Ce$^+$ and CeO$^+$ from a solution at 100 ng ml^{-1}.

the sample resides in the hottest part of the plasma, depends on the plasma operating parameters, particularly plasma power and central channel gas flow. Thus when a solution of a yttrium salt is introduced along the axis, the boundary beyond which most of the yttrium oxide molecules are dissociated is observed as the tip of the red initial radiation zone (see section 2.1.4). The orifice must therefore be placed beyond this point to avoid an excessive oxide response. This may be seen for a refractory species CeO$^+$ in Figure 2.7. Both Co and Ce were present in this solution and the Co$^+$ response is seen to fall steadily from 5 mm outwards. The Ce$^+$ response, however, is not at a maximum until about 7.5 mm beyond which it follows the Co$^+$ response down. At 5 mm the oxide response is relatively high at about 16% of the metal response but falls to below 1% after 7.5 mm. Thus a longer dwell time in the plasma is necessary to promote dissociation of CeO$^+$ into Ce$^+$.

The axial position of the extraction or sampling orifice must consequently by selected with some care. If the system is optimised without considering the yield of interfering oxide signals from matrix elements in the sample, quite misleading results may be obtained. This is particularly true if the full spectrum of a sample new to the operator is not checked for interferences prior to detailed analysis.

The ICP thus forms a very convenient ion source with a high yield of singly charged analyte ions, few doubly charged and oxide or other molecular and adduct ions and the facility for rapid interchange of samples. Its main drawbacks are the high temperature at which the ion reservoir must be maintained and the atmospheric pressure within it. The means of extracting ions into the mass spectrometer are discussed in the following section.

2.1.7 *Other plasmas*

Argon is by far the most common gas used to sustain the ICP. There has recently been substantial interest in mixed-gas ICPs to reduce the levels of certain polyatomic ions. For example, mixing $\sim 5\%$ N_2 into the outer gas flow can reduce the levels of ArO^+ and MO^+ substantially (Lam and Horlick, 1990a, b; Lam and McLaren, 1990b). Addition of $\sim 0.5\%$ Xe to the central gas flow can attenuate polyatomic ions like N_2^+, ClO^+, ArN^+ and ArO^+, although $\sim 90\%$ of the analyte signal is sacrificed (Smith *et al.*, 1991). Caruso and co-workers (Brown *et al.*, 1988; Mohamad *et al.*, 1989) have reported extensive studies with helium microwave induced plasmas, and initial experiments with a helium ICP have been described (Montaser *et al.*, 1987; Koppenaal and Quinton, 1988). Because of the high ionisation energy of helium, a plasma sustained in this gas may yield positive ions from non-metals more efficiently than an Ar plasma. At the present time, none of these alternate plasmas is in common use, although they may well prove useful analytically in the future.

2.2 Ion extraction

Extracting ions from the plasma into the vacuum system is obviously critical in ICP–MS. A schematic diagram of a typical extraction interface is shown in Figure 2.8. Ions first flow through a sampling orifice (approx. 1 mm diameter

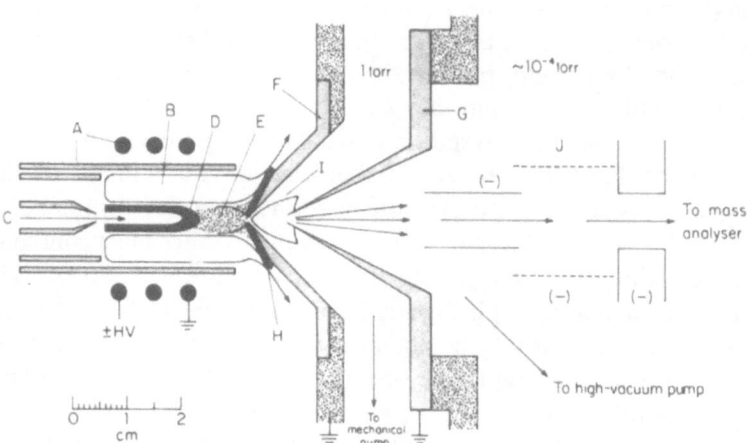

Figure 2.8 ICP and sampling interface for ICP–MS. A solution of Y is injected into the plasma with resulting red emission from excited YO and Y and blue emission from Y^+. A, torch and load coil (reverse geometry); B, induction region; C, aerosol gas flow; D, initial radiation zone; E, normal analytical zone (see Figure 2.1); F, sampler, G, skimmer; H, boundary layer of ICP gas deflected outside sampler; I, supersonic jet; J, ion lens. Reproduced with permission (Houk, 1986).

in a cooled cone) into a mechanically pumped vacuum system, where a supersonic jet forms. The central section of the jet flows through the skimmer orifice, which is also approximately 1 mm diameter. The extracted gas containing the ions attains supersonic velocities as it expands into the vacuum chamber and reaches the skimmer orifice in only a few microseconds (Douglas and French, 1988). Thus, the sample ions change little in nature or relative proportions during extraction. To a first approximation, they simply flow through the sampling orifice and then through the skimmer orifice. The discussion below covers the current state of understanding of ion extraction in ICP–MS.

2.2.1 *Boundary layer and sheath*

Two ways in which the plasma interacts with the sampling cone are shown in Figure 2.9. First, the plasma is deflected and cooled as it meets the metal cone. A boundary layer of gas forms between the plasma and the side of the cone. The temperature in the boundary layer is intermediate between that of the plasma and that of the cone. Chemical reactions such as oxide formation occur readily in the boundary layer, as can be seen by the red colour therein when a concentrated solution of yttrium is introduced into the plasma.

In modern ICP–MS instruments, the sampling orifice is large enough for the gas flow through it to puncture the boundary layer. Thus, the gas sampled is not cooled much while it is outside the sampling cone. There is still a thin, oblique boundary layer inside the lip of the orifice, and care must be taken to design the interface so that oxides formed in this layer do not pass through

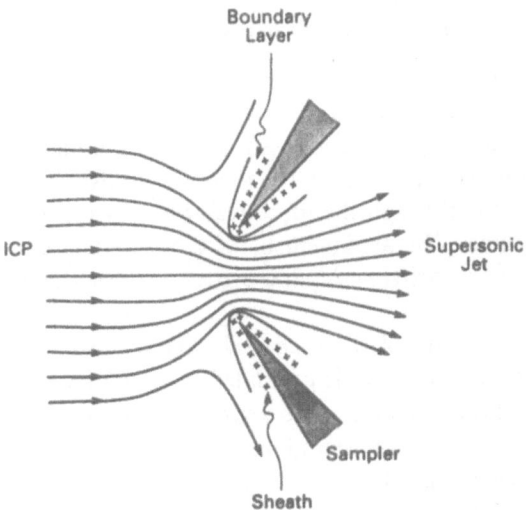

Figure 2.9 Interaction of ICP with sampling cone showing boundary layer, sheath and flow field into orifice. The thickness of the boundary layer has been exaggerated greatly for clarity.

the skimmer. The oxide response is minimised if the skimmer diameter is less than that of the sampling orifice, as shown by early work at Sciex (Douglas, 1985) and by experiments with small sampling orifices described by Gray (1986b) and by Vaughan and Horlick (1990a).

Figure 2.9 also illustrates an electrical interaction between the plasma and the conducting sampler. The plasma contains equal numbers of positive ions and electrons (neglecting doubly charged and negative ions) and is therefore electrically neutral. A surface immersed in the plasma collects both ions and electrons. The mobility of the electrons is much higher than that of the ions, so the electron flux (or current) to the surface is much higher than the ion flux. The potential of the surface becomes negative with respect to the plasma. The sheath region that forms over the surface is depleted in electrons, and positive ions are in excess (i.e. the + signs in Figure 2.9). The negative potential of the surface with respect to the plasma repels electrons and attracts ions to balance the two fluxes.

In ICP–MS, the surface contacting the plasma is the metal sampling cone, which is usually grounded. Since the potential of the sampling cone is fixed, the plasma appears to float at a positive potential, which is known as the plasma potential. The sheath co-exists with the boundary layer. The number of neutrals greatly exceeds the numbers of either ions or electrons, so the sheath is a dynamic region with particles constantly moving in and out via collisions. Although calculations indicate the sheath to be much thinner than the boundary layer (Houk et al., 1981a; Douglas and French, 1988), it still influences the ion extraction process by its interaction with the RF potential in the plasma, as described in the next section.

2.2.2 Plasma potential and secondary discharge

The plasma is maintained by RF energy (usually at 27 or 40 MHz) coupled to it by the load coil (Figure 2.8). Consequently, currents circulate at this frequency through the plasma, which is a good conductor. In addition to this inductive (magnetic) coupling, the load coil is also coupled capacitively (i.e. electrostatically) through the torch wall by the capacitance between the coil and the plasma. The arrangement of the electrical connections to the coil influences this capacitive coupling process, as described in the following discussion.

Traditionally, one end of the load coil is connected to the high voltage RF source while the other end is grounded. A potential gradient thus exists along such a coil, except at the moment when the field polarity reverses. When the plasma contacts the sampling cone and part of it is drawn through the orifice, the plasma is coupled to the cone through the very thin sheath. The impedance of this sheath is much lower than that of the capacitative coupling between the plasma and load coil. Thus, the plasma acquires an RF potential which is determined by the ratio of these two impedances which act as a potential divider. An RF current flows through this coupling from the load coil. However, the RF current flow to the grounded cone is modified by the different mobilities of ions and electrons in the sheath. During negative half cycles, the current is carried mainly by electrons, which can flow to ground far more readily than the positive ions that carry current during positive half cycles.

Because of these effects, the plasma assumes a net mean positive DC potential. This offset or bias potential may be considerably larger than the floating potential due to the sheath alone (Douglas and French, 1986; Gray, 1986c). The same effect is observed in low pressure RF discharges used for sputtering and production of thin films (Chapman, 1980). Numerous measurements with Langmuir probes illustrate both sources of plasma potential (Douglas and French, 1986; Gray *et al.*, 1987; Houk *et al.*, 1987; Jakubowski *et al.*, 1989; Lim *et al.*, 1989b).

If the plasma potential is high enough, it can cause an electrical discharge between the plasma and the sampling orifice. This phenomenon is manifest as a crackling discharge into the orifice. In the author's experience, the discharge tends to be most intense when the orifice is separated from the torch by ~ 10 cm. In the early work this discharge was referred to as the 'pinch' (Houk *et al.*, 1981a); it is now usually referred to as a secondary discharge. A severe discharge is characterised by bright white emission from the gas flowing into the orifice and is detrimental in that it erodes the orifice, generates multiply charged ions, and induces high kinetic energies and a wide spread of kinetic energy in the extracted ion beam (Houk *et al.*, 1981b; Douglas and French, 1986).

Minimising this discharge was a key step in the early development of ICP–MS. Modification of the load coil was one successful approach. The coil arrangement shown in Figure 2.8, often referred to as the reversed coil, attenuates the discharge to manageable levels. Grounding the end of the coil nearest to the interface reduces the electric field close to the sampling cone. The plasma potential can also be reduced by: (a) use of a low aerosol gas flow rate ($0.5–0.9 \, l \, min^{-1}$), (b) reducing the solvent load to the plasma (e.g. by desolvating aqueous aerosols before they are injected), (c) moving the sampling orifice close to the load coil and (d) substituting a two-turn load coil (instead of the original three turns). These measures weaken the discharge but do not eliminate it fully, because the ion kinetic energy (see below) and certain properties of the spectra such as the ratio M^{2+}/M^{+} vary with operating conditions in a fashion that is not consistent with the expected changes in the plasma. The fundamental reasons why the plasma potential is sensitive to sampling position and operating conditions are not clear at this time. Perhaps the plasma potential depends critically on the composition of the sheath, which forms the bottom of the potential divider chain. Conditions in the sheath are, in turn, affected by temperature and ion populations in the plasma, which depend on sampling position and operating parameters.

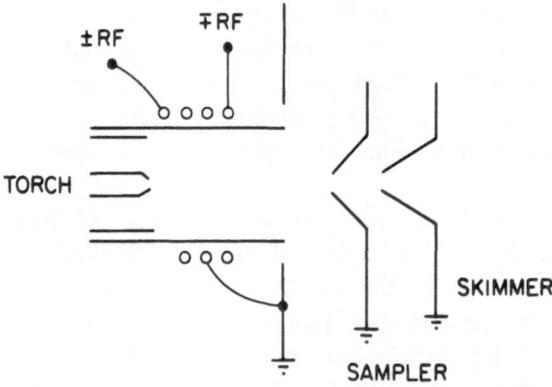

Figure 2.10   Centre-tapped load coil (Douglas and French, 1986).

The centre tapped load coil arrangement shown in Figure 2.10 seems to eliminate almost totally the secondary discharge. In this arrangement, the centre is grounded and voltages of equal magnitude but opposite polarity (i.e. opposite phase), are applied to either end. At any instant, the positive gradient from one end to the centre is balanced by the opposite phase, negative gradient from the other end, resulting in a low bias potential in the plasma from the RF coupling. Thus, the net potential is only a few volts (Douglas and French, 1986). This load coil was a key part of the early ELANs. The same effect is produced by the Colpitts oscillator circuit in the new ELANs (Figure 2.11). With these systems, small changes in plasma operating parameters induce little change in plasma potential (Douglas and French, 1986) and only modest changes in ion kinetic energies (Fulford and Douglas, 1986). In addition, performance characteristics such as the yield of M^{2+} and MO^{+} vary with plasma operating conditions in a manner consistent with the expected changes in the abundance of these species in the ICP. At one time, this relative independence of mass spectral properties from plasma operating conditions was considered by some to be a major advantage relative to the performance obtained with unsymmetrical coils. With the widespread use of aerosol desolvation (Hutton and Eaton, 1987; Zhu and Browner, 1988),

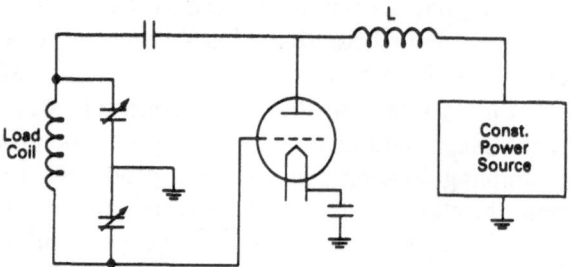

Figure 2.11   Colpitts oscillator circuit for generating ICP with low plasma potential. Reproduced with permission (Douglas, 1991).

dry sample introduction, and other empirical ways to minimise plasma potential with unbalanced load coils, the authors' view is that excellent results can be obtained from either type of coil arrangement, and this is not a major distinction between the various instruments as far as analytical performance is concerned.

2.2.3   Supersonic jet

The gas flowing through the sampling orifice expands into the first stage of the vacuum chamber, which is evacuated by a mechanical pump. The pressure ratio is roughly 1 bar/1 mbar, which is more than sufficient for a supersonic jet to be formed inside the first stage. This jet is shown schematically in Figure 2.12 and in two photographs in Figure 2.13. The jet consists of a freely expanding region often called the zone of silence surrounded by shock waves called the barrel shock and Mach disc. The barrel shock is the curved, faint white region at the top and bottom, and the Mach disc is the white region at the right of Figure 2.13a (Gray, 1989b). The barrel shock and Mach disc are caused by collisions between fast atoms from the jet and the background gas, which reheat the atoms and induce emission.

The position of the onset of the Mach disc is given by

$$X_M = 0.67 \, D_o (P_0/P_1)^{1/2} \tag{2.1}$$

where X_M = position of Mach disc from the sampling orifice along the central axis, D_o = diameter of sampling orifice, P_0 = pressure in ICP and P_1 = background pressure in extraction chamber (Olivares and Houk, 1985a; Kawaguchi et al., 1988b). To avoid losses of ions due to collisions and scattering, the skimmer is positioned with its open tip inside the Mach disc. Thus, the central core of the zone of silence passes through the skimmer into the second vacuum

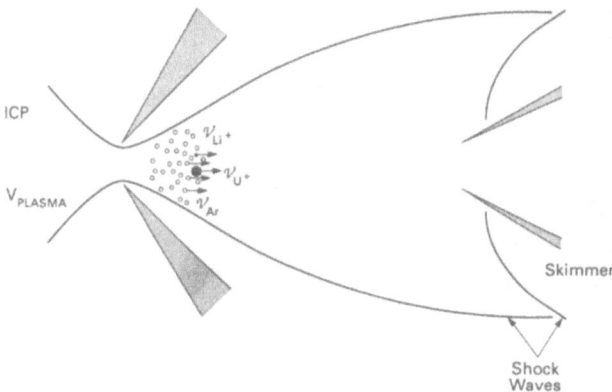

Figure 2.12   Illustration of sampler, skimmer, and supersonic jet showing how light ions (·) and heavy ions (●) are all accelerated to the same velocity (v) as neutral Ar(○) in the jet.

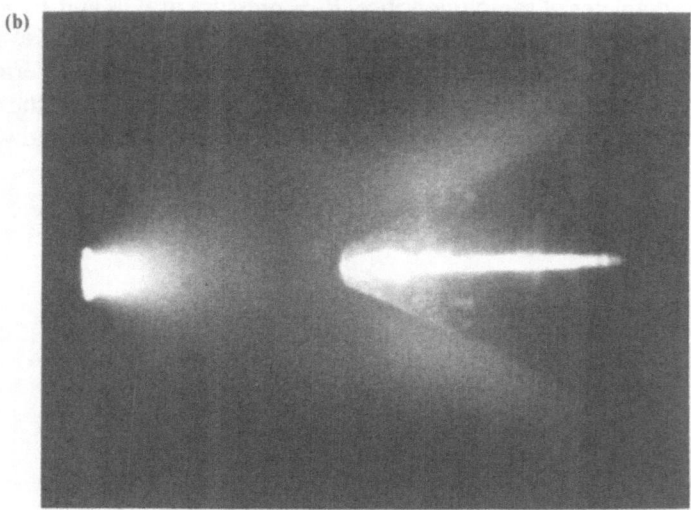

Figure 2.13 Photographs of supersonic jet: (a) without skimmer, (b) with skimmer present at usual position. The plasma is at the left and is blocked by the sampling cone. The white oval at the left is the sampling orifice. The extracted gas flows from left to right.

stage. The Mach disc is now replaced by a shock wave that forms outside the skimmer downstream from the tip, as shown in Figure 2.13b. A skimmer–sampler separation of roughly 2/3 of the distance to the onset of the Mach disc usually provides optimum ion transmission (Douglas and French, 1988; Lam and Horlick, 1990b).

The extraction interface used in ICP–MS is very similar to those used by chemical physicists for generating cold molecular beams (Beijerinck, 1985; Campargue, 1966; Milne and Greene, 1967) or for sampling ions from flames or other plasmas (Stearns et al., 1979; Sugden, 1965; Hayhurst and Telford, 1971). The major difference is that a much larger sampling orifice (~ 1 mm as compared to $\sim 100\,\mu$m) is used in ICP–MS to prevent plugging due to condensed solids (Douglas and Kerr, 1988).

2.2.4 Gas dynamics

The gas dynamic properties of the extraction process have been described in detail in an excellent paper by Douglas and French (1988) and in a recent book chapter by Douglas (1991). Some salient points from these papers are given below. We will assume typical ICP–MS conditions, namely, that the diameter of the sampling orifice is ~ 1 mm, the skimmer diameter is comparable to or slightly smaller than that of the sampler, the secondary discharge is not severe, and the skimmer is located as shown in Figure 2.12.

The sampler collects gas from a region of the plasma of cross-sectional diameter roughly eight times that of the sampling orifice. The gas flow rate for an Ar plasma at ~ 5000 K is $\sim 10^{21}$ atoms/s through the sampler. Of the gas that passes through the sampler, only $\sim 1\%$ ($\sim 10^{19}$ atoms/s) also traverses the skimmer. Furthermore, only the centreline flow gets through the skimmer, resulting in a spatial resolution in the plasma comparable to the skimmer diameter. The ion distribution plots in Figures 2.5 and 2.6, and other plots in the literature (Douglas and French, 1988), show structure of sufficient detail to lend credence to this estimate of spatial resolution.

With a large sampling orifice, the flow punctures the boundary layer cleanly (Figure 2.12), and the gas is cooled very little until it gets inside the orifice. The presence of the metal sampling cone has little effect on the upstream plasma unless the secondary discharge is intense. Once the sampled gas passes through the sampling orifice, collisions occur for the first few orifice diameters, after which the atoms continue to flow under essentially collisionless conditions. Douglas and French estimate that approximately 250 collisions occur between neutral Ar atoms and other species during extraction. Collisions involving other species are less frequent because they are far less abundant than neutral Ar. For example, the densities of both ions and electrons are less than that of neutral Ar by a factor of $\sim 10^{-3}$. The extraction process takes roughly 3 μs. Thus, there is little opportunity for ion loss by recombination between ions and electrons.

There is substantial experimental evidence for the lack of ion-electron recombination during extraction:

1. Ionisation temperature can be measured mass spectrometrically by observing ratios of signals from ions of widely different ionisation energy. Such values are generally 7000–9000 K when various thermometric species are used (Houk *et al.*, 1981b; Crain *et al.*, 1990; Wilson *et al.*, 1987c), which are close to those measured optically in the plasma. Thus, if ions and electrons recombined significantly, different ions such as Fe^+, Co^+, Ni^+, etc. would have to do so at essentially the same rate, which is unlikely.
2. Fluorescence from neutral Na was not observable inside the supersonic jet, indicating that Na is nearly completely ionised in the plasma and few neutral Na atoms were formed during extraction (Lim *et al.*, 1989a).
3. Measurements of electron density in the jet indicate this parameter to be given roughly by

$$n_{e,J} \sim (n_{e,ICP} n_{a,J})/n_{a,ICP}$$

where $n_{e,J}$ = electron density in jet, $n_{e,ICP}$ = electron density in ICP, $n_{a,J}$ = density of neutral Ar in jet and $n_{a,ICP}$ = density of neutral Ar in ICP. Thus, the electrons are simply rarefied by the same ratio as the other species in the jet (Lim and Houk, 1990) and are not greatly depleted by other processes. Assuming the plasma potential is small, the jet is still electrically neutral, with roughly equal numbers of positive ions and electrons. Collisions reduce the kinetic temperature of the heavy particles to ~ 300 K but do not cool the electrons nearly as much, their temperature remaining close to that in the source, as verified by Langmuir probe measurements (Lim and Houk, 1990).

The low number of collisions during and the short time duration of the expansion suggest that the sampling process is not complicated by extensive chemical reactions, and the ions extracted are more or less representative of those in the plasma (Douglas and French, 1988). The oxide ions (MO^+) observed are probably present mainly in the plasma, particularly if the sampling orifice is too close to the initial radiation zone (Figures 2.1 and 2.8). Some MO^+ ions can be formed from M^+ ions by reactions in the cool boundary layer in front of or inside the sampling orifice; use of a skimmer orifice that is smaller than the sampling orifice helps prevent these additional MO^+ ions from passing through the skimmer to the mass spectrometer (Gray, 1986b; Vaughan and Horlick, 1990a). The origins of the other polyatomic ions are less clearly established. The following statements represent the authors' opinions on this matter based on literature and experience but not on conclusive experiments. Strongly bound species like N_2^+ and O_2^+ are likely to be present in the plasma. Species like Ar_2^+, ArO^+, ArN^+ and ArM^+ may be formed by condensation reactions with Ar during extraction; many of these ions (or their neutral analogues) can be made deliberately in

supersonic beams or by ion–molecule reactions for studies of ion energetics
and chemistry (Ng, 1983, 1991). Chemical reactions involving adsorbed layers
of atoms and/or molecules inside the tip of the sampling cone may also be
important. Many of the background species observed from the ICP (e.g.
OH^+, H_2O^+, H_3O^+ and Ar_2^+) are also seen during mass spectrometric
sampling of various plasmas, flames and other high-temperature media.

2.2.5   Ion kinetic energies

As described by Fulford and Douglas (1986), the supersonic jet also influences
the kinetic energies of the extracted ions. If the plasma potential is small, the
ions are simply entrained in the flow of natural gas through the orifice
(Figure 2.12), and they reach the same speed as the terminal velocity of Ar
in the expansion. Since all ions have the same velocity (v), their kinetic energy is
$\sim 0.5\,mv^2$ and is thus proportional to mass (m). Thus, a plot of ion energy
versus mass should be a straight line with heavier ions having greater kinetic
energy.

The ion energies shown in Figure 2.14 were determined with the home-made
instrument in Ames. A positive stopping voltage was applied to the mass
analyser and the voltage necessary to attenuate the ion signal to the
background level was measured. These curves show that the ion kinetic
energy does indeed increase with mass. If the supersonic expansion were the
only phenomenon that influenced the ion energy, the y intercepts (corresponding to hypothetical ions with mass zero!) would be zero, and the ion
energies would not vary much with plasma operating conditions. In fact, the
y intercepts are not zero and the ion energies do vary with aerosol gas flow
rate. The instrument used for Figure 2.14 has a reversed load coil (i.e.
Figure 2.8) and thus still has a significant residual plasma potential. The ions
originate from a region in the plasma whose potential is positive with respect

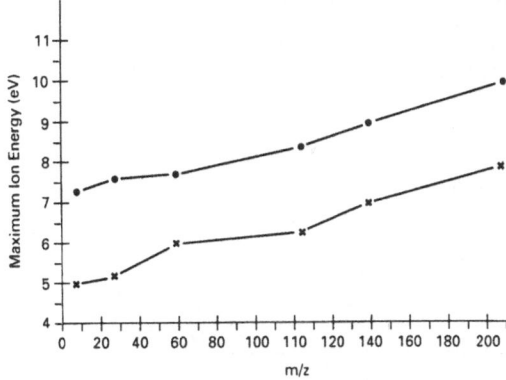

Figure 2.14   Ion kinetic energies for Ames instrument (Crain *et al.*, 1990) as functions of m/z
for two aerosol gas flow rates. ●, $1.0\,l\,min^{-1}$; ×, $0.73\,l\,min^{-1}$.

to the grounded sampling cone. As they enter the aperture they are accelerated across this potential difference and thus acquire extra kinetic energy that displaces the measured ion energies to higher values than would be obtained from the supersonic expansion alone. Since the plasma potential is sensitive to aerosol gas flow rate, so is the ion kinetic energy. Gray and Williams (1987b) report data that differ somewhat from Figure 2.14 in that the range of ion energies for ions of different m/z also varies with aerosol gas flow rate. At approximately $0.9 \, l \, min^{-1}$, the range of ion energies was narrowest.

The variation of ion energy with ion mass and, for some load coil geometries, with plasma operating conditions complicates ion focusing, as described below.

2.3 Ion focusing

After the ions leave the skimmer, they must be conveyed to the mass analyser. Ion lenses are used for this function. The basic principles of operation of ion lenses are the same as those for the electron lenses used in television sets and electron microprobes. Also, ion lenses have many characteristics in common with lenses for light optics. There are several general references on ion optics (e.g. Spangenburg, 1948; Busch and Busch, 1990). An excellent simulation program called SIMION is also available that runs readily on common personal computers (Dahl and Delmore, 1990). A recent paper describes the application of SIMION to the ion lens from an ELAN (Vaughan and Horlick, 1990b). The following discussion provides background about the basic principles of ion optics and some special points pertinent to ICP–MS.

2.3.1 *Operation of ion lenses*

Figure 2.15 illustrates the general problem of transmitting and focusing ions through an ion lens. Suppose a positive ion of charge z is formed in a region of potential $V_{initial}$. It has a potential energy $z \, V_{initial}$. This ion will travel through a given region as long as the potential in that region is below $V_{initial}$;

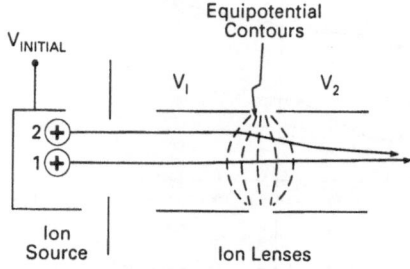

Figure 2.15 Schematic diagram illustrating operation of ion lens.

otherwise, it will turn around and go back toward the source. Thus, the ion in Figure 2.15 will travel to the right if the potentials V_1 and V_2 are less than $V_{initial}$. The situation is exactly analogous to the operation of a roller coaster, which is first hauled as high as possible, thus acquiring potential energy. It then rushes downhill upon release as this potential energy is converted to kinetic energy. However, the coaster (and the analogous ion) can never traverse a spot higher than the initial elevation. Stopping the ions by applying a positive voltage to an intermediate electrode or to the quadrupole rods is, in fact, the method used to estimate ion kinetic energy for Figure 2.14 (Olivares and Houk, 1985a; Fulford and Douglas, 1986).

When the ion moves into a region of potential V, the kinetic energy of the ion becomes $z(V_{initial} - V)$. The velocity (lower case v) of the ion is therefore $v = [2z(V_{initial} - V)/m]^{1/2}$. The potential of such a region may be established by an electrode such as a grid or cylinder suspended in the vacuum system along the path desired for the ion. Consider the first cylinder (labelled V_1) in Figure 2.15. Once inside the cylinder, the ion experiences a uniform potential and moves at constant velocity in the same direction as when it entered the cylinder, just as a roller coaster moves at constant velocity (neglecting friction) on a level stretch. By imparting a directed velocity to the ion, the lens draws it toward the mass analyser and retains it inside the vacuum system while the unwanted neutral particles flow to the pump. As the ion nears the exit of the first cylinder, it enters a new field region, which may be used as a lens to constrict or focus the stream of ions. This focusing action can improve the fraction of ions leaving the source that are transmitted downstream to the mass analyser.

Consider the region between the two cylinders in Figure 2.15. If $V_1 \neq V_2$, the potential between the cylinders varies with position. The figure depicts curved equipotential contours in this region. These curved equipotential surfaces between cylinders provide the focusing action. Ion 1, which leaves the source on centre, is acted upon symmetrically and passes straight through the lens. This ion is easily collected anyway. The real improvement lies for ion 2, which initially leaves the source displaced from the axis of the lens. The forces acting on ion 2 in the region between the two cylinders are unbalanced. If V_1 and V_2 are adjusted properly, ion 2 can be deflected closer to the axis. Once inside the second cylinder, the ion travels along a straight path, which may cross the axis and diverge again. More electrodes can be used downstream from V_2 to provide additional focusing action and to further adjust the ion path. For example, it may be desirable to position the cross-over point so that the ion beam is focused on a differential pumping aperture between separate vacuum stages.

As indicated by the preceding discussion, the focusing action of the curved equipotential surfaces inside the lens improves ion transmission and sensitivity, just as the curved surfaces of an optical lens cause a focusing effect by refraction. In fact, an ion lens may be thought of as analogous to an unusual optical lens in which the refractive index changes continuously with position inside the lens. Like optical lenses, ion lenses can be characterised by focal points, principal planes, etc. (Spangenburg, 1948; Busch and Busch, 1990). Unlike optical lenses, however, the transmission and focal properties of an ion lens can be varied externally simply by changing the voltages applied to the lens elements.

Ion motion through electrostatic fields such as those used in most ion lenses is governed by the Laplace equation (Spangenburg, 1948). The value of m/z does not appear in this equation. To a first approximation, the focal properties of ion lenses should not differ for ions of different m/z, and it should be easy to find a single set of lens voltages that transmit ions of all m/z values with uniform efficiency. The lens should also transmit ions of different m/z uniformly if they start at the same position with the same initial energy. This utopian behaviour is based on several assumptions that are not met in ICP–MS, however.

2.3.2   Ion lenses in ICP–MS

The ion lenses in the most common instruments are shown in Figure 2.16. In each lens, several electrodes are strung together to confine the ions on their way to the mass analyser. Each lens incorporates a central disc to prevent photons from the plasma from reaching the detector. The sampler

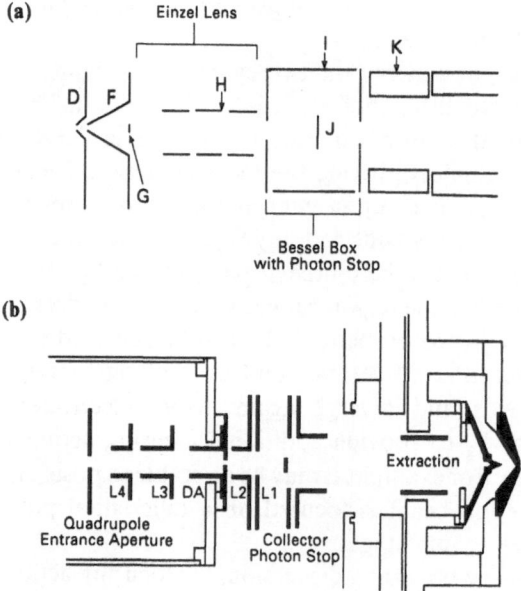

Figure 2.16   Ion lenses from ELAN (a) and PlasmaQuad (b). In the ELAN 5000, the Bessel box also serves as the differential pumping orifice. Note that the ELAN has two photon stops, the second of which is much larger than that on the PlasmaQuad.

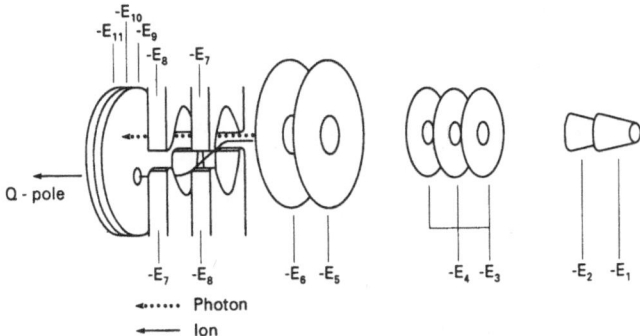

Figure 2.17 Ion lens from Yokogawa PMS 2000 showing bent ion path. Reproduced with permission from Yokogawa.

and skimmer stare into the heart of the ICP, which is a good source of vacuum ultraviolet radiation that can activate the detector (LaFreniere *et al.*, 1987). In some systems, the voltage applied to the centre stop may be adjusted to deflect some of the ions around it safely. Nevertheless, 50–80% of the ions are probably lost here, as shown by measurements of ion current with the stop removed.

Offsetting the entrance to the quadrupole from the line of sight through the lens and sampler is another effective way to minimise the background from photons. The PMS 2000 from Yokogawa accomplishes this with the lens arrangement shown in Figure 2.17. The ion count rates obtainable with this lens are very high ($\sim 10^8$ counts s^{-1} per ppm) with a typical background of 50 counts s^{-1}. This type of lens may prove superior to those using a central baffle to block photons.

Generally, the skimmer is grounded. However, as discussed in section 2.2.5, the ions gain kinetic energy during extraction from both the gas dynamic effect of the supersonic expansion and any plasma potential above that of the sampler. The ion beam also has an energy spread of a few electron volts. Thus, a quadrupole bias voltage of $+3$ to $+15$ V may be necessary to stop the ions to determine their energy (Figure 2.14). Also, ions of different mass have different kinetic energies and thus have different paths through the lens.

The juxtaposition of these non-ideal conditions in ICP–MS means that different ion optical conditions are required to transmit ions of different m/z and that the sensitivity for different elements is not as even across the mass range as the high ionisation efficiencies of the different elements (Figures 2.2 and 2.3) would indicate. Furthermore, the extent, and possibly even the direction, of the mass discrimination effect depend on ion lens settings and ion energy, the latter of which can be influenced by plasma potential and plasma operating conditions.

A common method of minimising these complications is illustrated by the plots of ion signal versus lens voltage shown in Figure 2.18. This figure shows

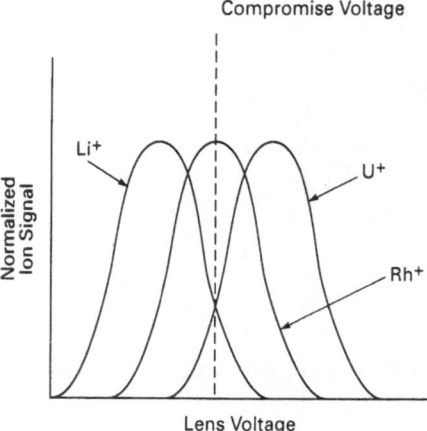

Figure 2.18 Typical plots of normalised ion signal as a function of lens voltage for ions of widely different m/z. The dotted line shows a reasonable compromise voltage for multi-element analysis.

that the voltage necessary to yield maximum signal for Li^+ differs from the optimum voltage for Rh^+ or U^+. Often, the ion lens voltages are selected to maximise the ion signal for an ion like Rh^+ in the middle of the mass range, as shown by the dotted line labelled 'compromise voltage'. Some transmission for Li^+ and U^+ is usually sacrificed compared to that obtainable if the lenses were reset to maximise signal for light or heavy ions. Thus, the uneven transmission across the mass range should be recognised but need not cause a severe loss of sensitivity when a single set of ion lens conditions are used for multi-element analysis.

It is possible to isolate the sampler and skimmer electrically from ground and apply the same voltage to both of them. In this case, the ion kinetic energies and the lens voltages necessary to transmit ions also track the offset voltage applied to the interface. This is one way to accelerate the ion beam to high kinetic energy and is adopted in the high resolution ICP–MS devices that use magnetic sector mass analysers (section 2.5).

2.3.3 Space charge effects

As mentioned in section 2.2.3, few ions are lost to recombination during the extraction process. Thus, the ion current through the sampler is quite high ($\sim 0.1\,A$), and the ion current through the skimmer is typically $1\,mA$. In the plasma and in the supersonic jet, this ion current is balanced by an equal electron current, so the beam acts more or less as if it were neutral (Douglas and French, 1988). However, as the beam leaves the skimmer, the electric field established by the lens collects ions and repels electrons. The electrons are no longer present to keep the ions confined in a narrow beam, the beam

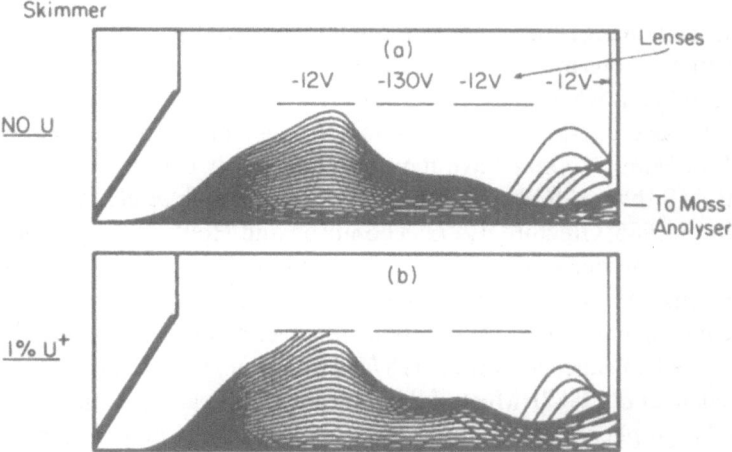

Figure 2.19 Calculated ion trajectories for Co$^+$ including effects of space charge. Only the upper half of the ion lens from Figure 2.16a is shown. Note that the beam is slightly wider and more Co$^+$ ions are lost at the entrance to the Bessel box when U$^+$ is present. Reproduced with permission (Gillson *et al.*, 1988).

suddenly is not quasi-neutral, and the ion density is still very high. The mutual repulsion of ions of like charge limits the total number of ions that can be compressed into a beam of a given size. Space charge effects should become substantial in ICP–MS at total beam currents of the order of 1 μA (Olivares and Houk, 1985b; Gillson *et al.*, 1988), roughly three orders of magnitude below the actual beam current cited above. The simple treatment of ion lenses described by the Laplace equation assumes that the ions do not interact while in the lens, so the high ion current causes space-charge effects that are further reasons for non-ideal behaviour in ion optics in ICP–MS.

As shown by the ion trajectory calculations of Gillson *et al.* (1988) depicted in Figure 2.19, the ion beam expands greatly due to space-charge effects. This expansion makes it difficult to collect all the ions leaving the skimmer and is probably a major source of ion loss in ICP–MS. Furthermore, if the same space-charge force acts on all the ions, the light ions are affected the most and are deflected most extensively. A greater fraction of light ions is deflected outside the acceptance volume of the lens than is the case for heavy ions, which could contribute to the generally poorer sensitivity for light elements and for the need for different focusing voltages for light and heavy ions (e.g. Figure 2.18).

Finally, the transmission of the ion lens now depends on the total beam current and the mass of the ions comprising the beam. Figure 2.19 also shows the ion trajectories for Co$^+$ in a beam that is mostly Ar$^+$ and O$^+$ with only 1% U$^+$ added. With the U$^+$ present, more Co$^+$ trajectories hit the wall at the right than when the beam is only Ar$^+$ and O$^+$. Thus, even a small change in the total ion current caused by addition of just a modest amount of matrix element can change the fraction of analyte ions that gets through the lens. Heavy matrix ions are themselves deflected less and stay closer to the centre of the ion beam where they can do the most damage.

These space charge effects are a major cause of matrix interferences in ICP–MS. Many workers have reported that matrix effects are more severe in ICP–MS than in ICP emission spectrometry (Houk *et al.*, 1980; Olivares and Houk, 1986; Gregoire, 1987a; Thompson and Houk, 1987; Vickers *et al.*, 1989). Also, the mass of both interferent and analyte are important. Heavy matrix ions suppress analyte signals more extensively than light matrices, and heavy analyte ions are suppressed less severely than light ones (Tan and Horlick, 1987; Beauchemin *et al.*, 1987a; Crain *et al.*, 1988; Gillson *et al.*, 1988; Kawaguchi *et al.*, 1987). Most of these observations can be explained via the space-charge phenomenon. Alleviation of these effects and their analytical symptoms would greatly improve the analytical capabilities of ICP–MS. Caruso and co-workers have described an interesting and potentially valuable scheme in which the ion lens voltages are adjusted to maximise the analyte signal with the sample matrix present (Wang *et al.*, 1990; Sheppard *et al.*, 1991). This procedure can reduce the extent of the matrix effect significantly.

2.4 Quadrupole mass spectrometers

The ion lens provides little or no m/z separation of the extracted ion beam. In most ICP–MS instruments, this function is performed by a quadrupole mass analyser. This section describes the operating principles and pertinent properties of these mass analysers. The discussion draws largely from a paper by Miller and Denton (1986). The detailed treatises by Dawson (1976, 1986) are highly recommended for those interested in the authoritative picture on quadrupoles.

2.4.1 *Quadrupole configuration*

A diagram of a typical quadrupole mass filter is shown in Figure 2.20. Four straight metal rods or metallised surfaces are suspended parallel to and equidistant from the axis. Ideally, the surfaces of these rods have a hyperbolic shape, although round rods that approximate hyperbola are usually used instead. The rods are manufactured and mounted to very high dimensional tolerances (10 μm or less). Opposite pairs are connected together. DC and RF voltages of amplitude U and V, respectively, are applied to each pair. As shown in Figure 2.20, the DC voltage is positive for one pair and negative for the other pair. The RF voltages on each pair have the same amplitude but are opposite in sign, i.e. they are 180° out of phase.

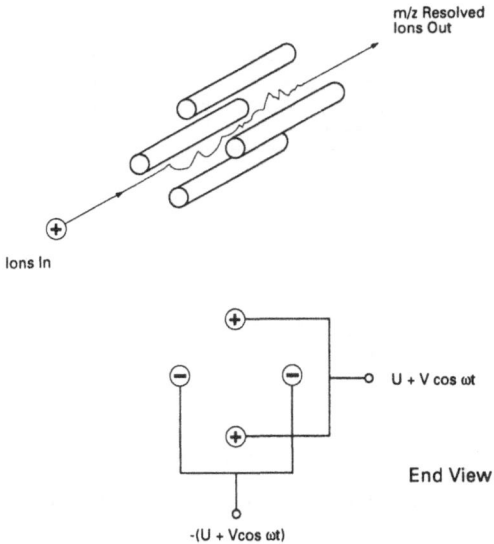

Figure 2.20   Schematic diagram of quadrupole rods showing ion trajectory and applied voltages.

The ions to be separated are introduced along the axis into one end of the quadrupole structure at velocities determined by their energy and mass. The applied RF voltages deflect all the ions into oscillatory paths through the rods. If the RF and DC voltages are selected properly, only ions of a given m/z ratio will have stable paths through the rods and will emerge from the other end. Other ions will be deflected too much and will strike the rods and be neutralised and lost there. Thus, the dimensions of the ion trajectories relative to the boundaries of the rods are critical.

2.4.2   Ion trajectories and stability diagram

For simplicity, the ion paths are considered separately in each of the two rod planes, as depicted in Figure 2.21. In the positive rod plane (Figure 2.21a), the lighter ions tend to be deflected too much and strike the rods, while the ions of interest and the heavier ions have stable paths. In this plane, the quadrupole acts like a high pass mass filter. In the negative rod plane (Figure 2.21b), the heavier ions tend to be lost preferentially, and the ions of interest and the lighter ions have stable paths. Thus, the quadrupole acts like a low pass mass filter in the positive plane. Of course, the positive and negative planes are superimposed physically, so these two filtering actions occur on the same ion beam at the same time. This juxtaposition of high pass and low pass mass filtering produces a structure that transmits ions only at the m/z value of interest. The quadrupole has a ready analogy in light optics, namely the use of two cut-off filters with different transmission characteristics to isolate a narrow spectral window.

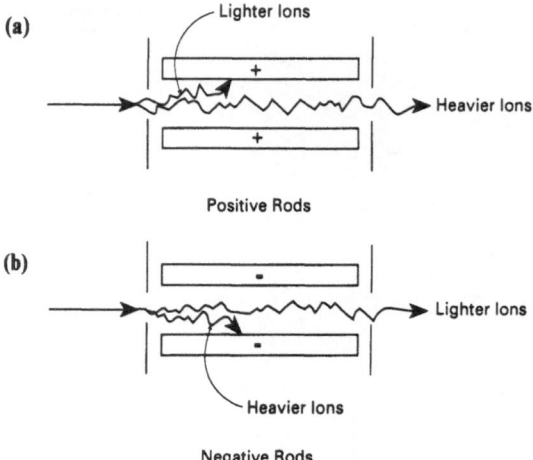

(a)

Lighter Ions

Heavier Ions

Positive Rods

(b)

Lighter Ions

Heavier Ions

Negative Rods

Figure 2.21 Side views of the ion separation processes in the two rod planes of a quadrupole.

The ion trajectories and transmission properties of a quadrupole can be calculated with reasonable accuracy (Dawson, 1976). Although such calculations are outside the scope of this book, these results can be used to provide a convenient and informative picture of the function of a quadrupole via the stability diagram, the most important part of which is shown in Figure 2.22a. The stability diagram is essentially a plot of the parameter $a \sim U/(m/z)$ as a function of $q \sim V/(m/z)$. Only the upper right quadrant of the entire stability diagram is shown. The ion trajectories are determined by $U, V, m/z$, and other fixed parameters such as the inscribed radius between rods (r_o) and RF frequency (ω). For most values of (a, q), the RF and DC fields displace the ion outside the rod boundaries, where it is lost. Such trajectories are termed unstable ones, because ions on these trajectories have unstable paths. Ions with stable paths are those that stay within the rods; these ions have (a, q) values that fall within the pyramidal enclosure of the stability diagram.

To provide the m/z separation, the scan line shown in Figure 2.22b is superimposed upon the stability diagram. At particular values of U and V, a given point on the scan line corresponds to one particular value of m/z. Thus, as the power supply changes U and V during the scan, a and q change as well, which corresponds to moving from one m/z to another along the scan line. With the scan line present the only ions that have stable paths are those that fall on the scan line and inside the sharply pointed tip of the stability diagram. As shown in Figure 2.22b, an ion with $m/z = M$ will have a stable path through the rods if a (i.e. the DC voltage) and q (i.e. the RF voltage) are selected so that the corresponding value (a, q) on the scan line stays under the tip of the stability diagram. At the same values of U and V, neighbouring ions at $M + 1$ and $M - 1$ have different (a, q) values that are

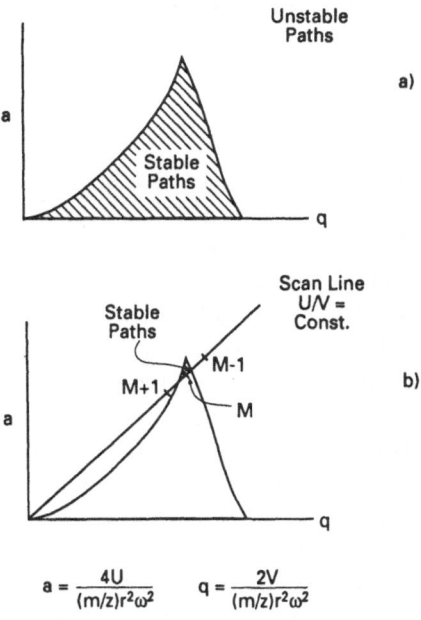

$$a = \frac{4U}{(m/z)r^2\omega^2} \qquad q = \frac{2V}{(m/z)r^2\omega^2}$$

Figure 2.22 Stability diagram for ion transmission through quadrupole. For fixed values of U and V, the (a, q) values for ions of different $m/z = M, M - 1$ and $M + 1$ are indicated along the scan line. Adapted from Miller and Denton (1986) with permission.

outside the stable region, so these ions are lost. Generally, as the m/z transmitted by a quadrupole is scanned, the voltages U and V are changed while the ratio U/V is kept constant. As U and V are changed, the operating points for ions of different m/z will move into the stable region, and a mass spectrum can be obtained by varying U and V.

Note that the degree of separation of M from $M - 1$ or $M + 1$ depends on the slope of the scan line. If the ratio U/V is increased, the scan line has a higher slope and passes closer to the tip of the stability diagram, the band of m/z values transmitted through the mass analyser is narrower, and adjacent ions are separated more extensively. Thus, increasing the ratio U/V increases the resolution. Unlike optical instruments, both the resolution ($\sim U/V$) and the m/z transmitted ($\sim$ magnitudes of U and V) are controlled electronically via the voltages applied to the rods. There are no moving parts, and the resolution and transmission are not very sensitive to the size of the entrance and exit apertures. However, the number of ions transmitted depends roughly on the area of the stability diagram enclosed. As resolution is increased (i.e. as the U/V is increased), the area at the tip of the stability diagram decreases, and the fraction of ions at the m/z value of interest that will get through the quadrupole also decreases. The usual compromise between resolution and transmission or sensitivity also holds for quadrupole mass spectrometers. Other regions of the stability diagram (not shown) or other scanning methods can be employed (Dawson, 1986), but the above discussion describes the method used in the vast majority of quadrupoles.

2.4.3   *Characteristics of mass spectra from quadrupoles*

The m/z scale is linear with the values of U and V. This feature simplifies computer control of the device. However, the peak shape is not precisely the same across the m/z range, and calibration or verification of the peak positions with a solution containing ions at regular mass intervals is advisable.

Typical peak shapes at different resolution settings are shown in Figure 2.23. The peaks are generally not quite symmetrical and are usually a little broader on the low m/z side. At low resolution, the peaks are gently curved at the top. They are usually not flat-topped. As resolution is progressively increased, the peak becomes sharper and the sides of the peak get closer together until the ion signal no longer reaches its maximum, as shown by the peak labelled high resolution in Figure 2.23. If resolution is increased further, the peak maximum declines further, which illustrates the compromise between resolution and sensitivity mentioned above. For ICP–MS, this means that calibration curves should be measured anew if resolution is changed.

The concepts of resolution and transmission have ready analogues in optical spectroscopy. The abundance sensitivity is an additional important figure of merit, as also shown in Figure 2.23. Suppose a peak is present only at $m/z = M$, and there are no peaks at $M - 1$ or $M + 1$. The abundance sensitivity is the ratio of the net signal at M to that at $M - 1$ or at $M + 1$. Thus, abundance sensitivity is a measure of the ability to measure a weak peak from a trace constituent at a m/z value adjacent to a major peak. A large numerical value for abundance sensitivity is desirable. The quadrupoles used for ICP–MS can achieve abundance sensitivities of 10^6 on the low mass side

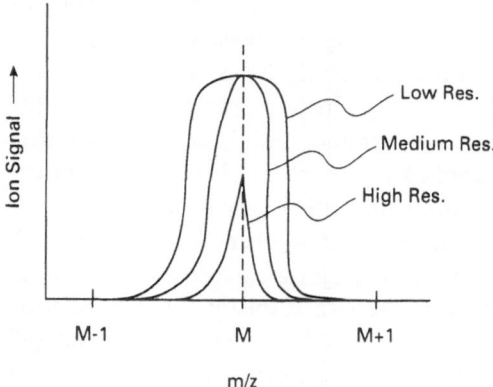

Figure 2.23   Typical peak shapes from a quadrupole mass analyser at different resolution settings.

and 10^8 on the high mass side, although values of 10^4 and 10^6 are more typical in routine use.

Several other important characteristics of quadrupoles should be mentioned. They can operate at fairly high pressures ($\sim 1 \times 10^{-4}$ torr), as long as the mean free path (see section 2.8) is greater than the length of the device and the pressure is low enough to prevent electrical discharges inside the rods. Since the ICP operates at atmospheric pressure, the gas load is high, and the tolerance of the quadrupole to relatively high operating pressures simplifies the vacuum requirements. The typical operating pressures in ICP–MS instruments are limited largely by the electron multiplier rather than by the mass analyser, as indicated by the recent production of the PQe by VG Elemental which uses only a Faraday cup detector and thus requires only two vacuum stages.

Quadrupoles perform best with ions having low kinetic energies. The resolution is not greatly sensitive to ion energy spread in the axial direction if the maximum ion energy stays below a certain value, usually roughly 20 eV. If the ion energy is too high, the ions pass through the quadrupole too quickly and do not experience enough RF cycles for proper resolution. The peaks resulting from such high-energy ions tend to be broad and/or split. To adjust the axial ion energy and ion transit time, an additional DC potential (often called 'pole bias') can be supplied to all four rods to either accelerate slow ions or to retard fast ones. This DC bias is of the same magnitude and polarity to all four rods. It merely offsets the mean potential inside the rod set and should not be confused with the other DC voltage (U in Figure 2.20) that is of different polarity on either pair of rods and that contributes to the mass filtering action. A spread of velocities in the radial direction does cause deterioration of peak shape and transmission. Use of RF-only quadrupole rods (section 2.4.4) at the entrance to the mass analyser is one way to avoid this problem.

Finally, only ions of a given m/z value are transmitted at any given time. The m/z value can be scanned or switched very rapidly, but ions of different m/z cannot be monitored simultaneously with a conventional quadrupole. Thus, a quadrupole ICP–MS instrument is analogous to a very fast and versatile sequential scanning ICP emission instrument. Quadrupole ICP–MS instruments do not perform simultaneous multi-element analysis, regardless of how fast or how long they are scanned, despite occasional claims to the contrary.

2.4.4 RF-only quadrupole

Quadrupoles used in ICP–MS often have short rods on the entrance and exit to the mass analyser (Figure 2.24). These rods are the same radius as the main rod set in the centre and have only RF and DC offset (i.e. pole bias) potentials applied. The value of U is zero, so the 'scan line' (a misnomer in

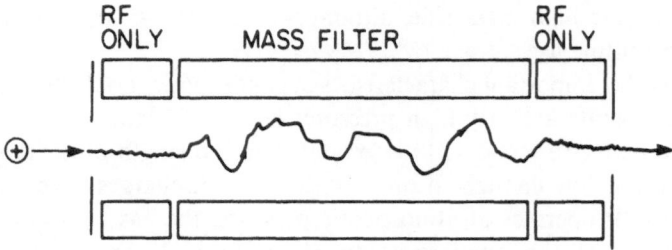

Figure 2.24 Side view of quadrupole showing function of RF-only rods.

this case) lies along the q axis in the stability diagram (Figure 2.22). Pivoting the scan line down to the horizontal axis is tantamount to operating at very low resolution. A wide range of m/z values now fall within the stable region of the stability diagram. Physically, the ions still oscillate through the RF-only quad, but with much smaller deflections than when they pass through a mass filter (Figure 2.24). It is often said erroneously that an RF-only quadrupole transmits ions of all m/z values, but in fact only ions of m/z above the value given by the intersection of the right leg of the boundary line with the axis are transmitted. Thus, an RF-only quadrupole acts as a high pass mass filter with very low resolution, and if used alone, it would interest us little.

When combined with a mass filter, as shown in Figure 2.24, RF-only quadrupoles substantially improve the performance of the latter. The small amplitude ion oscillations mean that an RF-only quadrupole actually can collimate divergent ions closer to the central axis. Thus, ions injected off-axis are focused closer to the axis in the RF-only quadrupole. The transmission efficiency can be better than if only the mass filter is used, particularly if the source and ion optical conditions are such that the ion beam is divergent if the RF-only rods are omitted.

The RF-only quad also alleviates the following problem. When an ion is injected into a mass filter, it passes through a region in front of the rods where the potentials are lower than those inside the rod structure. This region of intermediate potential is called a fringing field, and it generally causes problems in mass spectrometry. Another fringe field region exists between the downstream ends of the rods and the exit aperture. Such fringe field regions exist in Figure 2.21 between the rods and the entrance and exit apertures. In these regions, the potential would be somewhere between the ideal quadrupole potential (determined by U and V) and the DC potentials applied to each aperture.

First, consider the effect of the fringe field at the entrance to the quadrupole. If a and q are selected so that the ion has a stable path when inside the rods, the actual values of a and q in the fringing field are too low, and the ion briefly experiences (a, q) values that are outside the stability region. These ions in the fringe field are deflected before they get inside the rods and some may be lost. In addition, the extent of loss may differ for ions of different

m/z, thus causing mass discrimination. With the RF-only quadrupole at the entrance (Figure 2.24), the ions pass smoothly into the central mass filter with high efficiency. A fringe field still exists at the entrance to the RF-only quad but its defocusing effect is less severe because even a low value of (a, q) still remains within the stability region when the scan line is down on the horizontal axis. A more elegant expression is to say that the RF-only quad matches the emittance of the previous ion optics to the acceptance of the mass filter. An RF-only quad after the mass filter prevents ions from spraying out of the quad over a large volume and makes it easier to channel them through an exit aperture to a detector. On some instruments, a separate pole bias is provided for the RF-only rods to provide additional focusing action.

2.4.5 Scanning and data acquisition

As described in section 2.4.2, the m/z transmitted by a quadrupole mass analyser is determined by constant factors such as the dimensions of the rods, the frequency of the RF voltage and by the variables U and V. The power supply that regulates U and V can itself be controlled externally, e.g. by a DC voltage ramp from a computer-controlled digital-to-analog converter or function generator. Several data acquisition modes are possible.

If U and V are not changed, the mass filter transmits only one m/z value continuously. This mode is called selected ion monitoring or single ion monitoring. It provides a 100% duty cycle on the m/z value of interest but precludes acquisition of spectral information elsewhere. For multi-element measurements, the values of U and V can be changed continuously in one scan, as shown in Figure 2.25a. Alternatively, U and V can be changed under computer control rapidly between selected discrete values (peak hopping, Figure 2.25b), or the quadrupole can be scanned repetitively through the m/z region of interest (multi-channel scanning, Figure 2.25c). The latter two methods have the advantage that the hopping or scanning can be done repetitively and rapidly to average out some of the fluctuations in ion signal, thereby improving precision. The flight time of ions through the quadrupole ($\sim 20\,\mu s$) is the ultimate lower limit to the dwell time at each m/z position. In practice, the power supplies that drive the quadrupole need $50\,\mu s$ to 1 ms to settle when U and V are changed.

There have been numerous arguments as to the relative advantages of peak hopping versus multi-channel scanning. Peak hopping spends a greater fraction of the time on the selected peaks and is usually superior for repetitive determination of the same suite of elements. However, important peaks that are not included in the computer program may be neglected by peak hopping procedures. For example, unanticipated spectral interferences from polyatomic ions are easily overlooked in peak hopping methods. With multi-channel scanning, all peaks in the selected m/z region are recorded, so this method is often preferred for providing a full record of the elemental composition of

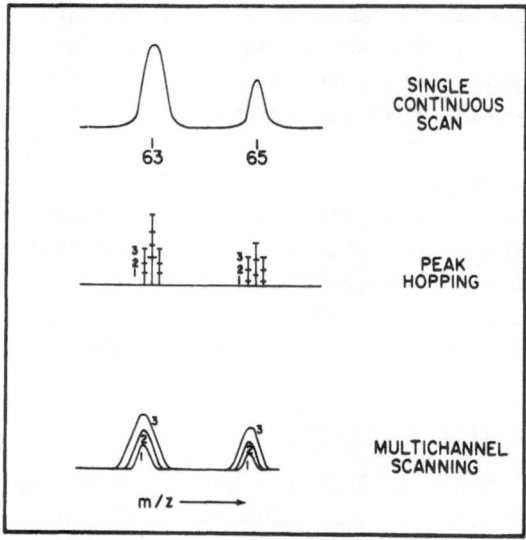

Figure 2.25 Scanning and data acquisition modes. Three successive cycles (indicated by 1, 2, 3) are shown for peak hopping and multi-channel scanning. In peak hopping, each m/z position is typically monitored for ~ 0.1 s. In multi-channel scanning, each scan takes typically 0.01 s per m/z unit in the window observed. Reproduced with permission (Houk, 1990).

the sample. In the authors' view, these methods should be viewed as complementary rather than competitive, and the ideal data acquisition system from ICP–MS would do a good job with either.

2.5 Other mass spectrometers

Quadrupole mass spectrometers are compact and convenient to use with a high pressure source such as an ICP, but the resolution is insufficient to separate chemically different ions at the same nominal m/z value. Polyatomic ions are not overly abundant in the mass spectra observed from the ICP but can in some cases cause serious overlap interferences with analyte ions. Some examples are given in Table 2.2. Note that a spectral resolution of $\sim 10\,000$ would alleviate most polyatomic ion interferences, but much higher resolution would be required to separate different atomic ions at the same nominal m/z. Naturally, the more intense the interferent is relative to the analyte peak, the higher the resolution required to separate the two.

Two groups have published initial reports of a different approach to mass analysis in ICP–MS with fairly high resolution for this purpose (Bradshaw *et al.*, 1989; Morita *et al.*, 1989). A double-focusing magnetic sector mass spectrometer is used, as shown in Figure 2.26. The ICP is not altered, but the sampling interface (both sampler and skimmer) is isolated electrically from the vacuum chamber (to prevent spurious discharges) and biased at a

Table 2.2 Resolution required to separate interferences

Analyte ion	Interfering ion	Resolution[a]
$^{28}Si^+$	$^{14}N_2^+$	960
$^{32}S^+$	$^{16}O_2^+$	1800
$^{56}Fe^+$	$^{40}Ar^{16}O^+$	2500
$^{80}Se^+$	$^{40}Ar_2^+$	9700
$^{159}Tb^{+\ b}$	$^{143}Nd^{16}O^+$	7800
$^{165}Ho^{+\ b}$	$^{149}Sm^{16}O^+$	8900
$^{58}Ni^+$	$^{58}Fe^+$	28000
$^{148}Nd^+$	$^{148}Sm^+$	77000
$^{87}Rb^+$	$^{87}Sr^+$	300000

[a] Resolution $= M/\Delta M$. The numerical values reflect the resolution required to separate peaks of similar intensity.
[b] Mono-isotopic elements.

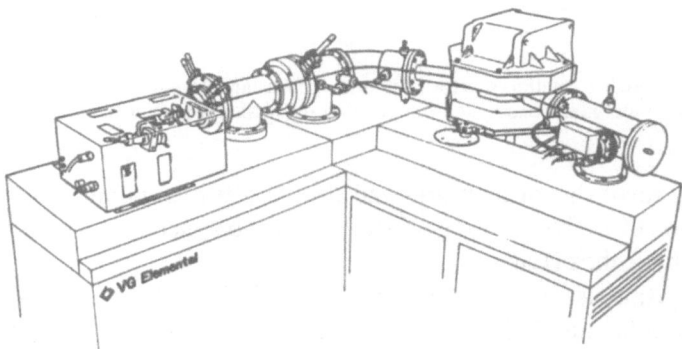

Figure 2.26 Double focusing ICP–MS device from VG. Reproduced with permission from VG Elemental.

high positive potential. This bias provides the necessary ion accelerating voltage (typically $+4\,kV$ to $+8\,kV$) for formation of a proper ion beam for the subsequent analysers. The ions from the ICP pass through the interface despite the high positive voltage on the latter, which indicates that the plasma potential also follows the bias voltage applied to the interface.

After the interface, lenses are used to change the shape of the beam from one with a circular cross-section to one with a slit-like profile that can be imaged on the slit system for efficient ion transmission. The beam then passes through an electrostatic energy analyser that selects only ions within a band of kinetic energies for transmission through the exit slit. These energy-resolved ions are then m/z analysed in a magnetic sector and detected. Both direction and velocity focusing are provided. Each stage (sampler, skimmer and ion lens, electrostatic analyser, magnetic analyser and detector) is separated by

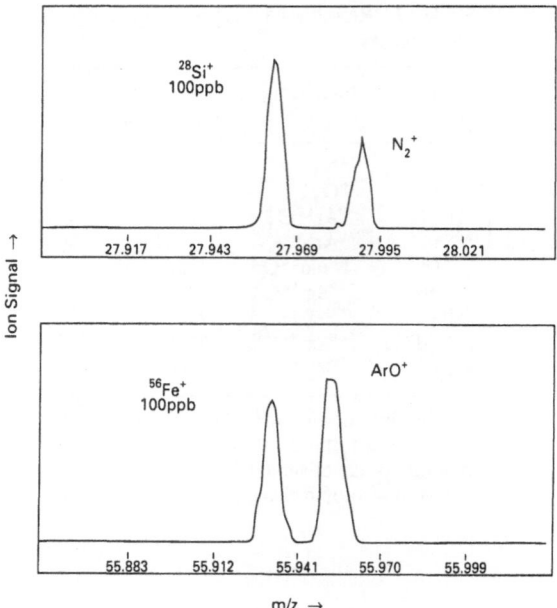

Figure 2.27 High resolution spectra from double focusing instrument illustrating separation of polyatomic ion interferences. Reproduced with permission from VG Elemental.

relatively small slits or apertures and thus has its own pump. A good basic description of double-focusing principles and of magnetic mass spectrometers is given by Roboz (1968).

Some examples of spectra obtained with this device are shown in Figure 2.27. Si^+ and Ni_2^+ and Fe^+ and ArO^+, yield separate triangular peaks, as do many other troublesome interfering pairs. In general, different atomic ions at the same nominal m/z (e.g. $^{87}Sr^+$ and $^{87}Rb^+$) cannot be separated by this instrument, as the resolution required is $\gtrsim 30\,000$ (Table 2.2). The device can also be operated in a low resolution mode (i.e. with the slits opened wide) to yield the flat-topped peaks shown in Figure 2.28, which may prove advantageous for isotope ratio measurements. The background is very low ($\lesssim 1$ count s^{-1}); no photon stop is necessary because of the bent flight path and the narrow slits. Because the beam is focused efficiently, the ion count rates are comparable to or higher than those obtained with quadrupoles, so detection limits can be better with the double focusing instrument.

As yet, only a few of these instruments have been delivered, so it is premature to elaborate on their merits relative to quadrupole devices, but the better resolution and detection limits with similar sensitivity are very impressive. Because of the high accelerating voltage, space-charge effects may possibly be less severe with the magnetic instruments, although extensive studies of matrix effects have not been reported. The scan speed is slower with the high resolution device, and it is naturally more cumbersome and expensive than

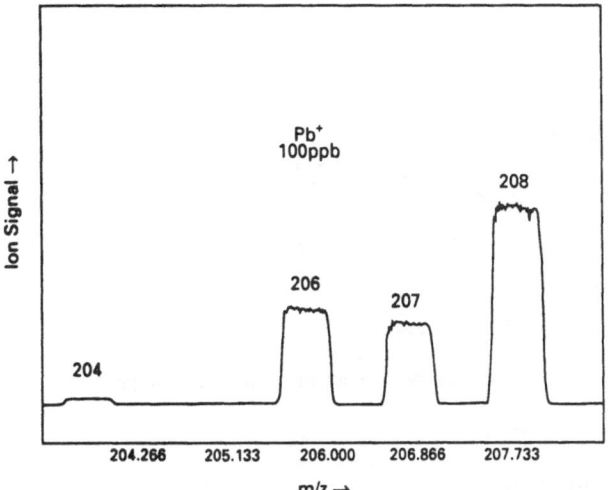

Figure 2.28 Low resolution spectra from double focusing instrument showing flat-topped peaks. The vertical scale is less sensitive (i.e. the peak count rates are greater) than for Figure 2.27.

a quadrupole instrument, but these are the only disadvantages apparent at this time.

There are many types of mass analysers, and most could be adapted for use with the ICP. Perhaps the most likely candidates are the ion trap for a possible 'benchtop' device, the ion cyclotron resonance-Fourier transform instrument for very high resolution, or time-of-flight devices for moderately high resolution without high cost and complexity. These three devices are usually used with pulsed, low pressure ion sources, but methods are being investigated for their use with continuous, high-pressure sources similar to the ICP. Since most workers in ICP–MS have been converts from ICP emission spectrometry, there has not been much work on alternate mass analysers for ICP–MS as yet. As in GC–MS, it is not likely that any other analyser will displace the quadrupole in the great majority of instruments.

2.6 Ion detection

2.6.1 *Channeltron electron multipliers*

These are the most common ion detectors used in ICP–MS instruments. Several good references are recommended for more detailed information (Kurz, 1979; Timothy and Bybee, 1978).

Figure 2.29 shows a diagram of a Channeltron electron multiplier. The operating principles are similar to those of a photomultiplier, except that there are no discrete dynodes. Instead, an open glass tube with a cone at one end is used. The interior of the tube and cone are coated with a lead oxide

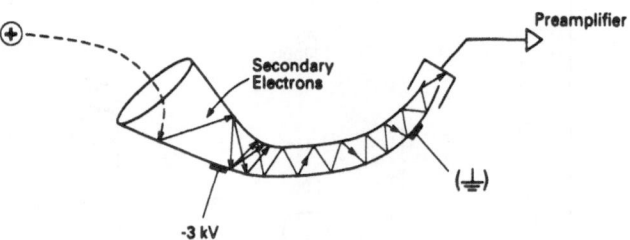

Figure 2.29 Channeltron electron multiplier.

semiconducting material whose exact composition is proprietary. Electrical connections are made to the semiconducting coating through metal strips. For detection of positive ions, the cone is biased at a high negative potential ($\sim -3\,\mathrm{kV}$) and the back of the tube near the collector is held near ground. Relative to either end, the resistance of the interior coating varies continuously with position. Thus, when a voltage is applied across the tube, a continuous gradient of potential exists with position inside the tube.

Suppose a positive ion leaves the mass analyser and is attracted to the high negative potential at the cone. When the ion hits this surface, one or more secondary electrons are ejected. Inside the tube, the potential varies continuously with position, so the secondary electron(s) move further into the tube to regions closer to ground. These secondary electrons hit another section of the coating and more secondary electrons are emitted. This process is repeated many times as the secondary electrons pass down the tube. The result is a discrete pulse containing as many as 10^8 electrons at the collector after an ion strikes the mouth of the detector.

Some pertinent properties of Channeltrons are as follows. They can be vented to air repeatedly without damage provided, of course, that the high voltage is switched off first. This feature makes them convenient for systems that have to be dismantled frequently for cleaning or other maintenance. They require a pressure below approximately 5×10^{-5} torr during operation or spurious discharges form in the detector chamber. A brand new multiplier often has a very high and erratic gain at first, which usually degrades and stabilises gradually with time. A Channeltron has a limited lifetime that is determined by the total accumulated charge, i.e. (input ions) × (gain). After this lifetime is exceeded, the interior surfaces are exhausted and the multiplier should be replaced. The lifetime of an ageing multiplier can be extended somewhat by applying a more negative voltage to boost the gain, but the need to raise the applied voltage is usually a sign that the end is near for a particular Channeltron.

2.6.2 Signal measurement by pulse counting

In the usual mode of detection, the applied voltage is high enough to saturate the gain of the detector. An ion yields one pulse of $\sim 10^8$ electrons at the collector. This pulse is sensed and shaped by a fast pre-amplifier. The output pulse from the pre-amplifier goes to a digital discriminator and counting circuit, which counts only pulses with amplitudes above a certain threshold level. The threshold level is chosen high enough to discriminate against low-amplitude pulses caused by spurious emission of electrons from inside the tube. The dark current count rate can be very low with a Channeltron, typically 1 count s^{-1} or less. Very weak ion signals can be sensed in this way. Photons from the plasma can also activate the detector and must be blocked, e.g. by the baffle in the centre of the ion lens in Figure 2.16 or by offsetting the quadrupole from the centreline of the ion optics (Figure 2.17).

Despite the use of pulse counting and careful blocking of the optical axis between the detector and plasma, the actual background during operating is generally 10–50 counts s^{-1}, substantially higher than the dark count rate of the multiplier. The phenomena responsible for this background are not known precisely (Kawaguchi et al., 1988a). Perhaps a few photons from the plasma or the extraction interface scatter into the detector chamber despite the best efforts to exclude them. In some cases, the high sensitivity may yield very small peaks at many m/z values from impurities in the blank solution. In the authors' experience, the background tends to increase with the total ion signal, which could indicate that the background is related to or caused by ions in some way. A few ions may pass directly through the very centre of the quadrupole (where ions are not deflected because they experience the same net force in all directions). Conceivably, photons may be emitted when fast-moving ions with unstable trajectories strike the rods. Some of these photons could escape from the rod housing into the detector.

The upper end of the dynamic range is limited by two phenomena. First, the multiplier can only sustain a given maximum current flow. At a gain of 10^8 and an ion count rate of 10^6 ions s^{-1}, the current flow at the collector is $16 \mu A$. At high count rates, the gain decreases because the electron current exceeds the standing current in the semiconducting coating. The fraction of ion pulses that are too low to be counted now increases. Pulse pileup, i.e. arrival of two pulses in a time interval shorter than the response time constants of the counting circuit, is a second problem that could also limit linearity, although the gain loss described above probably begins at lower count rates with the Channeltrons used in ICP–MS.

Both these effects cause the calibration curve to droop toward the horizontal axis at high count rates. A peak that is split in the centre is another symptom of count rates that exceed the linear range. Early ICP–MS investigators often remarked that the technique was actually too sensitive to measure minor sample constituents. There are several ways to extend the linear range. One is to attenuate the ion beam by a known and stable fraction. On the Perkin-Elmer Sciex instrument, the ion beam is defocused by altering the voltage on some of the exit lenses between the rods and the detector. Another approach is to change the gain of the detector by reducing the applied voltage. Standard analog methods are then used to measure the current produced. It should be noted that the actual dynamic range in each detection mode is still roughly 10^6 whether the ion beam is defocused or the gain is reduced; the improvement arises because the upper limit of this range is shifted to higher ion arrival rates.

2.6.3 Other detectors

Several other detectors have been tried with ICP–MS. These include discrete dynode electron multipliers (Russ, 1989), the Daly detector (Huang et al., 1987), and the Coniphot detector in the Nermag instrument. Modest improvements in linear range and precision have been reported with these alternate detectors. For some applications, the sensitivity is high enough that a simple Faraday cup (i.e. a metal electrode with no gain mechanism) is sufficient. Detection with a Faraday cup can also be considered another way to extend the upper end of the dynamic range. The time constant of the DC amplifier used to measure current limits use of the Faraday cup to relatively low scan rates.

2.7 Vacuum considerations

The ICP is at atmospheric pressure, whereas mass spectrometry requires movement of ions without collisions. Therefore, the mass spectrometer is contained within a vacuum system, and a working knowledge of basic vacuum principles is essential. Space does not permit a full discussion, and the interested reader is referred to several other references (Dushman and Lafferty, 1962; Chambers et al., 1989; Leybold-Heraeus, 1990).

2.7.1 Properties and flow of gases

The concept of mean free path is central to vacuum considerations. Mean free path (λ) is essentially the average distance a gas atom travels between collisions with another gas atom. Thus, λ varies with collision cross-section for the gas of interest and with temperature and gas density. A useful formula for air at 25°C is

$$\lambda(\text{cm}) = 5/P(\text{mtorr}) = 0.0066/P(\text{mbar}) \qquad (2.2)$$

where P denotes pressure and $1\,\text{atm} = 760\,\text{torr} = 10^3\,\text{mbar} = 101\,000\,\text{Pa}$.

The concept of flow is also important. An accurate way to visualise the function of a pump is to consider it to be a one-way hole connected to the vacuum system. Ideally, gas can flow from the vacuum system into the hole but cannot get back out into the vacuum system. The pump is actually just a gas trap, and gas must be able to flow to the pump for it to function. Thus, the ability of a pump to evacuate a chamber is related to the flow characteristics of gas, which will in turn depend on the gas pressure and dimensions of the chamber.

Three flow regimes can be readily identified. At high pressures, the flow is turbulent, which we shall not consider. At pressures of a few millibars, viscous or continuum flow prevails. Here, the mean free path is much less than the diameter and length of the vacuum chamber and connecting pipes. Thus, collisions between gas atoms are much more frequent than collisions of atoms with the walls. Atoms leave the chamber by being entrained in a stream of gas flowing into the pump. In an ICP–MS system, the gas flow through the lines connecting the mechanical pump to the interface is more or less in the viscous flow regime. Electrical discharges between high voltage components and ground occur readily at these pressures.

The third flow regime prevails in the high vacuum stages of an ICP–MS system. At pressures of 10^{-4} mbar or less, the gas density is low, and the mean free path is generally much greater than the dimensions of the vacuum system. In this case, collisions between gas atoms are much less frequent than collisions of atoms with the walls of the chamber. Atoms bounce off the chamber walls and thus diffuse out of the chamber into the pump. This is the molecular flow regime prevalent in the ion lens and mass analyser chambers.

The ability of a pump to evacuate a chamber to a given pressure depends on the pumping speed and the size of the pipe connecting the pump to the chamber. Gas flows more readily through a short, fat connecting pipe than through a long, thin one. For example, in the molecular flow regime, the conductance of a pipe for air at 25°C is given by

$$C(l/s) = 12D^3/L \tag{2.3}$$

where D = diameter (cm) and L = length (cm).

In the viscous flow regime, the conductance of a pipe is given by a more complicated expression (Chambers *et al.*, 1989; Leybold-Heraeus, 1990)

$$C = (aD^4 \bar{P})/L + b \tag{2.4}$$

where $\bar{P}$ = average pressure and a and b are constants for a given gas and temperature. Note that, in the viscous flow regime, the conductance of a pipe is dependent on pressure, whereas the conductance of a pipe in the molecular flow regime is not dependent on pressure.

If a pump with pumping speed S_{pump} is connected to a chamber by a line of conductance C_{line} the net pumping speed in the chamber is S_{net}:

$$1/S_{net} = 1/S_{pump} + 1/C_{line} \tag{2.5}$$

Ideally, the chamber and connecting line are designed so that the conductance of the line is larger than the speed of the pump. In this case, the conductance of the line does not greatly restrict the net pumping speed of the apparatus. For the same reason, a 'large' diffusion pump with high pumping speed is one with a relatively large inlet throat. Conversely, there is little advantage to having C_{line} enormously greater than S_{pump}. Some compromise between the pumping speed and the overall size and cost of the vacuum system and pumping components is usually necessary.

2.7.2 A vacuum system for ICP–MS

A general diagram for a typical three-stage vacuum system for ICP–MS is shown in Figure 2.30. Some general points of interest are described below.

An enormous pump would be necessary behind the sampler to reduce the pressure immediately to the $\sim 10^{-5}$ mbar value necessary for the MS and detector. The alternative is to reduce the pressure gradually in separate stages by a technique called differential pumping. Hence, the apparatus has three orifices (sampler, skimmer and differential pumping orifice), each with its own sealed chamber and pump.

A mechanical pump is used on the first stage. The bulk of the extracted gas is removed here. The pumping speed is low, so a relatively long, thin connecting line can be tolerated without great loss of pumping speed in the chamber. The interface volume must be large enough not to restrict the pumping speed. The pressure can be estimated roughly in this chamber with a thermocouple gauge. More accurate measurements require a better gauge such as a capacitance manometer.

In the second stage housing the ion lens, the pressure is $\sim 10^{-4}$ mbar. Virtually all this gas comes from the plasma through the skimmer. The total

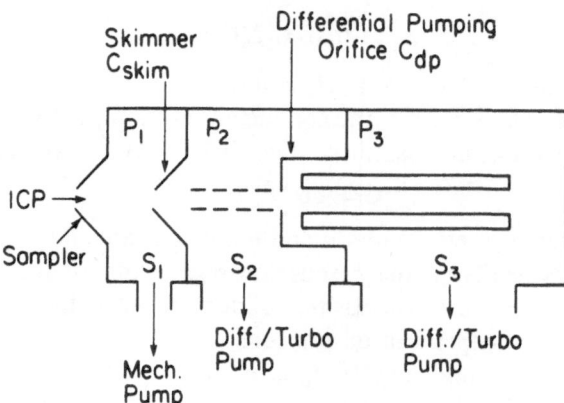

Figure 2.30 Three stage vacuum system for ICP–MS. P, S, and C denote pressure, pumping speed and conductance, respectively.

gas load and pressure can be estimated from

$$P_1 C_{skim} = P_2 S_2 \qquad (2.6)$$

where S_2 is the net pumping speed in the second chamber (i.e. $S_2 \leqslant$ the speed of the pump on the second chamber). The other parameters are identified in Figure 2.30. A pumping speed of $500\text{–}900 \, \text{l} \, \text{s}^{-1}$ is necessary, hence this is a large chamber with a sizeable diffusion pump or turbomolecular pump. The mean free path is about 25 cm, so ions can be transported roughly this distance before they are scattered by collisions with neutral gas atoms. The entrance to the next stage should therefore be closer to the skimmer than one mean free path. Since the second stage is fairly large, the wall containing the differential pumping orifice is often in a top hat that protrudes into the second stage. Ionisation gauges are generally used to monitor pressure here and in the third stage.

The third stage pressure is 4×10^{-5} mbar or less for the well-being of the quadrupole and detector. The mean free path at this pressure is roughly 5 m, which is substantially longer than the ion path. A smaller pump and connecting line are adequate here. The necessary pumping speed is given by

$$P_2 C_{dp} = P_3 S_3, \quad S_3 = C_{dp}(P_2/P_1) \qquad (2.7)$$

2.7.3   Pumps used in ICP–MS

We begin by reiterating the adage that the term 'pump' is a misnomer in that all pumps used in ICP–MS are best thought of as gas traps that serve as one-way holes. Gas flows from the chamber into the pump, where it is trapped or exhausted and thus cannot move back into the chamber. Different pumps differ primarily in the means used to generate the one-way hole.

The three types of pumps used in the present commercial ICP–MS devices are depicted in Figure 2.31. Mechanical pumps (Figure 2.31a) are used to evacuate the interface chamber and also serve as backing pumps to carry off the exhaust from the high vacuum pumps. Mechanical pumps operate by spinning a metal cam (called a rotor) inside a hollow metal cylinder (called a stator). The rotor contains two grooves that hold spring-loaded vanes that press against the inner wall of the stator. As the rotor spins, the vanes move around with it and also compress or expand the volume enclosed by them and the stator. In this fashion, gas from the chamber can flow into the pump, be compressed and then exhausted. The oil helps lubricate the vanes and seals them as they move within the stator. With mechanical pumps, proper oil level should be maintained and the oil must not be contaminated. Water in the pump oil is particularly troublesome, as it can rust the stator and cause formation of grooves that prevent proper sealing. If the power fails during operation, oil from the mechanical pump can be drawn from the oil reservoir into the chamber. The oil level should be checked and the oil

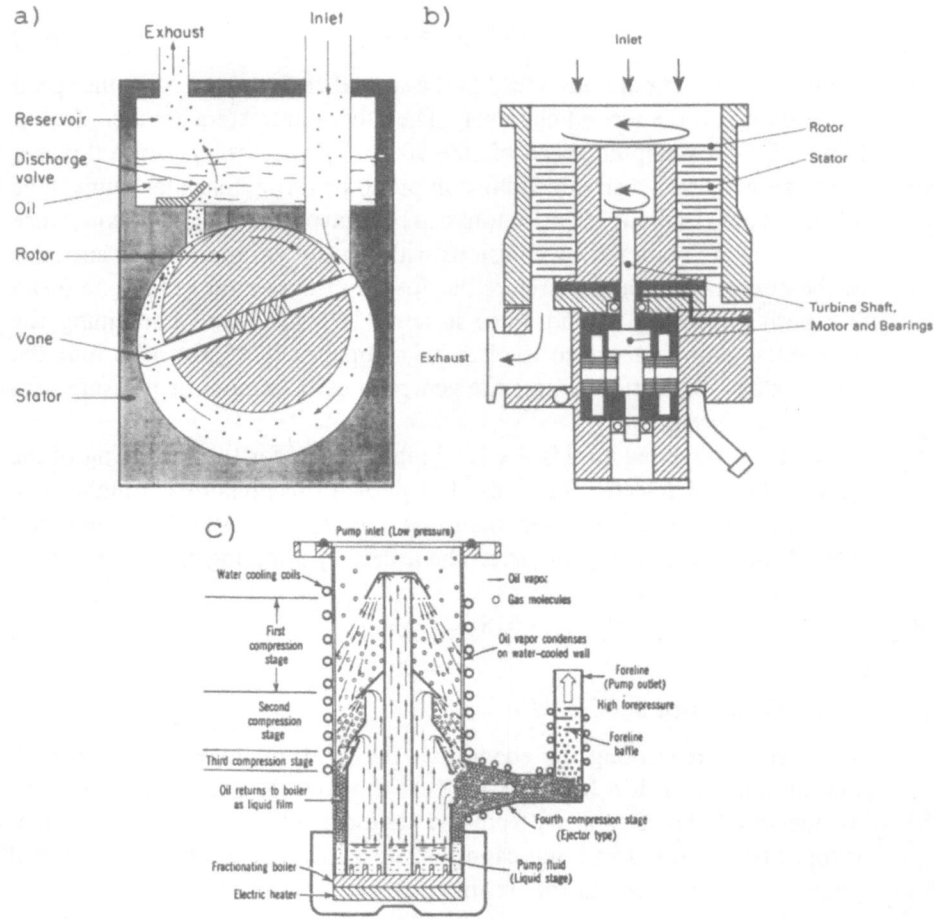

Figure 2.31 Diagrams of rotary pump (a), turbomolecular pump (b) and diffusion pump (c). (a) and (c) reproduced with permission (Roboz, 1968).

changed frequently (every few months), particularly for the pump on the interface, as the high pressure therein can cause oil to be lost gradually. A trap filled with molecular sieve or some other sorbent material can be inserted in the pumping line to prevent oil from reaching the chamber.

The minimum pressure achievable with a rotary pump is generally 10^{-4} mbar, so different methods are needed to pump the high vacuum chambers. At the present time, commercial ICP–MS instruments are moving more toward the use of turbomolecular pumps (Figure 2.31b) for this purpose. A turbo pump is a large metal cylinder with a wide inlet (so gas can flow out of the chamber easily) and a narrow outlet. A turbine is spun at high speeds ($\sim 30\,000$ rpm) inside the cylinder. The turbine blades intermesh with stationary blades mounted on the inner wall of the cylinder. All of these blades are angled so that gas atoms diffusing from the chamber into the turbine are deflected further into the canister to the exhaust, where they then flow into a mechanical pump. In this fashion, gas from the chamber flows into the pump, is compressed by the turbine and is then exhausted by the mechanical pump. Cooling and lubrication are necessary for the turbine bearings.

Diffusion pumps (Figure 2.31c) can also be used for the high vacuum chambers. Diffusion pumps have no moving mechanical parts. The pump consists of a metal tube that is water-cooled at the top wall and electrically heated at the bottom. A structure of metal jets and a viscous oil are inside the pump. The oil is volatile at the heater temperature but involatile at the wall temperature. Vaporised oil molecules pass up through the jet and are ejected at high velocity down toward the cool wall, where they condense and run back down into the boiler. Gas atoms from the vacuum vessel diffuse into the pump and collide with the high-speed jet of gaseous oil molecules. These collisions scatter the atoms from the chamber further down into the pump. Eventually, the gas atoms from the chamber flow into the exhaust port, where they are removed by a mechanical pump.

A turbo pump and diffusion pump with the same inlet diameter will have similar pumping speeds. The main distinction between turbo pumps and diffusion pumps is the mechanism used to generate the 'one-way hole' that allows gas atoms to be removed from the chamber. With diffusion pumps, some of the oil molecules from the pump can flow back surreptitiously into the chamber, where they coat and contaminate the ion lens, quadrupole rods and detector with an undesirable film of oil. This phenomenon, called backstreaming, becomes worse at higher inlet pressures. The pressure of the second stage of an ICP–MS device (i.e. $\sim 10^{-4}$ mbar) is about the maximum tolerable by a diffusion pump without severe backstreaming. Thus, diffusion pumps are usually used with cooled baffles between the chamber and the pump inlet to trap pump oil and prevent it from reaching the chamber. If the pressure is too high, oil from the mechanical pump can also get through the pump into the chamber. Oil contamination is particularly likely during a power failure because the mechanical pump shuts off immediately but the boiler on the diffusion pump takes a few minutes to cool. During this time, the oil inside the diffusion pump is still boiling but the pressure on the exhaust (often called the foreline) rises to intolerable levels (above ~ 0.5 mbar), 'pushing' oil out of the pump into the vacuum chamber.

Turbo pumps do not employ oil vapour in the pumping mechanism and are therefore much less prone to oil contamination, although oil from the mechanical pump can conceivably get into the chamber if the pressure is excessive. With either pump, care must be taken to ensure that the flow of coolant (usually water or air) is not interrupted.

The earlier ELANs use a different pumping mechanism altogether. A series of cryogenically cooled baffles surround the ion lens. Gas from the skimmer strikes the baffles and is condensed and trapped efficiently on the cold surfaces therein. This cryopumping provides very high pumping speed ($\sim 10\,000\,\mathrm{l\,s^{-1}}$) close to the gas source, so that only one high vacuum stage is necessary after the skimmer. Since there is no exhaust, the baffles gradually fill with trapped gas. Thus, it is necessary to regenerate the cryogenic surfaces periodically, typically after ~ 3 days of use, but this is easily accomplished under software control and poses little problem. One possible limitation of cryogenic pumping is that it does not pump helium effectively, and recent improvements in helium microwave plasmas (Brown *et al.*, 1988; Mohamad *et al.*, 1989) and ICPs (Montaser *et al.*, 1987; Koppenaal and Quinton, 1988) could make this capability desirable.

Note added in proof: Several recent studies of basic aspects of ICP–MS should be mentioned briefly here. Chambers *et al.* (1991a, b, c) have published a series of fundamental studies of the ion extraction process. In contrast to statements in this chapter and elsewhere in the literature, these authors consider charge separation during the sampling and skimming processes to be significant. The beam leaving the skimmer has been characterised with axial spatial resolution by Langmuir probe measurements (Niu *et al.*, 1991). The electron density behind the skimmer increases when matrix elements are added, particularly for heavy matrix elements such as uranium. These measurements may permit characterisation of the fundamental reasons for matrix interferences. Finally, Ross and Hieftje (1991) have described a way to mitigate matrix interferences by adjusting the solvent load and interface pressure.

3 Instrument options

J.G. WILLIAMS

3.1 Introduction

This chapter is mainly directed at the reader using ICP–MS systems with solution introduction. It is not intended as a detailed survey of all 'add-ons' for ICP–MS instrumentation nor as a detailed description of the theory of operation of these devices. The inductively coupled plasma source, its associated devices (nebuliser, spray chamber etc.) and the interface components can be readily modified by a user with some experience. Modification of the mass spectrometer, instrument electronics or system software is however, generally not a possibility for most ICP–MS users and for this reason will be omitted. It is advisable to consult the manufacturers before making any modifications to instrumentation, as this may invalidate the guarantee and can cause expensive damage.

The items considered will be:

(i) nebulisers, types in use, principles of operation, free running or pumped, analytical characteristics;
(ii) spray chambers, purpose of a spray chamber, types in use, temperature stabilised;
(iii) plasma torches, types in use, principles of operation;
(iv) sampling cones and skimmer cones, different designs, orifice diameters, materials of manufacture.

3.2 Nebulisers

3.2.1 *Introduction*

The sample introduction into the ICP, including nebulisation, is a critical part of the analytical process in ICP–MS. The great majority of ICP–MS analyses are carried out on liquid samples, however a gas stream is needed to get the sample into the plasma. The most convenient method for liquids to be introduced into the gas stream is as an aerosol from a nebuliser. A variety of methods can be used to produce an aerosol, such as the action of a high speed gas jet across the tip of a small orifice or the use of an ultrasonic transducer.

During the development of the nebuliser for optical spectroscopy it was found that the addition of a spray chamber (section 3.3) placed after the nebuliser could remove some of the large droplets produced, and thereby improve the stability of the spectral emission. So effective were the early crude devices that apart from manufacturing refinements, they are still used today in flame emission and atomic absorption spectrometers.

The most commonly used nebuliser designs for analytical atomic spectrometry are 'pneumatic' and 'ultrasonic' (Sharp, 1988a and references 3–15 therein), although other types such as electrostatic (Bailey, 1984), thermospray (Vestal and Ferguson, 1985), jet impact (Doherty and Hieftje, 1984) and mono-dispersive generators (Berglund and Liu, 1973; Lindblad and Schneider, 1965) have been used. Solid samples may be introduced as aerosols, using sputtering devices, and some work has been carried out using fluid beds for powder sample introduction (Ng *et al.*, 1984). The introduction of solid samples is discussed in chapter 10. A detailed description of the mechanism of operation and operating characteristics of pneumatic nebulisers for analytical atomic spectrometry has been given by Sharp (1988a).

Very few of the more novel devices for solution sample introduction have been used on ICP–MS instruments, with pneumatic nebulisers in almost universal use for routine analysis. To date there are no nebulisers which have been designed specifically for use on ICP–MS instruments and this situation is likely to remain since the liquid sample introduction requirements of ICP–MS and ICP–AES systems are essentially the same. Equally the practical problems associated with the operation of the nebulisers such as unstable operation, blockage and poor efficiency, apply equally to both ICP systems.

Essentially three types of pneumatic nebuliser are used in ICP–MS; the concentric flow (Meinhard, 1976), the cross flow (Kniseley *et al.*, 1974) and the Babington type nebuliser (Suddendorf and Boyer, 1978). These are shown schematically in Figure 3.1.

3.2.2 *Concentric nebulisers*

Probably the most widely used ICP nebuliser is the one-piece Meinhard glass concentric nebuliser (Meinhard, 1976), shown in Figure 3.2. It can be obtained as a general purpose nebuliser with limited salt tolerance, in a high salt tolerance design, and for applications that require a low nebuliser gas flow. For ICP–MS, a nebuliser gas flow of $0.75-1.00 \, 1 \, min^{-1}$ is typical, which is produced from a line pressure of 165 kPa (24 lb per square inch) at which the 'free running' uptake of water is about $0.6 \, ml \, min^{-1}$.

3.2.2.1 *Mode of operation* Most concentric nebulisers are self-actuating and self priming, i.e. solutions are drawn up by the pressure drop generated as the nebuliser gas passes through the orifice. This mode of operation is called 'free-running'. The viscosity of the solution and the vertical distance through

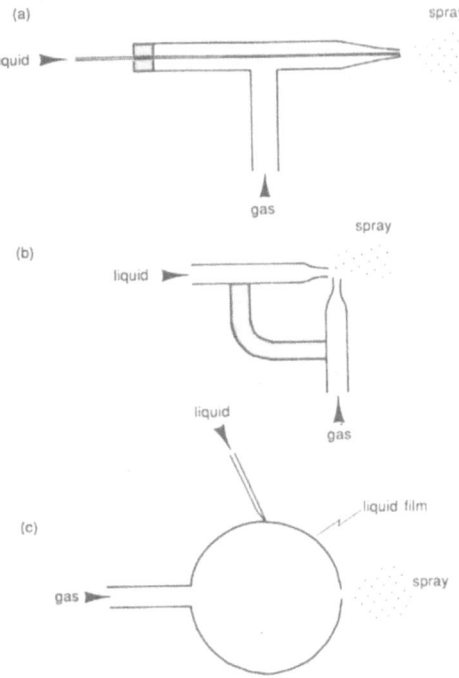

Figure 3.1 Schematic diagram of three types of pneumatic nebuliser. (a) Concentric flow, (b) cross flow, (c) Babington type. After Sharp (1988a).

which the liquid is lifted (the 'head') affect the rate of liquid transfer. The alternative and more common method is to meter the solution to the nebuliser using a peristaltic pump. This has a number of advantages which are described by Thompson and Walsh (1989).

(i) The head effect is eliminated.
(ii) Viscosity effects are reduced.
(iii) Metering uptake with a pump limits air which can be introduced and reduces the risk of the plasma becoming unstable.
(iv) The liquid uptake may be varied separately from the gas flow rate.
(v) The pump speed may be increased between samples in order to reduce flush-out time.

The main disadvantage of using a peristaltic pump is that it may contribute to poor precision. To achieve the flow rate required it is best to use a narrow bore pump tube and fast rotation of the rollers. The narrow bore tubing reduces the size of the successive liquid pulses and the high pump speed is needed to obtain the required sample uptake.

The effect of two different tube sizes to achieve the desired flow of $0.5\,\text{ml}\,\text{min}^{-1}$ is shown in Figure 3.3, where the ICP-MS response for

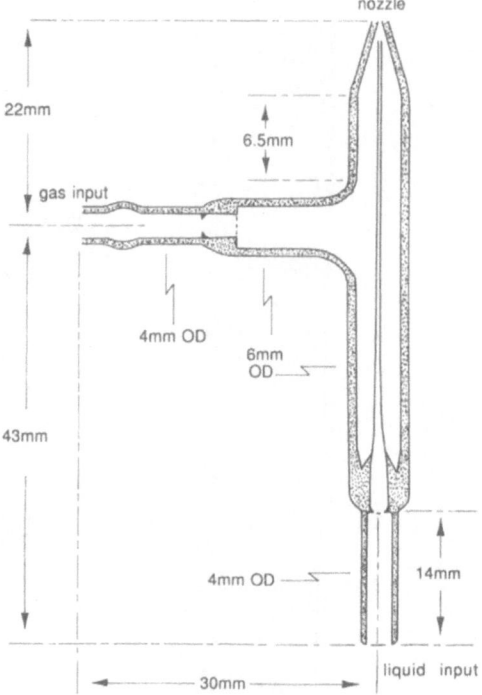

Figure 3.2 Meinhard concentric glass nebuliser. From Meinhard (1976).

$100\,\mathrm{ng\,ml^{-1}}$ In was monitored at $115\,m/z$. With the 1 mm tubing (Figure 3.3a) regular spikes in the analyte signal are produced at a frequency of about 1 Hz, corresponding to the pump roller period rate. A smaller bore tubing of 0.38 mm shows only a general noise level (Figure 3.3b) without the pulse and this is similar to that obtained if the nebuliser is allowed to aspirate freely (Figure 3.3c). An additional advantage of using fine bore tubing is that it reduces uptake and wash-out time. Pump tubing is normally made of PVC and linked to the nebuliser by a length of PTFE tubing. PVC tubing is suitable for most aqueous applications, but organic solvents require tubing made from much more expensive materials such as Viton.

The variation of analyte signal response as a result of pumping a Meinhard nebuliser at different flow rates can be seen in Figure 3.4. Signal increases linearly with uptake rate up to about $0.3\,\mathrm{ml\,min^{-1}}$. Beyond this point response changes little and it will be seen that the free aspiration or natural uptake occurs midway along the plateau. As a consequence flow rates as low as $0.2\,\mathrm{ml\,min^{-1}}$ can be used, which is particularly useful if sample volumes are limited. Where very little sample is available, a recirculating system of the type described by Hulmston (1983) or Isoyama *et al.* (1990) may be used. Recirculating systems may enrich the sample and give memory effects. In

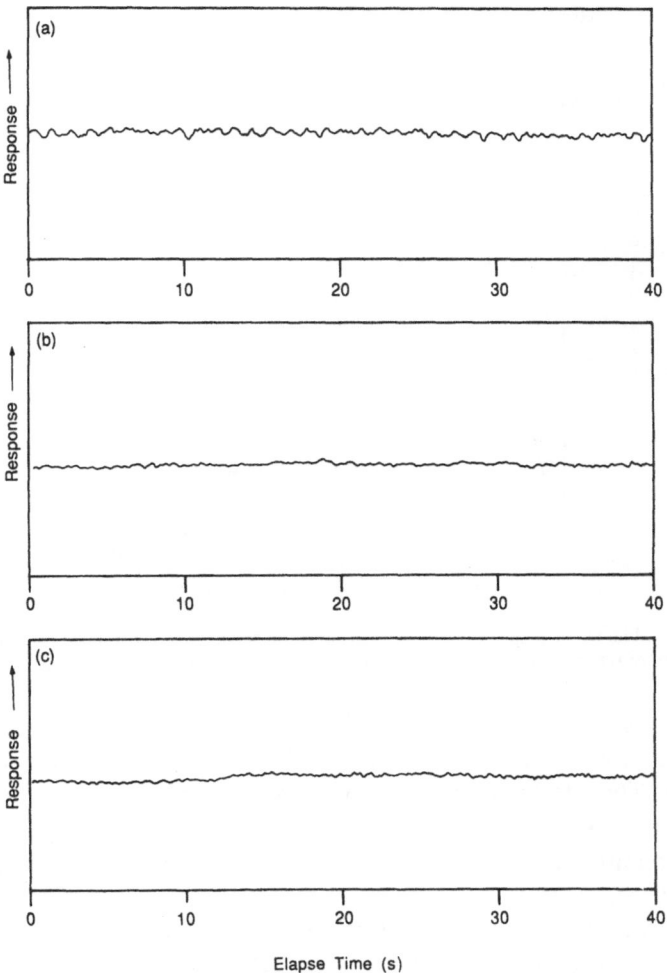

Figure 3.3 Comparison of short term signal stability with a Meinhard nebuliser for 100 ng ml^{-1} In solution, aspirated at 0.5 ml min^{-1}. Full scale deflection 25 000 counts s^{-1}. (a) 1 mm i.d. tube pumped, (b) 0.38 mm i.d. pumped, (c) free running.

general they have been little used, but may be worth pursuing for isotope ratio measurements.

3.2.2.2 *Blockage* Nebuliser blockage may occur for two reasons. The first may be caused by suspended solids getting stuck in the narrow (about 0.3 mm diameter) central sample uptake capillary. This is likely in samples which contain significant amounts of suspended solids, but the risk can be minimised by filtering or centrifuging the sample. Nebuliser blockage may be prevented by using a fine uptake tube, similar in diameter to the nebuliser capillary, so that material collects in the tubing, which can be readily replaced. Blockages can be dealt with by a number of methods such as forcing liquid or gas back-

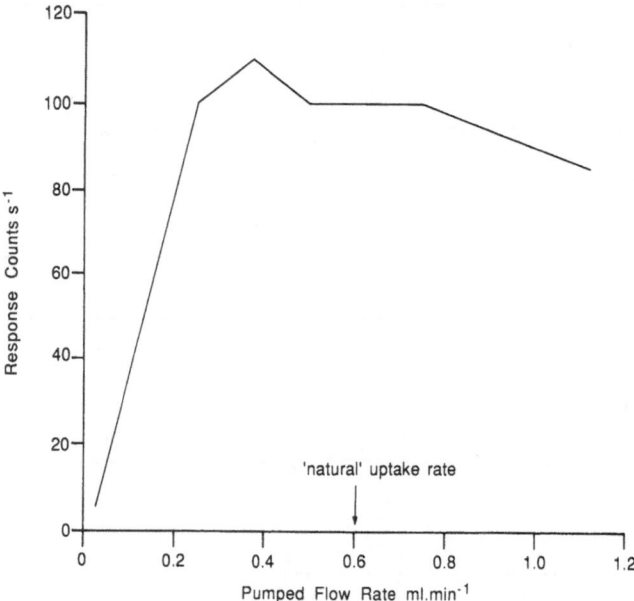

Figure 3.4 The effect of pumping a Meinhard nebuliser at increasing flow rate. The natural uptake rate of $0.6\,ml\,min^{-1}$ does not exactly correspond to maximum response.

wards along the capillary (backflushing), chemical dissolution or mechanical treatment, where a fine wire is pushed down the capillary to dislodge the blockage. These techniques have to be used with extreme caution as it is very easy to inflict permanent damage on the device. Ultrasonic cleaning of the nebuliser must not be carried out as this will usually damage the glass capillary.

The second type of blockage is caused by dried solute from a sample of high dissolved solids content partly obstructing the orifices. This interferes with the operation of the nebuliser and may reduce the signal. This effect has been explained by Sharp (1989a). Some droplets fall back on to the face of the nebuliser adjacent to the gas annulus. If the deposited solution contains salt close to saturation, the cold dry gas from the nebuliser causes sufficient cooling and evaporation to initiate crystal formation along the gas annulus of the nebuliser. This eventually stops the nebulisation. However, before this point is reached, introduction of distilled water can gradually reverse the salting process, but once nebulisation has stopped, mechanical removal or washing of the nebuliser is required. A number of authors (Burman, 1981; Gustavsson, 1979; Walton and Goulter, 1985) have proposed wetting of the gas stream to reduce evaporation. This can be effective for ICP–AES, however, it could lead to problems in ICP–MS where high aerosol water content causes high levels of polyatomic ions (see section 3.3). Regular washing of the gas annulus can prevent salt deposits building up, but the most effective method is to use the high salt tolerance nebuliser (Baginski and Meinhard, 1984).

Blockage of the annular gas orifice can occur if dust particles or fibre particles get into the gas supply lines. This type of blockage is almost impossible to correct, so it is imperative that the gas lines are disturbed as little as possible.

3.2.3 Cross-flow nebulisers

Much of the general behaviour of concentric nebulisers is also shown by cross-flow designs. They are less prone to blockage and salting up, but these can still occur as the sample solution is pumped through a capillary. Early ICP–MS work used the cross flow, but the lack of a readily available commercial type led to manufacturers supplying concentric, and more recently, high solids nebulisers.

The operation of the cross-flow nebuliser is similar to that of a scent spray, where a horizontal jet of gas passes across the top of a vertical tube. The reduced pressure that is generated draws liquid up the tube where at the top, it is disrupted into a cloud of fine droplets. These devices are capable of 'free running' but may not be self-priming. For this reason, it is normal to use a peristaltic pump.

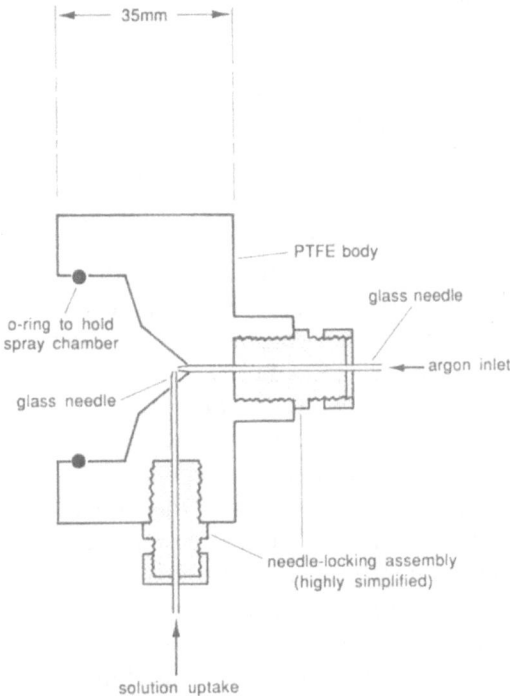

Figure 3.5 Simplified diagram of an adjustable, glass capillary cross flow nebuliser. From Kniseley *et al.* (1974).

Among the early models was the nebuliser designed by Kniseley *et al.* (1974), which had adjustable capillaries for both gas flow and sample uptake (Figure 3.5). These models were, however, prone to misalignment. Current models of cross-flow nebulisers have bodies constructed of Ryton or polypropylene and fixed capillaries of glass or sapphire. A 'fixed geometry' cross-flow nebuliser made of a hard, dense plastic material such as Ryton, with sapphire capillaries, is useful for introducing corrosive liquids that would damage an all glass nebuliser. During the later stages of development of ICP–MS in the United Kingdom, Gray and Date (1983) used a fixed geometry cross-flow nebuliser manufactured by the Jarrel-Ash Company (now Thermo-Electron). This type of cross-flow nebuliser has been in use in the authors' laboratory for a number of years in preference to a Meinhard nebuliser, because of the rugged construction and relative ease with which they can be cleaned. It is a very satisfactory alternative to concentric design. If blocked it can be cleaned by pushing a fine wire down the liquid capillary, or forcing the gas flow up the liquid capillary by blocking the exit face of the nebuliser. Fixed cross-flow nebulisers are not damaged if cleaned in an ultrasonic bath.

Gray and Williams (1987b) and Williams (1989) found that the performance of the ICP–MS instrument was very similar when using either a cross-flow or Meinhard nebuliser. Other types of cross-flow nebuliser are available such as the MAK (Anderson *et al.*, 1981). These, however, have not found widespread use.

3.2.4 *Babington type nebuliser*

The original device developed by Babington (1973) allowed a film of water to flow over the surface of a sphere. Gas forced through an aperture beneath the film produced an aerosol. The essential feature of this type of nebuliser is that liquid flows freely over a small aperture, rather than passing through a fine capillary and thus has a great tolerance to high dissolved solids. Since the delivery of the sample is not constrained by a capillary, slurries can be nebulised (Williams *et al.*, 1987; Ebdon *et al.*, 1988). They are not self-priming or self-aspirating and therefore must be pumped.

The original Babington nebuliser design showed extensive memory effects as the solution was allowed to wet the entire face of a sphere. Suddendorf and Boyer (1978) proposed constraining the liquid in a V-groove and introducing gas from a small hole in the bottom of the groove (Figure 3.6). This made a significant improvement to memory effects, although Babington type nebulisers do still have worse memory effects when compared to concentric or cross-flow designs. There are now a number of similar systems available, however, the two most commonly used in ICP–MS are probably the V-groove nebuliser produced by Van der Plas products (Ripson and de Galan, 1981) and the Ebdon nebuliser (Ebdon and Cave, 1982).

The main advantage of the V-groove nebuliser is its resistance to blockage.

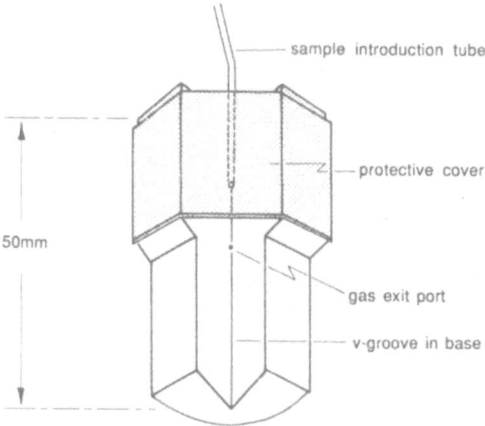

Figure 3.6 Suddendorf and Boyers' (1978) Babington type nebuliser.

However, it is not an optimum geometry for aerosol generation, which is evident from the coarser particle size distribution and lower efficiency (Sharp, 1988a). These disadvantages are not necessarily translated into a degradation of analytical performance, since the spray chamber (section 3.3) is the principal determinant of particle size reaching the plasma. If necessary an increased supply of liquid can be used to overcome inefficiency in nebuliser performance.

3.2.5 *Frit type nebuliser*

Concentric and cross-flow nebulisers only produce about 1% of droplets of the correct size to conduct to the plasma. This source of inefficiency has attracted some attention. An alternative design is the frit nebuliser (Layman and Lichte, 1982) which produces droplets of mean size 1 μm. The design is essentially the same as the Babington types, but with a fine glass frit matrix in place of a V-groove. The V-groove device optimises resistance to blockage at the expense of gas-liquid mixing efficiency (particle size distribution), whereas the frit provides excellent gas-liquid mixing but at the expense of poor sample handling characteristics, i.e. the nebuliser is prone to clogging and salting up and exhibits poor clean out characteristics.

A variation on the frit type nebuliser is the Hildebrand grid nebuliser (Figure 3.7) which uses a fine mesh grid positioned in front of the gas stream. Sample flows over the grid and nebulisation takes place from the wetted surface. Brotherton *et al.* (1989) used this nebuliser for analysis of high matrix solutions with ICP–MS and compared the performance of the nebuliser to that of a Meinhard. They concluded that the limiting factor in introducing salt solutions was not the grid nebuliser itself, but the ICP–MS interface, which begins to block up with the introduction of high salt solutions.

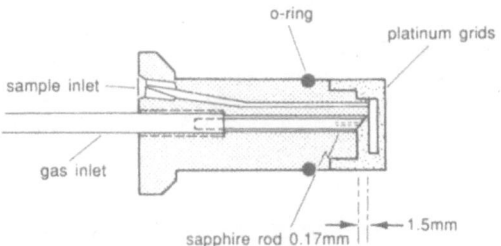

Figure 3.7 Simplified diagram of Hildebrand grid nebuliser. Front platinum grid is 100 mesh × 0.061 mm thickness. Rear grid is of the same mesh size and 1.5 mm thickness. After Brotherton *et al.* (1989).

3.2.6 *Ultrasonic nebuliser (USN)*

This method of aerosol generation was used in the development of both ICP–AES (Wendt and Fassel, 1965) and ICP–MS (Houk *et al.*, 1980). Solution is fed to the surface of a piezoelectric transducer operated at a frequency of between 0.2 and 10 Mhz. The longitudinal wave, which is propagated at right angles to the surface of the transducer towards the liquid-air interface, produces pressure that breaks the surface into an aerosol. There is a very high efficiency of production with this type of device, which is independent of gas flow. Thus more analyte can be transported to the ICP at a slower nebuliser gas flow rate than that used with pneumatic nebulisers. This potentially gives a considerable increase in sensitivity, and improvement of detection limits, since the analytes have a longer residence time in the plasma. However, the increased efficiency does allow a greater quantity of water to enter the plasma, producing a cooling effect, which is counter-productive for analyte ionisation. For ICP–MS it is essential that the aerosol is desolvated since high solvent loadings lead to high levels of interference ions (see section 3.3). This is normally achieved using a heated tube to evaporate the water droplets and a condenser to remove the resulting water vapour. The remaining solute particles then enter the plasma.

For simple aqueous matrices, detection limits with USNs are usually an order of magnitude better than with pneumatic devices. Tingfa *et al.* (1990) compared the multi-element detection limits for ICP–MS using a USN with those of a pneumatic nebuliser, and showed an improvement factor of between 8 and 200, depending on the element. In samples with a complex matrix, the background signal from the matrix also grows and leads to increased levels of spectral and non-spectral interferences. They generally have the longest wash-out times of most nebuliser types.

Apart from the analytical limitations of the USN, the high cost of commercial systems is prohibitive for most laboratories (approximately £10 000) as USNs require a separate RF source. In addition, the devices tend to be unreliable due to their complexity. Despite these drawbacks, one

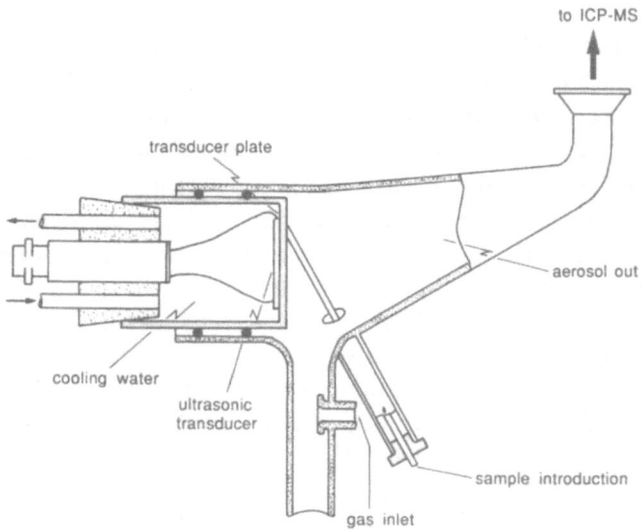

Figure 3.8 Schematic diagram of a continuous-flow ultrasonic nebuliser. After Fassel and Bear (1986).

research group at Ames laboratory uses a USN (Figure 3.8) developed by Fassel and Bear (1986) for much of their ICP–MS investigations (Jiang *et al.*, 1987; Jiang and Houk, 1988; Jiang *et al.*, 1988).

3.3 Spray chambers

3.3.1 *Principles*

Aerosol transport efficiency is defined as the percentage of the mass of nebulised solution that actually reaches the plasma. For this percentage to be high and for rapid desolvation, volatilisation and atomisation of the aerosol droplets when they reach the plasma, a nebuliser must produce droplets of $< 10\,\mu$m. However, pneumatic nebulisers produce aerosols with a broad distribution of droplet diameters up to $100\,\mu$m. The primary task of the spray chamber is to remove large droplets (i.e. $> 10\,\mu$m) from the gas stream and deliver them to waste. As the gas flow carrying the aerosol enters the spray chamber, it undergoes sharp changes in direction which the larger droplets cannot follow. These droplets strike the walls and subsequently run to waste. The spray chamber ensures that only droplets small enough to remain in suspension in the gas flow are carried into the plasma. With most pneumatic nebulisers this means a loss of about 99% of the sample solution. Readers interested in the theory and dynamics of spray chambers are directed to the excellent review on the subject by Sharp (1988b).

3.3.2 *Operation*

The search for an ideal spray chamber has led to a variety of designs being proposed and evaluated. The most common formats are those employing flow reversal cyclonic action and impact beads. Droplets leaving the 'ideal' spray chamber should be $< 10\,\mu m$ in diameter and have a small size distribution. They should also have a high analyte mass flux and transport efficiency, a low wash-out time and good pressure-temperature stability. These quantities are dependent on the sample and gas flow rates. Some of these criteria are met by the double pass (flow reversal) system (Figure 3.9) developed by Scott *et al.* (1974). Various versions of this basic design are widely used in ICP–MS instruments. A number of alternative spray chamber designs (cyclonic or impact bead) are available and used in ICP–AES, but their use on ICP–MS instruments has been limited. Spray chambers may be constructed of glass, polyethylene, PTFE or Ryton. Allenby (1987) described the use of a PTFE spray chamber on an ICP–MS instrument, used to determine trace elements in uranic materials which had been dissolved in hydrofluoric acid. Solutions containing up to 2% of fluoride were aspirated into the chamber with no deleterious effects to the spray chamber. Commercially produced inert spray chambers are widely available for analysis of solutions which are corrosive to glass.

Removal of waste solution has to be carried out in a way that retains a small positive pressure in the spray chamber to drive the aerosol through the injector tube. This can be achieved simply by dipping the drain tube into the waste container, however, the pressure will change slightly as the container fills with this arrangement. Drains should be designed so that the liquid runs away smoothly through devices such as fritted glass discs or capillaries. Liquid build-up in the chamber must be avoided, as this causes pressure changes resulting in signal drift and leads to extended memory effects. Pumping away

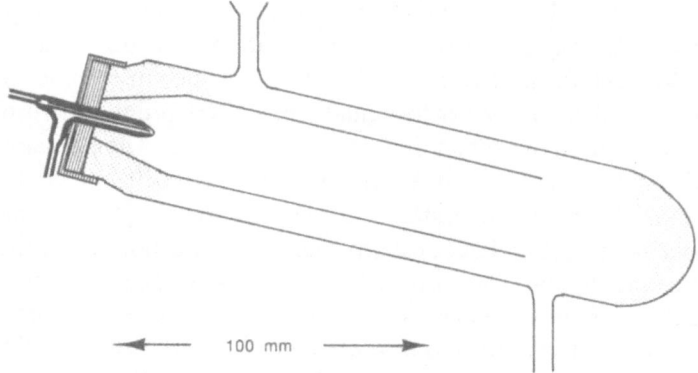

100 mm

Figure 3.9 Double pass spray chamber as described by Scott *et al.* (1974). The shaded area is dead volume where aerosol remains after the sample solution has been removed. Excessive dead volume can lead to increased wash-out times.

waste solution is a convenient method, particularly if a second channel is available on the pump used to supply the nebuliser.

The wide dynamic response, high throughput of samples and excellent sensitivity of ICP–MS are of little value if the sample introduction system is subject to memory effects. The time taken for sample wash-out between samples may have a major influence on the total time required for analysis, particularly in some isotope ratio determinations. Certain elements are very troublesome, particularly those which are volatile, such as Br and Hg. These are retained on the glassware, particularly on the injector tip of the torch, for a long period of time, even if they are introduced at low levels ($10 \, ng \, ml^{-1}$). Most elements, however, do not present such a problem and typical clean out times are between about 60 and 180 s.

Different designs of nebuliser can critically affect wash-out time. A comparison of wash-out times for a Meinhard and a de Galan nebuliser under the same operating conditions is shown in Figure 3.10. The response for $100 \, ng \, ml^{-1}$ In in 1% HNO_3 was monitored using the 115 m/z setting. The sensitivity was about 1.5×10^6 counts s^{-1} per $\mu g \, ml^{-1}$. After a steady signal rate was obtained, the uptake tube was placed in a washing solution of 2% HNO_3. With the Meinhard nebuliser the signal decayed to about 1% of the original rate after about 8 s, 0.1% in 35 s and has returned to background levels in about 1 min (Figure 3.10a). The wash-out time for the de Galan V-groove is much longer than that of the Meinhard (Figure 3.10b) of the order of 165 s to reach 1% of the original signal. This is probably because

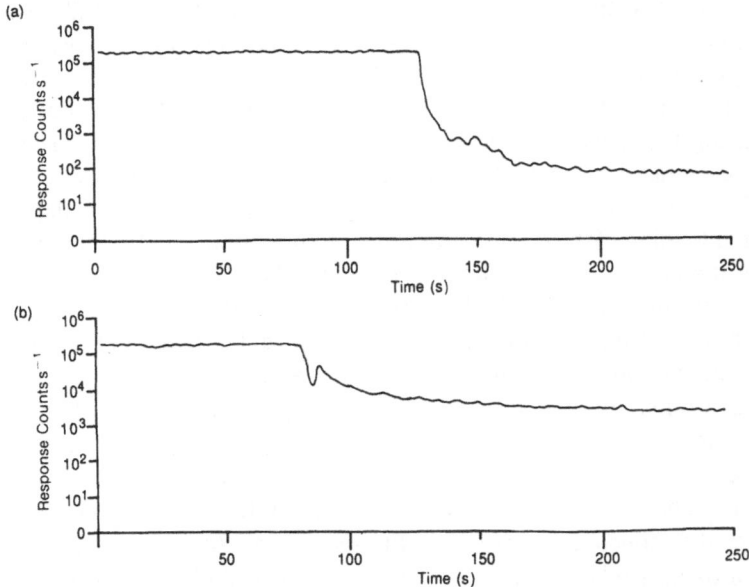

Figure 3.10 A comparison of wash-out characteristics for $100 \, ng \, ml^{-1}$ In for (a) Meinhard nebuliser and (b) de Galan V-groove nebuliser.

liquid is retained on the face of the nebuliser and drawn back into the groove. If high throughput is likely for an instrument, the wash-out characteristics must therefore be a major consideration during instrument operation and appropriate measure taken to assess wash-out times for individual elements. During an analysis run, it is advisable to monitor wash-out by choosing one of the elements in a suite under investigation, selecting the appropriate mass, and monitoring the signal decay until a suitable background level is achieved. This will depend on the level of the element(s) under investigation. The time period for the required signal decay to occur can then be used as a pre-set wash-out time. This is particularly important if an auto-sampler is to be used and time delays have to be programmed into the system to allow for wash-out. Increasing the pump speed to maximum between samples will not improve wash-out times of the spray chamber, but will reduce the time taken for washing solution or for the next sample to reach the nebuliser.

3.3.3 Thermally stabilised spray chambers for ICP–MS

During the early development of ICP–MS, water was identified as a major source of ions for the formation of polyatomic species. The advantages of carrying out analysis in the absence of water, with techniques such as laser ablation (Gray, 1985b), electrothermal vaporisation (Whittaker et al., 1989) or arc nebulisation (Jiang and Houk, 1986) have been reported. Gray (1986a) calculated the population levels of H^+, O^+ and Ar^+ in the plasma from the Saha equation along with those ions which are derived from common mineral acids such as N^+, S^+, Cl^+ and C^+. Because of the high levels of these ions in the plasma it was suggested that only very low significant levels of polyatomic ions can be formed. Reducing the amount of water or solvent introduced into the plasma should reduce the levels of polyatomic ions. Stabilising the delivery of water vapour into the plasma should lead to a more reproducible signal for polyatomic ions. In addition, Gray et al. (1987) and Hutton and Eaton (1987) showed that ion energies may also be dependent on water loading of the plasma. For ICP–MS it is important that these remain stable, and at a low level, as they can critically affect instrument performance.

The variation of the $^{40}Ar^{16}O^+$ response with time and spray chamber temperature, using a pneumatic nebuliser is shown in Figure 3.11. Figure 3.12 shows the total amount of water from both vapour and aerosol entering the plasma at different temperatures assuming a 1% sample transport efficiency. For an instrument operating with the spray chamber temperature at > 25°C, more water enters the plasma as vapour, than in the form of aerosol. Cooling the spray chamber causes much of the vapour to condense on the walls, significantly reducing the water input to the plasma. At about 10°C the levels of both oxygen related (e.g. ArO^+) and other (e.g. $ArAr^+$) polyatomic ions are greatly reduced (Hutton and Eaton, 1987; Williams, 1989). In addition refractory oxide ions, doubly charged ions, energies and any corrosion

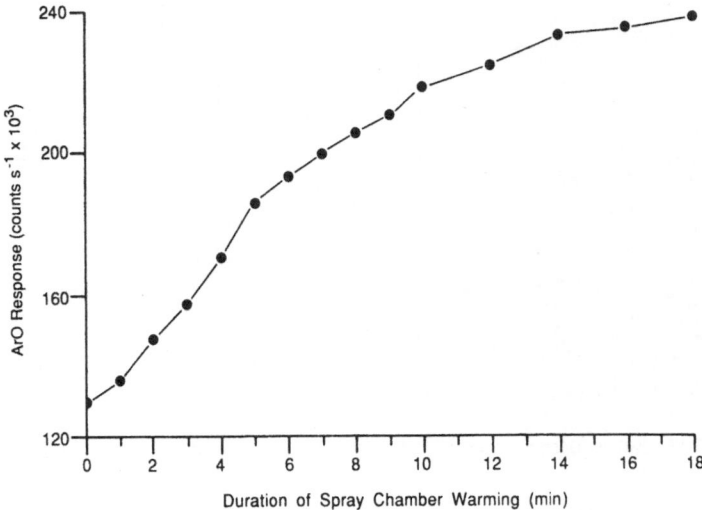

Figure 3.11   Variation of $^{40}Ar\ ^{16}O^+$ response with time and spray chamber temperature (spray chamber at 25°C at 0 min and 35°C after 18 min). From Williams (1989).

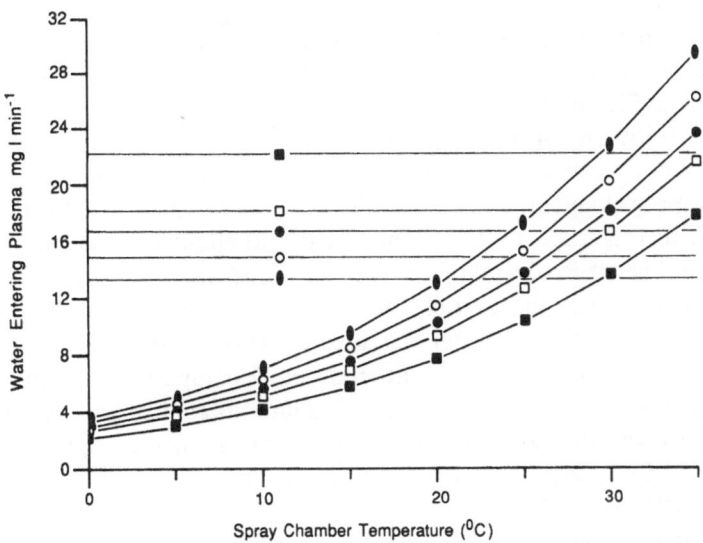

Figure 3.12   The mass of water entering the plasma is shown for five carrier gas flows. Horizontal lines represent aerosol component (assumed at 1% with no temperature effect) and curves represent vapour component. ■, 0.45 l min^{-1}; □, 0.55 l min^{-1}; ●, 0.6 l min^{-1}; ○, 0.67 l min^{-1}; ◗, 0.75 l min^{-1}. After Williams (1989).

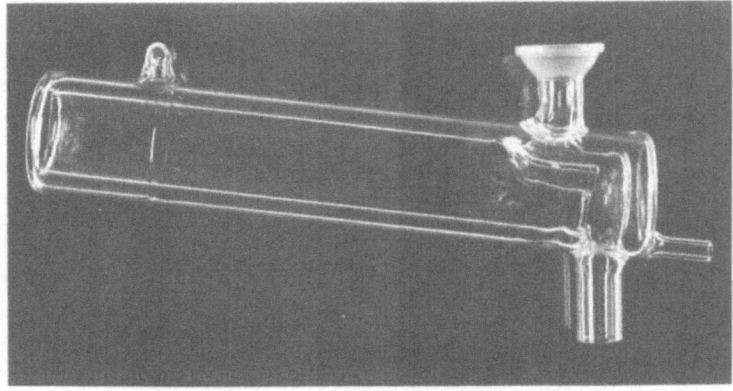

Figure 3.13 Surrey design water cooled single pass spray chamber. From Williams (1989).

products from sampling and skimmer cones are all reduced and maintained at a constant level.

The Scott double pass spray chamber is the type most commonly used with ICP–MS and ICP–AES instruments. A water jacketed version of this design is available and fitted to most ICP–MS systems. The only chamber designed specifically for an ICP–MS instrument has been described by Williams (1989) and is shown in Figure 3.13. This is a single pass type 125 mm long with a volume of some 60 ml. A single pass design has fewer internal surfaces and less dead space exposed to sample solution which reduces memory and wash-out times. A small baffle is placed directly below the outlet to prevent any large droplets accumulating at the outlet. A dripline path from the baffle to the waste outlet prevents pulsing from excess solution running to waste. A large aerosol outlet (10 mm) was found to avoid obstruction by condensation after a period of time. Wash-out times on Co for this spray chamber, using a Meinhard nebuliser, are < 10 s for the signal to decay to 1%, < 35 s to 0.1% signal decay and about 1 min to background.

The on-line effect of cooling the spray chamber from 35 to 11°C is shown in Figure 3.14. Here the level of $^{40}Ar^{16}O^+$ is reduced by more than a factor of 2, where ordinary tap water was used for cooling. In addition the levels of most other polyatomic species are reduced. A comparison of some oxygen based species found when introducing 1% solutions of nitric and hydrochloric acid is shown in Table 3.1. Concentrations are calculated relative to $10\,ng\,ml^{-1}\,Co^+$. Depending on the species, the interferences were reduced by between 10% and 60%.

Thermally stabilised spray chambers are essential where organic solvents are used. The higher vapour pressure produces increased solvent loading in the ICP resulting in plasma instability. Temperature fluctuations in the spray chamber may have a significant effect on signal stability. Only small

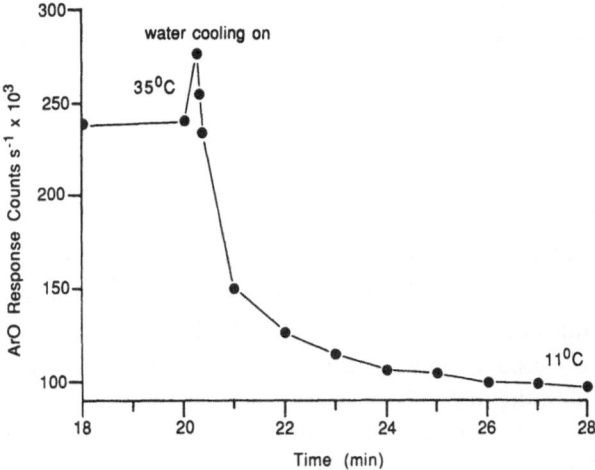

Figure 3.14   Variation of $^{40}Ar\ ^{16}O^+$ response with decreasing spray chamber temperature over a period of 10 min. The temperature was initially stabilised at 35°C before cooling to 11°C. From Williams (1989).

Table 3.1   Effect of cooling spray chamber temperature on oxygen containing polyatomic ion peaks[a]

	1% HNO₃		1% HCl	
Mass (m/z)	Uncooled	Cooled	Uncooled	Cooled
31	20.0	9.78	19.7	9.44
32	8030	2560	8270	2630
48	1.40	0.64	2.18	1.17
49	0.08	0.08	1.42	1.86
50	0.21	0.12	0.36	0.21
51	0.17	0.13	19.9	12.7
52	0.90	0.49	2.23	1.11
53	0.22	0.08	7.00	4.63
56	64.4	16.9	61.6	14.8
64	1.60	0.51	1.82	0.49
65	1.70	1.30	1.84	1.88
66	0.25	0.17	0.23	0.12
67	0.27	0.08	0.39	0.07
69	0.49	0.14	0.49	0.15

[a] Values are expressed as equivalent concentrations (ng ml⁻¹), referenced to 10 ng ml⁻¹ Co. From Williams (1989).

temperature changes can drastically alter organic solvent loading of the carrier gas (Hutton, 1986). Operation of the chamber at $< 0°C$ for the analysis of samples in volatile organic solvent such as white spirit, xylene, ethanol, acetone or chloroform is often necessary. To prevent freezing of recirculated cooling water, anti-freeze should be added. Maintaining temperatures of

$< 0°C$ when introducing aqueous solvents is inadvisable, as it may cause the aerosol and waste to freeze in the spray chamber and block it.

3.4 Torches

3.4.1 *Construction*

Torches should be constructed to a high degree of accuracy using good quality quartz, an important feature for the stability and formation of the plasma. The injector tube diameter is generally about 1.5 mm. Narrower bore tubes are rarely used as they block easily, an important consideration when solutions containing high dissolved solids are to be analysed. They do, however, offer better plasma penetration. For high dissolved solids or slurry analysis, wider bore injectors (typically 3.0 mm) are often used, with little degradation in instrument performance.

There are various types of injector tube in use (see Thompson and Walsh, 1989) including tapered injector tip, capillary injector tip and complete capillary injector tip. Individual suppliers tend to favour a particular design for production reasons, unless a specific design is requested. The capillary tip injector is probably most commonly used, where the final 25 mm (approximately) of 3–4 mm diameter tube are at a diameter of 1.5 mm. A tapered injector tip tends to degrade in performance as the tip devitrifies and silica is lost.

3.4.2 *Demountable torches*

The plasma torch does not have to be of fixed configuration. It can be obtained partly demountable (i.e. with a removable injector tube) or fully demountable. Integral torches generally have better long term consistency although when they do finally degrade, replacement of the coolant tube has to be carried out by a glass blower. Replacement of injectors is generally not possible.

Fully demountable torches provide an easier (and cheaper) means of replacing devitrified parts, although they can be difficult to reconstruct. Partly demountable torches offer a good compromise and are used routinely on this author's ICP–MS instruments. This arrangement allows a range of injector diameters to be used (i.e. 1.5 mm for solution analysis and 3.0 mm for high dissolved solids and slurries), without the need to remove the torch from its assembly. In addition, materials other than quartz can be used for the injector tube. A particularly useful alternative is an alumina injector which allows HF solutions to be analysed.

Apart from overcoming the problems of injector blockage, when high dissolved solids or slurries are analysed, it has been found that a wider bore injector (e.g. 3.0 mm) allows high dissolved solids solutions to be nebulised for longer periods of time before cone blockage occurs, compared with a standard 1.5 mm injector (Williams and Gray, 1988).

3.4.3 *Alignment*

The torch is normally mounted horizontally with the centre of the injector aligned on the axis with the sampling cone aperture. The final 25 mm of the coolant tube should be inside the load coil, such that the distance between the turn of the coil edge furthest from the sampling cone is 3–5 mm distant from the edge of the auxiliary tube. This positions the plasma correctly in the torch. The distance between the torch and sampling cone is usually between 10 and 15 mm as measured from the edge of the first turn of the load coil to the tip of the sampling cone. This static alignment method is generally satisfactory, however, a dynamic alignment method, such as the one described in chapter 2, is preferable.

3.4.4 *Specialised torches*

3.4.4.1 *Torches for hazardous gases*   The trace element content of gases used in the micro-electronic industry has to be accurately determined to very low levels, as impurities can adversely affect the properties of the final products. These gases are extremely hazardous and specialised handling techniques must be employed with them. Some, such as trimethyl gallium, trimethyl aluminium, dimethyl zinc and silane are pyrophoric, i.e. they will spontaneously combust in air.

To allow these dangerous gases to be introduced into the ICP a special four gas inlet torch has been designed (Streusand *et al.*, 1990) which allows the addition of water *in situ* to hydrolyse the gas safely. In addition calibration standards can be added to the gas via the water.

3.4.4.2 *Low flow torch*   Gordon *et al.* (1988) carried out a preliminary investigation of a torch externally cooled by water, which allowed coolant gas flows of only 2–3 l min^{-1} to be used. Reasonable performance was found to be achievable, however, interference species (oxides, doubly charged etc.) were significantly higher.

A torch with a 9 mm diameter coolant gas tube has been characterised by Ross *et al.* (1990). An analytically useful plasma was supported by 850 W forward power (40.68 MHz RF frequency) and 8.7 l min^{-1} total argon flow rate. Although the 9 mm source produced similar sensitivities, detection limits, doubly charged ion ratios and oxide ion ratios to a conventional plasma, the plasma sampling depth and sampling position had to be adjusted individually for each element in order to obtain maximum sensitivity.

3.4.4.3 *Mixed gas and sheath gas torches* A few workers have reported that the presence of a second gas in the Ar flow changes the fundamental properties of the plasma (Murillo and Mermet, 1989; Evans and Ebdon, 1989). The addition of a few percent of hydrogen as a sheath around the injector gas flow enhances the ionisation process in ICP–AES (Murillo and Mermet, 1989), and is attributed to the higher thermal conductivity of this diatomic gas. This leads to a more efficient transfer of energy with the plasma, resulting in improved desolvation, volatilisation and dissociation processes.

The use of mixed gas plasma, produced by adding a sheathing gas around the nebuliser gas, has been considered by Beauchemin and Craig (1990). Modifications were not made directly to the torch. To introduce the gas (H_2 or N_2) a sheathing device is inserted between the spray chamber and injector as shown in Figure 3.15. The addition of either gas reduced sensitivity but improved stability, resulting in generally improved detection limits. In addition, mass bias was decreased when mixed gases were used and lower nebuliser gas flow could be used. This final point shows potential for coupling a gas chromatograph to an ICP–MS instrument. The use of sheathing gases is not widespread at the moment as performance improvements are marginal, optimisation is complex and expensive extra gas flow control equipment is required.

A sheathing gas of argon was used by Lichte *et al.* (1987) in a specially designed four gas inlet torch. This allowed adjustment of the sample injection velocity into the plasma without affecting the performance of the nebuliser or the rate of total sample delivery. In addition this torch was water cooled to improve its lifetime.

To determine trace elements in organic based solution it is necessary to add a small amount of oxygen to the nebuliser gas flow to prevent the condensation of particulate carbon on the sampling cone orifice. Hausler (1987) analysed xylene solutions of organically bound metals using an addition of 2% oxygen to the argon flow, introduced between the spray chamber and injector. Hutton (1986) introduced 5–7% oxygen to the nebuliser

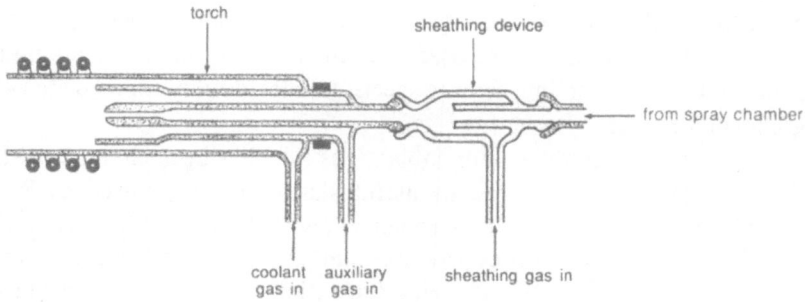

Figure 3.15 Schematic of the glassware arrangement used for the addition of a sheathing gas. After Beauchemin and Craig (1990).

gas for the analysis of NIST 1634a (residual fuel oil) diluted in white spirit. Generally limits of detection were comparable to those obtained with the analysis of aqueous solutions, except for some elements below 80 m/z where there were interferences from polyatomic ions.

3.5 Interface

3.5.1 *Introduction*

The sampling and skimmer cones are critical components of the interface—the heart of an ICP–MS instrument. The function of the interface is to extract gas from the plasma that is representative of the original sample. The design of the interface is highly specialised. There is very little independent research into cone and skimmer design, most of it is carried out by the instrument companies. Figure 3.16 shows the sampling cone and skimmers used in the author's instruments. Much of the early work in interface design was carried out by Douglas *et al.* (1983b) and Bray (1982). Basic cone and skimmer design has changed little since those original concepts.

3.5.2 *Sampling cones*

3.5.2.1 *Construction*   At the interface, the plasma flares out around the side of the sampling cone, in the tip of which the aperture is drilled. Most of the injector flow from the plasma is extracted by the aperture while the majority of the gas that passes over the sides of the cone comes from the annulus. Aperture diameters of 0.75–1.2 mm are most commonly used. The section of the cone tapers towards the centre, so that its diameter/length ratio is approximately unity and the base of the cone provides good heat transfer. A variety of metals may be used such as aluminium, copper, nickel and platinum, although nickel is usually found to give the best compromise between cost and durability. The material used to construct cones must have a high thermal and electrical conductivity. In the analysis of organic materials the use of a platinum cone is advisable as oxygen is added to the nebuliser

Figure 3.16 Sampling cone and skimmers used in the author's instruments.

gas flow to aid the dissociation of organic compounds, and platinum cones are less susceptible to degradation than nickel in this highly reactive environment. Unfortunately highly oxygen resistant materials are usually those which have poor thermal conductivity.

3.5.2.2 *Operation* Sampling cones are replaceable and are usually retained by screws on a water-cooled plate which forms the front wall of the vacuum system. Alternatively the cones can have threaded bases which allows them to be screwed directly to the front wall of the interface. Nickel sampling cones are very durable and will usually operate for many months provided that samples containing high acid concentrations or salt content are not introduced into the plasma and good thermal contact is made between the cone base and the water-cooled plate. However, repeat analysis of 10% sulphuric acid, for example, can reduce cone lifetimes to a matter of days. Careful attention to acid type and concentration is important if sampling cone lifetimes are to be maximised.

Complete cone failure is rarely sudden. Cone blockage due to the introduction of samples with high dissolved or suspended solids is readily alleviated by removal of the cone (after switching off the plasma) and cleaning. Solutions containing up to 0.2% solids can be tolerated for long periods, which can be extended by the use of large bore (3 mm) injector tubes (Williams and Gray, 1988). Wear makes itself apparent by surface pitting, roughness and even cracking. Near the end of its useful life, the aperture diameter tends to increase, the lip profile of the hole becomes rounded and spectral interferences become more prominent. Monitoring the first stage vacuum may give an indication of the condition of the aperture. Replacement is straightforward and can usually be postponed until a convenient moment arises in an analytical procedure. This problem should never arise as the good operator will be aware of slow cone deterioration. Cleaning and inspection should be carried out daily.

3.5.2.3 *Cleaning* The sampling cone should be cleaned on the face and rear using a proprietary fine abrasive metal polish powder ('Polaris' stainless steel reviver) made into a paste with deionised water and a soft cloth. All polish must be completely removed before refitting and water may be dried off using acetone washes. An ultrasonic bath will facilitate the washing. Coarse abrasives of whatever type or dilute acids washes should not be used as these methods introduce pits and grooves into the surfaces of the sampling cone. Distortions in the surface material cause disruptions of the plasma gas flow as it is sampled and can lead to higher levels of interfering ions being formed. In addition, material can be more easily deposited on the surface of the cone. Apart from the aperture itself, the most important area to clean is the inside surface of the cone, closest to the aperture. It is also the most difficult area to clean. It is in this area that the free jet begins to form after the sampled plasma has passed through the aperture. Therefore the surface must be as clean and smooth as possible so that gas flow is not disrupted.

3.5.3 *Skimmer cones*

Mounted directly behind and on axis with the sampling cone at a distance of 6–7 mm, is the skimmer which is also usually made of nickel. The condition of the skimmer tip has a direct bearing on the sensitivity of the ICP–MS instrument and the levels of some polyatomic ions, notably the main argon dimer at $80\,m/z$. It is very important that the skimmer tip is inspected on a daily basis. In normal operation, even with samples containing high levels of dissolved solids, it is rare for the skimmer to become blocked. Generally deposition occurs on the exterior of the cone, at least 1 mm downstream from the tip, which remains clean. However, this is not a universal occurrence as it has been noted that the skimmer can block in some situations (Douglas and Kerr, 1988). The skimmer is mechanically less robust than the sampling cone, particularly at the tip, the edges of which must be sharp and regular. In use the skimmer tip runs hot and becomes annealed. Therefore it must be handled with care especially when cleaning, which should be carried out using the method for sampling cones.

4 Sample introduction for liquids and gases

J.G. WILLIAMS

4.1 Introduction

The selection of the best sample introduction procedure for an analysis by ICP–MS requires a number of points to be considered. These include:

  (i) the form of sample (e.g. solid, liquid or gas);
  (ii) the sample matrix;
  (iii) the concentrations and ranges of the elements to be determined;
  (iv) the accuracy required;
  (v) the precision required;
  (vi) the amount of material available;
  (vii) the number of determinations required per hour;
  (viii) additional requirements, such as information on speciation;
  (ix) spectral and non-spectral interferences that may have to be overcome;
  (x) the hazards of handling the sample;
  (xi) any instrument damage or corrosion that may occur.

This chapter will concentrate specifically on sample techniques, although in any real analysis this is an extension of sample preparation. Therefore the selection of a suitable technique can depend on available and effective preparation procedures. Preparation will not be discussed here, except where it is intimately linked to sample introduction. Solid sample techniques are discussed in detail elsewhere, together with the methods of preparation required for these techniques. Direct introduction of solutions using a 'conventional' nebuliser and spray chamber configuration are discussed in chapter 3.

Irrespective of the detection system (e.g. mass spectrometry, atomic emission spectrometry or atomic absorption spectrometry), the goals of sample introduction are the reproducible transfer of a representative portion of sample material to the atomiser cell (e.g. ICP) with high efficiency and with no adverse effects. As has been shown in chapter 3 and in some later chapters, several of these criteria are mutually contradictory. It is in the pursuit of these goals and in consideration of the points outlined at the beginning of this chapter that a whole range of 'alternative' sample introduction techniques have been developed for atomic spectrometry and are in use with ICP–MS systems.

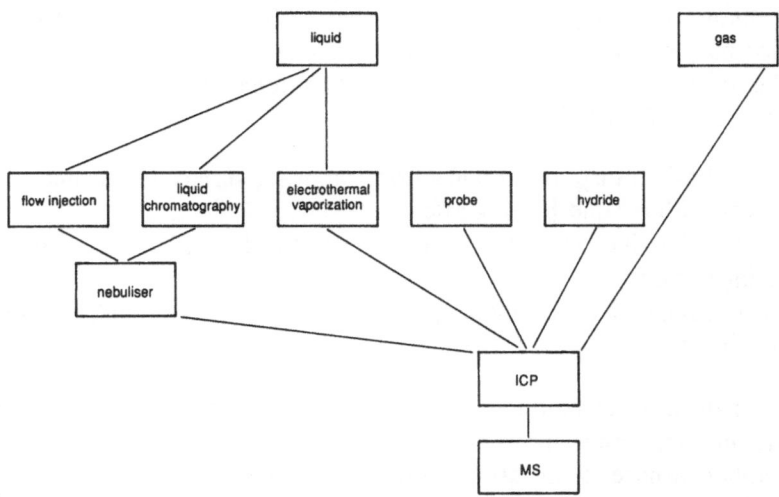

Figure 4.1 Summary of alternative methods of introducing a liquid or gas sample into the plasma of an ICP–MS system.

Some techniques, such as flow injection or liquid chromatography, are used in conjunction with a nebulisation system to provide more information about a sample. Others, such as electrothermal vaporisation or hydride generation, replace the nebuliser and spray chamber and can offer some advantages like simpler transport mechanisms or better transport efficiency.

Figure 4.1 summarises the many and varied alternative methods of introducing a liquid or gas sample into the plasma of an ICP–MS system.

4.2 Electrothermal vaporisation

4.2.1 *Principles*

The analytical chemist is often asked to analyse samples in micro-volume amounts, with very low elemental concentrations. In this situation electro-thermal vaporisation (ETV) may be a useful method of sample introduction offering advantages of low volume requirement, high transport efficiency (compared to pneumatic nebulisation) and easy removal of solvent. A microlitre amount (typically $5–100\,\mu l$) of sample is deposited into an electrically conductive vaporisation cell. Initially a low current is applied to the cell, causing resistive heating to occur, which dries the sample. This is then followed by a high current for a short period of time (typically 5 s) to completely vaporise the sample. An optional 'ash' stage may be used to remove some of the matrix prior to the analyte vaporisation stage. A continuous stream of gas (usually Ar) is passed through the unit and carries the sample vapour into the plasma. A variable current supply is required for controlling cell heating, although the sequence of the heating steps is generally carried out using a suitable electronic control system.

The ideal ETV device should include operational characteristics such as (Park *et al.*, 1987a):

  (i) near 100% transport efficiency from vaporiser surface to plasma;
 (ii) a cell with a rapid heating rate;
(iii) a cell that has zero reactivity with chemicals and air even at high temperatures;
 (iv) a wide range of operating temperatures, with a relatively high maximum ($\geqslant 3000°C$).

So far the most common application of the electrothermal device is as an atom source for atomic absorption (L'vov, 1984) in electrothermal atomisation or graphite furnace atomic absorption spectroscopy (ETA–AAS or GFAAS). It is also used as a sample introduction method for inductively coupled plasma atomic emission spectroscopy (ICP–AES). The first studies of its use with ICP–MS were carried out by Gray and Date (1983) and Gray (1986a). Absolute detection limits of between 1 and 12 pg were obtained for As, Cd, Pb, Se and Zn. In addition, the level of oxide ions was found to be between one or two orders of magnitude lower than when a nebuliser was used. In this application the device simply dries and vaporises the sample and the ICP atomises, excites and ionises it. In some cases common operating conditions can be used for many elements, but the separation of vaporisation and excitation/ionisation permits independent optimisation.

The major advantage of ETV as a means of sample introduction is the dramatic improvement in analyte transport efficiency over pneumatic nebulisation, i.e. from < 4% to > 60%. This results in improvements of sensitivity of at least an order of magnitude. In addition, the use of ETV allows difficult samples such as those in organic solvents or having high total dissolved solids (TDS) levels to be analysed. By optimising the thermal cycle, the matrix can be removed prior to analysis, which prevents it from reaching the plasma. This is particularly attractive for ICP–MS where high levels of TDS can lead to sampling cone blockage, spectral and non-spectral interferences. The use of an ETV device also enables initial removal of the solvent, drastically reducing the intensity of the background spectrum, thereby permitting determination of elements, like Fe (Whittaker *et al.*, 1989), which have major isotopes isobaric with background species such as ArO.

4.2.2 *Instrumentation*

4.2.2.1 *Vaporisation cells*   The vaporisation cell or 'furnace' should be small, so that samples may be reproducibly deposited at the same spot and so that rapid heating may be provided to vaporise the sample in one 'pulse'. However, it must be large enough to permit precise metering of samples from liquid dispensers and micropipettes.

A variety of materials have been used in the construction of vaporisation cells, most are similar to those which have been used for ETA–AAS. Generally these are either metals (Ta, W, or Re) or graphite. Cells in the form of a graphite rod (Darke *et al.*, 1989; Date and Cheung, 1987; Gray and Date, 1983; Whittaker *et al.*, 1989; Williams, 1989), graphite platform (Gregoire, 1988; Hall *et al.*, 1988b; Park and Hall, 1987, 1988), graphite tube (Hall *et al.*, 1990b; Newman *et al.*, 1989), rhenium filament (Park *et al.*, 1987a, b), tungsten filament (Park and Hall, 1988), tungsten ribbon (Tsukahara and Kubota, 1990) and a graphite tube with a tantalum and tungsten rod (Shen *et al.*, 1990) have all been used for ETV–ICP–MS. Although graphite is most commonly used, it is recognised from ETA–AAS work that metals do have some advantages. They have a rapid heating rate, freedom from carbide forming problems, and do not absorb sample into the furnace material, giving low blank levels. However, metal furnace materials evaporate, partly from oxidation by impurities in the argon or by the formation of volatile compounds with the sample matrix. These may give interferences and shorten the lifetime of the furnace. The addition of hydrogen to the argon carrier gas, in order to reduce oxidation, has been shown to be necessary to prolong the life of furnaces (Suzuki *et al.*, 1981).

Park and Hall (1988) reported that tantalum was preferable to tungsten for metal furnaces as it was extremely resistant to acid attack. Tungsten was found to degrade quickly and became brittle and an isobaric interference on ^{203}Tl resulted from the formation of ^{186}W^{16}O^{1}H. In contrast graphite was found to give about half the signal obtained with a tantalum filament. In this work hydrogen was not used. However, where it had been added, tungsten was found to be superior because of less interaction between it and analytes (such as Co and Ni), less oxidation and better reproducibility (Tsukahara and Kubota, 1990).

Despite the advantages that metal furnaces may offer, graphite is often preferred. Porosity can be overcome by using pyrolytic or pyrolytically coated graphite, metal carbide coated graphite or a ceramic material, pyrolytically coated with graphite. However, the performance of coatings is dependent upon the analytes under investigation and the operating conditions used. Carbide formation can be overcome by the additon of reactive gases to the carrier gas flow. The ETV device used by Hall *et al.* (1990) in the determination of picogram (10^{-12} g) levels of U and Pu, was a commercial system (VG Elemental Ltd, UK), based around a graphite tube. In this work about 1% trifluoromethane (Freon 23) was added to the argon carrier gas to prevent the actinide elements reacting with the graphite furnace at high temperatures (2600°C) to form carbides. Their formation leads to severe memory effects and poor analyte sensitivity. However, in the presence of Freon, the elements react with the liberated fluorine radicals to form volatile fluorides instead of

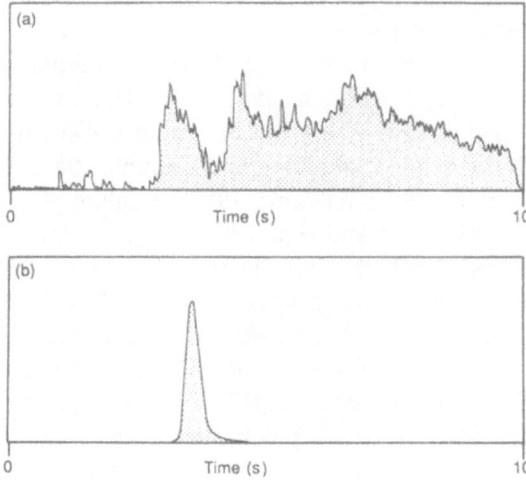

Figure 4.2   Effect of introducing 0.1% Freon into the argon carrier gas flow. 30 μl of 0.1 ng ml⁻¹
U solution were injected in each case. (a) No Freon, 2760 peak area counts; and (b) with Freon
addition, 108 453 peak area counts. After Hulmston and Hutton (1990).

forming carbides. This improves the transport efficiency, reduces memory
effects and increases sensitivity (Figure 4.2).

4.2.2.2   *Electrothermal vaporiser designs*   The first studies of ICP–MS with
ETV sample introduction (Gray and Date, 1983) were carried out using an
ETV system similar to that described by Gunn *et al.* (1978), which is shown
in Figure 4.3. The glass chamber volume was 1 litre and the furnace was a
carbon rod with a 5 μl capacity dimple at the centre. This design has been
used by other workers (Darke *et al.*, 1989; Date and Cheung, 1987; Whittaker
et al., 1989; Williams, 1989). A similar commercial system (Seiko instruments
Inc., Japan) with a glass chamber volume of 300 ml and a tungsten furnace
has been characterised by Tsukahara and Kubota (1990).

The first system designed specifically for ICP–MS (Figure 4.4) also used
the glass chamber concept, however the volume of envelope above the furnace
was only about 5 ml (Park *et al.*, 1987a). It was suggested that larger volumes
can introduce an excessive dilution of the sample vapour by the carrier gas,
which leads to an increase in duration and a reduction in the level of the
transient signal. However, smaller volumes may result in appreciable vapour
condensation on the walls of the chamber.

In the system using an open furnace the carrier gas is allowed to flow over
the furnace and entrains the vapour that is evolved above it during heating.
The vapour is carried to the ICP by a length of plastic tubing. Although the
analyte vapour is capable of being transported large distances (10 m) without
serious loss or dilution (Ng and Caruso, 1985), all ETV–ICP–MS work to

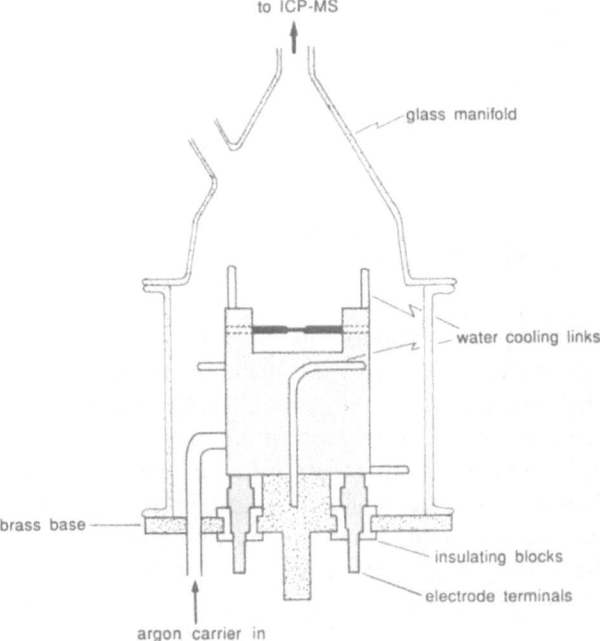

Figure 4.3   Electrothermal vaporisation device described by Gunn *et al.* (1978).

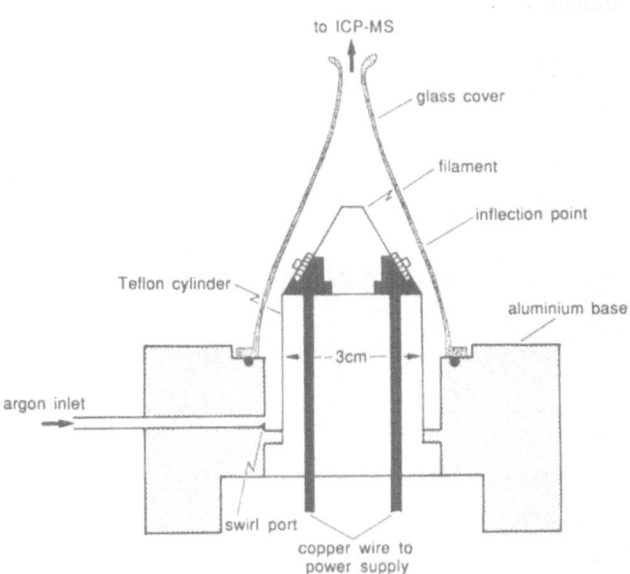

Figure 4.4   Electrothermal vaporisation device described by Park *et al.* (1987a).

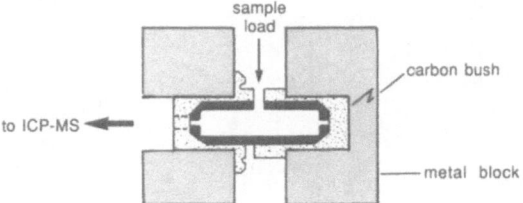

Figure 4.5 Simplified diagram of modified GFAAS module for electrothermal vaporisation, available from Perkin Elmer Corporation.

date has used a vapour transport tube of < 1 m in length and 5–7 mm in diameter. For convenience the ETV device should be placed reasonably close to the ICP–MS instrument, although remote sampling is a possibility if the vapour can be transported down 10 m of tubing.

The commercial system available from Perkin Elmer Corporation (USA), is a modified GFAAS module, which can be seen in cross-section in Figure 4.5. The analyte vapour is generated in the graphite tube and is forced out through the end of the tube by argon carrier gas flow and into the ICP via a length of plastic tubing. The commercial system from VG Elemental Ltd (UK) developed specifically for ICP–MS, is of similar design. It consists of a graphite tube mounted in carbon bushes inside a quartz sleeve (Figure 4.6). The volume inside the quartz sleeve is about 17 ml and the internal volume of the graphite tube is about 0.8 ml.

No systematic comparisons between the various designs have been reported making it difficult to assess which is optimum for ICP–MS. Successful work

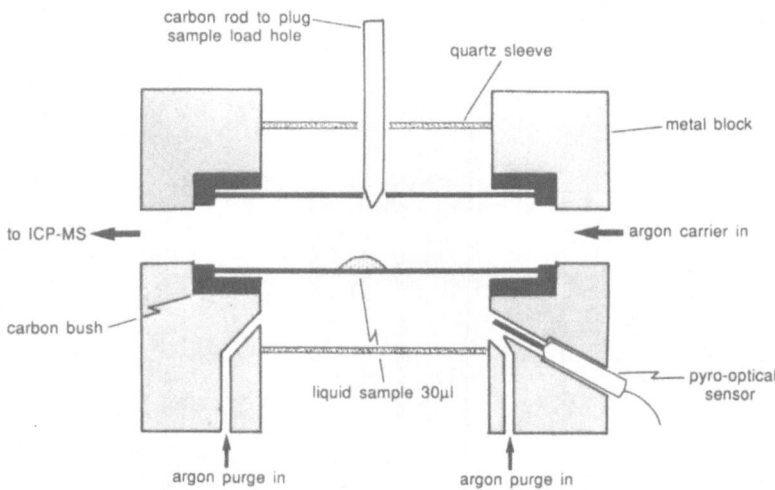

Figure 4.6 Commercially available electrothermal vaporisation device, manufactured by VG Elemental Ltd.

Table 4.1 Summary of ETV–ICP–MS operating conditions for elements determined

Elements determined	Matrix type	Ar carrier gas flow (l min^{-1})	Additional gas flow (l min^{-1})	Vaporisation temperature (°C)	Furnace material	Reference
Te	Biological fluids	4.5–5.0	O$_2$, 1.5	2650	Graphite coated ceramic tube	Newman et al. (1989)
Pu, U	Aqueous solutions	0.6–0.8	Freon, 0–0.005	2600	Graphite tube	Hall et al. (1990b)
Fe	Aqueous solutions	0.8	H$_2$, 0.05	2490	Tungsten ribbon	Tsukahara et al. (1990)
Co	Aqueous solutions	1.0	H$_2$, 0.05	2410	Tungsten ribbon	Tsukahara et al. (1990)
Pb	Aqueous solutions	1.0	H$_2$, 0.05	2410	Tungsten ribbon	Tsukahara et al. (1990)
Cd	Aqueous solutions	3.0	—	1250	Rhenium filament	Park et al. (1987a)
Pb	Aqueous solutions	2.0	—	2200	Rhenium filament	Park et al. (1987a)
Cu	Aqueous solutions	1.0	—	1900	Rhenium filament	Park et al. (1987a)
Fe	Aqueous solutions	2.0	—	2200	Rhenium filament	Park et al. (1987a)
Ni	Aqueous solutions	2.0	—	2200	Rhenium filament	Park et al. (1987a)
Mo, W	Geological materials	1.5	Freon, 0.003	3000	Graphite platform	Park and Hall (1987)
Tl	Geological materials	1.5	Freon, 0.003	3000	Tungsten filament	Park and Hall (1988)
Ir, Pd, Pl, Ru	Geological materials	1.0	—	3000	Graphite platform	Gregoire (1988)
Pb	Geological materials	1.0	—	Not given	Graphite rod	Date and Cheung (1987)
Fe	Serum	0.7	—	Not given	Graphite rod	Whittaker et al. (1989)
Pb	Fly ash	0.55	—	1400	Graphite rod	Darke et al. (1989)

has been reported for the determination of a number of elements in a variety of sample types (Table 4.1).

4.2.3   Operating parameters

4.2.3.1   *ETV system optimisation*   In addition to the ICP–MS parameters commonly optimised, such as carrier gas flow, forward power and sampling distance, the use of an ETV system requires optimisation of additional parameters such as the temperature, rate and duration of the drying, ashing and vaporisation stages. A number of studies have been carried out to investigate operating conditions.

An examination of the effect of carrier gas flow rate and (Re) filament temperature on As and V signals (Park *et al.*, 1987a) showed that signals were very sensitive to both. It is reported that only small changes in carrier gas flow rate seriously degrade analyte signals, as this probably affects the system optimisation (Gray and Williams, 1987b; Williams, 1989). This work also showed that for a given flow rate, the signal initially increased with vaporisation temperature and then declined. The signal for $1\,\mu g\,ml^{-1}$ Fe at a series of increasing graphite furnace vaporisation temperatures is shown in Figure 4.7 for a carrier gas flow of $0.71\,min^{-1}$. This observation has also been made for Co and Pb on a tungsten furnace (Tsukahara and Kubota, 1990). The loss of analyte signal at higher furnace temperatures may be due to the rapid expansion and contraction of the argon in the chamber caused by the hot furnace. Thus the carrier gas flow into the plasma is changing at a critical time, as the analyte vapour reaches the plasma, and is therefore no longer optimised during the integration. It is suggested that the carrier gas

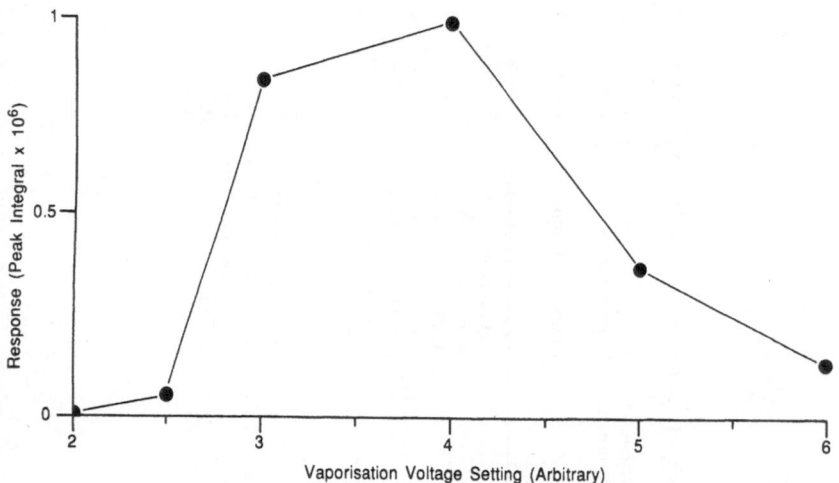

Figure 4.7   Signal for $5\,\mu l$ of $1\,\mu g\,ml^{-1}$ Fe at a series of increasing graphite furnace vaporisation temperatures.

Table 4.2 A typical ETV heating cycle or furnace programme[a]

Stage	Temperature(°C)	Ramp period[b](s)	Hold period[c](s)
Drying	100	15	45
Ashing	1700	30	10
Vaporisation	3000	0	5

[a] From Park and Hall (1987).
[b] Ramp period = time taken to reach set temperature from previous set temperature.
[c] Hold period = time period ETV cell is held at set temperature.

flow rate should be set below that at which the system was optimised to compensate for the transient effect. This implies that the optimum flow and temperature combination has to be found experimentally for each element. A list of elements, carrier gas flow rates and vaporisation temperature used for analysis of a variety of sample types is shown in Table 4.1.

In modern ETV systems the furnace power supply can be programmed to dry the sample after injection, ash it at an intermediate temperature (e.g. 500°C) and vaporise it. The temperature and duration of each of these steps can usually be controlled over a wide range. Optimising the operating conditions of the furnace is a vital step in the development of ETV–ICP–MS analytical methods. In the drying phase, the solvent must be driven off slowly to avoid spitting, caused by the boiling. With an open furnace, sample will be lost resulting in a reduced signal. With a tube furnace, sample will be sprayed around the inside of the tube, which will result in inconsistent vaporisation from sample to sample. The matrix or organic component of a sample may be removed by ashing it at the highest furnace temperature that does cause analyte loss. This can be determined experimentally.

In the vaporisation stage, the lowest temperature that is a compromise between rapid response and maximum analyte signal must be chosen. Too high a temperature may damage the furnace or distil off contaminants, but too low a temperature may result in incomplete analyte vaporisation and lead to analyte memory effects. A tube clean (i.e. a high temperature step) can be included in the analysis programme. A typical furnace programme can be seen in Table 4.2.

4.2.3.2 *ICP–MS system optimisation* There is some debate as to the best method of optimising the ICP–MS system (ion optics, plasma-sampling cone alignment). Some workers maximise signal response using a conventional nebuliser system, then switch to the ETV device (Park *et al.*, 1987b; Tsukahara and Kubota, 1990). This may not be advisable where the optimum ion lens settings change with the water content of the plasma (Hutton and Eaton, 1987; Williams, 1989). The system may be optimised by first monitoring the signal for ^{12}C normally present as a CO_2 impurity in the argon. Alternatively a volatile element such as Cd or Hg may be vaporised slowly from the furnace (Date and Cheung, 1987; Whittaker *et al.*, 1989; Williams, 1989; Gregoire, 1988; Hall *et al.*, 1990b).

4.2.3.3 *Data collection*   Data collection, or analyte ion counting, for ETV–ICP–MS may be carried out in two ways. In the first, the mass spectrometer is set to transmit only one isotope (single mass monitor) of the analyte to be determined. This mass can be monitored throughout the ETV cycle (i.e. drying, ashing and vaporisation stages). A transient signal such as the one shown in Figure 4.8 is produced and is analogous to the signal received in ETA–AAS. This method of data acquisition is useful if information on only one element is required or the vaporisation process is being investigated. However, this method does not make use of the quasi-simultaneous multi-element capability of ICP–MS. Alternatively data can be collected by scanning or peak hopping rapidly across a series of isotopes during the vaporisation step. This can provide both multi-element and isotopic information about a sample.

   In either, the period over which ions are counted is an important parameter. If single mass monitoring is used, either transient peak height or peak area can be determined. The peak area is generally recommended as peak height of a transient signal can show large fluctuations. Each element is vaporised from the surface of the furnace at a different temperature and the rate at which this occurs will depend on the physical and chemical characteristics

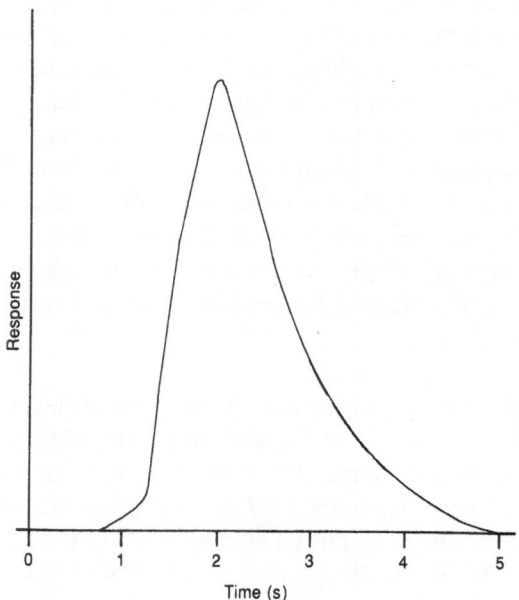

Figure 4.8   Idealised electrothermal vaporisation transient signal.

of the element, as well as the rate of heating of the furnace. The design characteristics of the system, such as the furnace material and chamber volume will also affect vaporisation. Thus, the total measurement time for an analysis is dependent on the element(s) being determined and the physical and chemical environment encountered during vaporisation.

The transient signal for a single element may have a duration of between 0.5 s and 10 s, but the occurrence of the peaks of transient signals, from a single vaporisation, for different elements may not be coincident. Therefore, for multi-element work, vaporisation and data collection must be correctly inter-related in order that the signals for all elements are collected.

Both scanning or peak hopping methods can provide multi-element or isotope ratio information on a sample. However, unlike single mass monitoring where the mass spectrometer transmits the same mass continuously, the total integration time is divided between the masses that are scanned or peak hopped. This results in a significant reduction in sensitivity. If sensitivity is an important criterion, data should be acquired using single mass monitor.

4.2.3.4 *Calibration strategies* Chapter 6 deals in detail with different calibration strategies for ICP–MS. These also apply to ETV–ICP–MS, however several key points must be considered when choosing a calibration method.

The analyte signal from ETV is transient, with a duration of a few seconds. The magnitude of the signal for an analyte of a given concentration in a matrix free solution with an optimised system, will depend mainly on the volume and position of each sample deposited in the furnace and the condition of the furnace material. If any one of these factors is not the same for each analysis, the transient peak size and shape will not be reproduced. Even with automated sample injection, the location and volume of the sample deposited can vary. This is compounded by continual deterioration in the condition of the furnace material surface with successive vaporisations. The net result is that precision is often significantly poorer than that obtained by nebulisation ICP–MS.

Precisions of 20% are typical for a manually loaded ETV system, and this affects the precision of analysis using any external calibration method, including standard additions. An internal standard can only be applied if either scanning or peak hopping of several masses is used. Such a standard has to be carefully chosen for an analysis, as it must be close in mass as well as having very similar chemical and physical properties to the element(s) of interest.

If more than one isotope of an element is monitored during vaporisation, and the precision for isotope ratios is calculated, it is generally about an order of magnitude better than the precision of individual isotope determinations. Table 4.3 shows the counts for ^{54}Fe, ^{56}Fe and ^{57}Fe, integrated over 10 s, obtained from ten consecutive analyses of 5 μl aliquots of a standard solution containing 1 $\mu g\,ml^{-1}$ Fe. Clearly, the precision is greatly improved

Table 4.3 Comparison of precision for isotope peak integrals and isotope ratios of natural iron by ETV–ICP–MS[a]

Peak integrals

	^{54}Fe	^{56}Fe	^{57}Fe
	28246	439324	11968
	27681	432494	11891
	33535	520067	14361
	30326	475534	13153
	18766	296968	8133
	35754	549414	15516
	22909	358506	10254
	23623	378846	10617
	23279	376682	10666
	21551	349739	9900
Mean	26567	417757	11646
SD	5196	75993	2101
RSD (%)	19.56	18.19	18.04

Isotope ratios

	^{54}Fe:^{56}Fe	^{57}Fe:^{56}Fe
	0.0643	0.0272
	0.0640	0.0275
	0·0645	0.0276
	0.0638	0.0277
	0.0632	0.0274
	0.0651	0.0282
	0.0639	0.0286
	0.0624	0.0280
	0.0618	0.0283
	0.0616	0.0283
Mean	0.0634	0.0279
SD	0.0011	0.0004
RSD (%)	1.75	1.59

[a] From Williams (1989).

by measurement of the isotope ratio rather than integrated counts at one mass.

The only calibration method that can exploit the use of isotope ratios is isotope dilution and it has been used to good effect in a number of applications (Park and Hall, 1987, 1988; Gregoire, 1988). It does have some drawbacks, and these are detailed in chapter 6.

4.2.4 *Applications and analytical performance of ETV–ICP–MS*

Few applications have been reported for ETV–ICP–MS. The majority have been for the analysis of geological materials and the remainder of biological or environmental samples. It is particularly useful where small sample volumes are available and it may have useful potential in the reduction of matrix and interference effects.

4.2.4.1 *Geological materials*   The ETV system first described by Park *et al.* (1987a) has been used for a series of geological applications. In the first, Mo and W were determined in 15 sediment, soil and rock reference materials using isotope dilution as a method of calibration (Park and Hall, 1987). In order to determine these elements at low levels by solution ICP–MS, an alkaline fusion method was devised to decompose the samples. A subsequent separation procedure, using activated charcoal was found to be necessary, as the leachate of the fusion had a high dissolved salt content ($\leqslant 4\%$). The separation step was time consuming and contributed to the 'blank' level, due to impurities in the reagents. In an attempt to eliminate the separation procedure, the fusion leachate was analysed directly for Mo and W by ETV–ICP–MS.

The furnace used for this study was a graphite platform. In order to prevent the formation of carbides, Mo_2C, MoC, W_2C and WC, 0.1% Freon was added to carrier gas. The ETV operating conditions used in the determination of Mo and W are shown in Table 4.2. The optimum conditions for the heating cycle were established following an investigation to identify the temperatures required to remove Na from the sample on the furnace and obtain the maximum Mo and W signals.

Much of the inferior precision obtained with ETV can be attributed to pipetting microlitre volumes of solution on to the furnace and to inconsistencies of vaporisation between individual analyses. Park and Hall used isotope dilution for Mo and W, and obtained precisions of about 1–2% RSD compared to 20% RSD with external calibration.

The determination of W proved difficult due to the loss of signal during vaporisation and memory between samples. These are symptoms of carbide formation, which appeared to occur in spite of the presence of 0.1% Freon. The use of an alternative furnace material may circumvent these problems although there are no reports in the literature of this for W determination. Despite these difficulties good W results were obtained on reference materials down to $0.06 \mu g\,g^{-1}$. Conversely, compared to nebulisation ICP–MS, the determination of Mo by ETV–ICP–MS resulted in much improved productivity (no matrix separation required) and a superior limit of detection of $0.03 \mu g\,g^{-1}$ ($0.08 \mu g\,g^{-1}$ by nebulisation).

In another application of this system, Tl was determined in 17 sediment and rock reference materials (Park and Hall, 1988). Two decomposition procedures were tested, both using HF and oxidising acids, one also incorporating fusion of the decomposition residue with $LiBO_2$. Good accuracy and precision were achieved in the direct analysis of the acid leachates, as demonstrated by comparison of the results with the certified values of the reference materials. A detection limit of $9\,ng\,g^{-1}$ was obtained, with an average precision of 4% RSD for Tl concentrations in the range $5-1400\,ng\,g^{-1}$. Isotope dilution was used as a method of calibration.

Sample introduction by ETV also permitted direct determination of Tl in solutions which contained about 1% dissolved solids. However, a spectral interference caused by a high concentration of Pb necessitated separation of Tl by extraction into isobutyl methyl ketone (IBMK) prior to analysis for those samples whose Pb contents were greater than $500\,\mu g\,g^{-1}$. A 5000-fold excess of a 'matrix' element should not cause spectral overlap, and such a separation would not be necessary if better mass analyser resolution had been available (Gray and Williams, 1987b; Williams, 1989). With adequate resolution a difference of concentration of 10^5 can be achieved. Initially in this study a graphite platform furnace was used, although problems were encountered in locating each $5\,\mu l$ aliquot of solution on the platform. Later studies with a tantalum filament showed it to be a very suitable furnace material, preferable to graphite.

In a similar system a pyrolytic graphite furnace was used by Gregoire (1988) for the determination of Ir Pd, Pt and Ru in geological materials. The use of ETV–ICP–MS offered better limits of detection than other techniques such as neutron activation analysis (NAA) or ETA–AAS and extended platinum group element determinations to very small samples such as mineral separates and sea-floor hydrothermal vent particulates.

Rock powders were decomposed using the nickel sulphide fire assay technique. Samples of $10\,g$ were pre-concentrated to a final solution volume of $0.5\,ml$. Isotope dilution was used as a method of calibration and the addition of $500\,\mu g\,ml^{-1}$ of Ni as nitrate to the samples resulted in a ten-fold enhancement in the analyte ion signal. Quantitative data for a series of ultramafic rock samples determined by NAA, nebulisation ICP–MS and ETV–ICP–MS compared well with one another over concentration ranges of $5-800\,ng\,ml^{-1}$ for Pd and Pt, $2-300\,ng\,ml^{-1}$ for Ru and $0.6-25\,ng\,ml^{-1}$ for Ir.

The application of ETV–ICP–MS to the determination of lead isotope ratios in geological materials has been described by Date and Cheung (1987). The ETV device was the graphite rod system first described by Gray and Date (1983). The accuracy and precision of the technique were assessed by analysing $5\,\mu l$ of $1\mu g\,ml^{-1}$ NBS 981 (calibrated lead isotopes) in scanning mode. Isotope ratio precisions of about 1.5% were obtained, close to those expected from counting statistics. Accuracy was within the limits of precision, except for $^{208}Pb : ^{204}Pb$ which showed a bias of 5.5% from the recommended value. Lead isotope ratio data obtained by ETV–ICP–MS and nebulisation ICP–MS for a galena concentrate and an altaite concentrate were compared and showed good agreement. This feasibility study showed that ETV–ICP–MS could be used as a rapid survey technique for lead isotope ratios in very small sample solution aliquots.

4.2.4.2 *Biological materials* The system used by Date and Cheung (1987) was later used by Whittaker *et al.* (1989) to determine iron isotope ratios in 5 μl aliquots of blood serum, without prior sample preparation. The organic matrix was removed with an ashing step. Following oral administration of 5 mg of enriched $^{54}FeSO_4$ and intravenous administration of 200 μg enriched $^{57}FeSO_4$ to non-pregnant women, the $^{54}Fe:^{56}Fe$ and $^{57}Fe:^{56}Fe$ isotope ratios in serum were measured, requiring 20 min per sample in quintuplicate. Changes in the fractional absorption of iron during human pregnancy could therefore be assessed.

The determination of the entire range of iron isotope ratios by ICP–MS with sample introduction by conventional nebulisation is prone to significant error as ^{54}Fe and ^{56}Fe suffer from severe polyatomic ion interferences by $^{40}Ar^{14}N^+$ and $^{40}Ar^{16}O^+$, respectively. Sample introduction with ETV significantly reduces the levels of these interferences since there is no accompanying solvent. Figure 4.9 shows the blank spectrum of the mass region 54–57 m/z in the absence of water, with peaks at 54 and 56 m/z reduced to background levels. The use of ETV–ICP–MS offers excellent scope for iron stable isotope tracer studies and the existence of polyatomic ions need not restrict these kinds of investigations.

Some tellurium containing compounds have demonstrated immuno-stimulatory and anti-tumour effects. Since extremely low doses have been proposed for use in laboratory animals and humans, it is necessary to detect picogram levels of these metal-based therapeutic agents. The Te content of biological fluids was assessed using ETV–ICP–MS by Newman *et al.* (1989). A Te containing drug (AS101) was prepared and used to spike blood plasma and urine samples. These were prepared by combining 0.1 ml of plasma, urine or a phosphate buffer with an equal volume of hyamine hydroxide matrix modifier. A 10 μl volume of prepared drug-hyamin hydroxide solution was

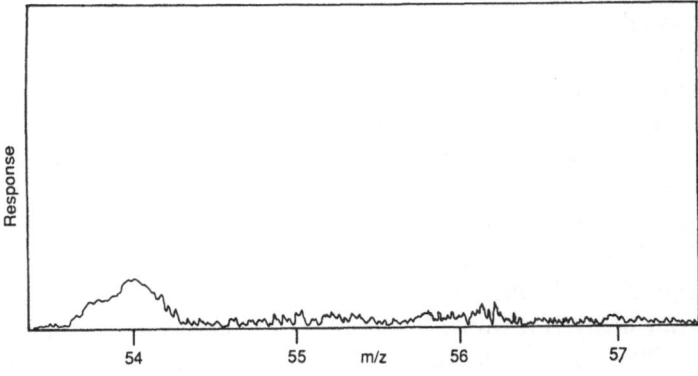

Figure 4.9 Blank spectrum of the mass region 54–57 m/z in the absence of water, with peaks at 54 and 56 m/z reduced to near background levels.

placed in the graphite furnace, a ceramic tube pyrolytically coated with graphite. Although the action of the matrix modifier was not understood its presence was essential to optimise rapid release of Te from the graphite-coated tube. The lower limit of detection of ^{130}Te was found to be 0.34 ng ml^{-1} (3.4 pg absolute) in the phosphate buffer, although it was degraded to 5.7 ng ml^{-1} (57 pg absolute) in the blood plasma matrix. However, this was still almost an order of magnitude better than limits of detection reported for other techniques and suggests that the technique was suitable for monitoring Te in biological samples.

4.2.4.3 *Environmental samples* A detailed procedure has been outlined by Park *et al.* (1987b) for the analysis of environmental and biological reference materials. Results obtained for the determination of As, Ag, Cu, Mn, Pb, Rb, V and Zn in orchard leaves (NBS 1571) and oyster tissue (NBS 1566) were in good agreement with the certified values with precisions between 2% and 13%. Studies on the effect of 1 μg of various elements on the signal of 1 ng of As, Cd and Cu showed that in the presence of these levels of matrix, both vaporisation and plasma loading interferences are possible, which has implications for real sample analysis. Absolute detection limits at the picogram level were obtained that were an order of magnitude better than those obtained by nebulisation ICP–MS.

In the wake of the recent nuclear accident at Chernobyl in the USSR and with increased public awareness of environmental issues, it is imperative that the nuclear industry is seen to maintain a high level of environmental monitoring. It is particularly important to carry out biological monitoring of the personnel working in nuclear installations. One necessary procedure is the determination of plutonium in urine at the femtogram level, which is normally carried out with alpha particle counting following chemical separation of the plutonium from the urine. The limit of detection by radiochemical methods is approximately 80 fg of Pu. However, at this level a counting time of about 20 days is required (Hall *et al.*, 1990b).

The feasibility of using ETV–ICP–MS to determine femtogram levels of Pu and U was investigated by Hall *et al.* (1990). It was found necessary to add about 1% Freon to the carrier gas to inhibit the formation of uranium and plutonium carbides. Absolute limits of detection of 0.8 fg for ^{235}U and 2.1 fg for ^{244}Pu were obtained in single mass monitoring mode. Signal response was found to be linear with concentration over the range 5–500 fg, with a precision for Pu of 8% RSD at an injected (50 μl) concentration of 50 fg. The ^{242}Pu:^{244}Pu isotope ratio was determined in scanning to a precision of 9% RSD and an accuracy of 2.2%.

For urine analysis it would be necessary to determine the concentration of both ^{230}Pu and ^{240}Pu to calculate the isotopic composition of Pu and therefore the total alpha activity. Unfortunately the measurement of these masses in scanning mode results in a significant reduction in sensitivity compared to single mass monitoring. Despite this current drawback ETV–ICP–MS has tremendous potential for this type of ultra-trace analysis.

4.3 Vapour generation and gas phase introduction

4.3.1 *Introduction*

Vapour sample introduction methods have been successfully used to introduce selected analytes into many types of atomic spectrometric instrument. It would be more widely practised if methods were available by which large suites of elements could be volatilised in a convenient manner. The principal benefit of using vapour introduction is that it avoids the use of a nebuliser. A pneumatic nebuliser can only offer about 1% transport efficiency, and some designs are prone to blockage from samples containing suspended solids or high levels of dissolved solids. Vapour introduction offers the potential for 100% transport efficiency, no blockage problems and the possibility of matrix separation and analyte pre-concentration. An additional benefit is that the plasma is often operated in the absence of water, which significantly reduces the levels of many polyatomic ion species.

Studies of vapour generation sample introduction for ICP–MS have so far been concentrated in two areas; (i) hydride and mercury vapour generation and (ii) osmium tetroxide vapour generation. The former have been used with a variety of atomic spectrometric techniques, but osmium tetroxide vapour generation is a relatively new technique which, used in conjunction with ICP–MS, can provide useful osmium isotope information. The purity of gases used in the manufacture of electronic devices can also be determined by ICP–MS.

4.3.2 *Hydride generation*

4.3.2.1 *Principles*   The elements As, Bi, Ge, Pb, Sb, Se, Sn and Te form hydrides, which are gaseous at ambient temperatures. These can be generated easily from aqueous solutions in a reducing environment. Mercury is reduced to its volatile elemental form in the same reaction. The most frequently used method of hydride generation is the acid-borohydride reaction

$$NaBH_4 + 3H_2O + HCl \rightarrow H_3BO_4 + NaCl + 8H \xrightarrow{\;E^{m+}\;} EH_n + H_2$$

where E = hydride forming element of interest and m may or may not equal n.

A number of experimental variations have been utilised to effect this reaction, such as alternative reducing agents or acids. Although generation of the hydride itself is relatively straightforward, it is essential that it is transported quantitatively and reproducibly, and introduced with the minimum disturbance of the plasma. In practice, the process can become quite complex, although the technique has progressed to the point that commercial systems compatible with various plasma spectrometers are available. For those interested in the background to the generation of gaseous hydrides for atomic spectrometry, reviews by Robbins and Caruso (1979), Nakahara (1983) and Thompson and Walsh (1989) can be recommended.

4.3.2.2 *Instrumentation*   Both continuous and batch processes have been used for hydride generation. Continuous hydride generation is the only method to have been used for ICP–MS. In this procedure the reagents are continuously pumped, usually by a multi-channel peristaltic pump, into some type of mixing chamber, where the acid-borohydride reaction takes place. The volatile hydrides and the main gaseous product, hydrogen, flow into a gas-liquid separator, where the liquid passes to drain and argon carrier gas is introduced to sweep the hydride and hydrogen gas mixture into the ICP. In the batch method, an aliquot of acid solution containing the analyte elements is mixed rapidly with an aliquot of borohydride reagent. The resulting volatile hydrides are swept into the ICP directly (Pickford, 1981) or as a plug following condensation in a liquid nitrogen trap (Hahn *et al.*, 1982). This allows venting of the copious amounts of hydrogen formed, rather than introducing it into the plasma.

Continuous hydride generation from ICP–MS   Several of the commercial continuous hydride generation systems used for ICP–MS are based on the work of Thompson *et al.* (1978a, b) for ICP–AES. Although there are differences in detail, the basic layout of the glass phase separator systems is the same and may be seen in Figure 4.10. The main requirements are: (i) efficient mixing of the acid sample solution and the borohydride; (ii) a short period of time to complete the reaction; (iii) separation of the product gases from the liquids; (iv) uniform mixing of product gases with the argon carrier gas and (v) a small positive pressure to sweep the gases into the ICP.

These requirements are achieved by having a high speed three channel peristaltic pump used with narrow bore silicone rubber tubing. Flows of between 3 and 10 ml min^{-1} are required for the liquids. A high flow rate minimises the time between insertion of the uptake tube into the sample solution and stabilisation of the analytical signal. Narrow bore tubes minimise the dead volume of sample required to fill the system. The four port valve should also have a fine bore and preferably be constructed in a hard wearing plastic such as Kel-F. This valve allows virtually instantaneous switching between the acid sample solution and the blank acid, which ensures there is no interruption in hydrogen production during the change-over of test solutions, when a new sample is introduced. If a four port valve is not available, manual interchange of sample and blank solutions is necessary. This is less direct than using a valve, and will allow a short interruption in

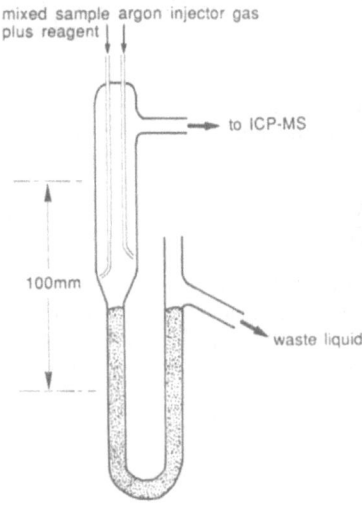

Figure 4.10 Configuration of a glass phase separator for hydride generation. After Thompson *et al.* (1978a).

the stream of hydrogen entering the ICP. This may result in an increase in reflected power in the ICP, due to mis-match between the plasma and generator, which can extinguish the plasma.

The connecting tube between the mixing point and the phase separator should be short, 10 cm or less, so that the reaction is almost complete when the liquid falls into the separator. Longer connecting tubes or delay coils can improve signal stability, however certain chemical interference effects are made worse by the delay coils.

The effervescent reaction in the phase separator can cause a certain amount of spray. This may be removed from the gas flow by a loose plug of glass wool in the gas line between the separator and the plasma torch.

The phase separator may be almost any device that allows separation of the product gases from the liquids. A porous PTFE tube separator (Wang *et al.*, 1988) exhibits faster hydride transfer and a lower signal fluctuation that a Thompson U-shaped separator (Thompson *et al.*, 1978a). A tubular membrane gas liquid separator suppresses peristaltic pump noise and prevents chloride vapour entering the plasma, enabling As and Se to be determined free from polyatomic interference (Branch *et al.*, 1991).

4.3.2.3 Operating parameters

Hydride generation system optimisation The concentrations of acid (usually HCl), reducing agent (usually $NaBH_4$) and oxidants (e.g. H_2O_2) are critical for efficient hydride generation. Table 4.4 is a summary of the optimum operating conditions for determination of lead by hydride generation

Table 4.4 Optimum operating conditions for the determination of lead by hydride generation ICP–MS[a]

ICP conditions	
Forward power	$1.2\,kW$
Reflected power	$< 5–15\,W$
Carrier gas flow	$1.46\,l\,min^{-1}$
Auxiliary gas flow	$1.4\,l\,min^{-1}$
Coolant gas flow	$11\,l\,min^{-1}$
Hydride generator	
Sample solution uptake	$9.4\,ml\,min^{-1}$
Blank solution uptake	$9.4\,ml\,min^{-1}$
$NaBH_4$ solution uptake	$4.7\,ml\,min^{-1}$
Reagent concentration	
HCl	0.6%
H_2O_2	1.0%
$NaBH_4$	5.0%
NaOH	0.1%

[a] From Wang et al. (1988).

ICP–MS. The gas flow rates and plasma forward power are similar to those used in pneumatic nebulisation, except that the carrier gas flow is higher. The higher gas flow, (i) maximises analyte sensitivity; (ii) sweeps the hydrides into the plasma faster, so that there is less opportunity for unstable hydrides to decompose in the phase separator and connecting tubing; and (iii) increases the dilution of the hydrogen, which makes tuning the ICP easier.

The correct wash-out time may be determined by monitoring the analyte signal after switching from the sample to the blank, although a period of about 20 s would be sufficient for most elements with the arrangement shown in Table 4.4. However, high concentrations of bismuth may cause problems as the element tends to 'plate out' on the connecting tubing and injector, which leads to prolonged memory effects.

Like any chemical reaction hydride generation is temperature sensitive. It is advisable to have all samples and reagents equilibrated to the temperature of the laboratory where the analysis will take place. Although all of the hydride forming elements can be reduced by $NaBH_4$ in acid solutions, the efficiency of the reduction depends critically on the acid concentration and varies between different acids. The concentration in Table 4.4 is optimum for lead hydride generation, but not for multi-element hydride generation. Concentrations of HCl of between 10% and 40% are required to effect maximum reduction of As, Bi, Sb, Se and Te. However, a concentration of $< 1\%$ of HCl or a weak organic acid such as 1% m/v tartaric acid is required to effectively reduce Ge and Sn. Optimum reduction of Hg is achieved with 0.08% HCl.

The minimum $NaBH_4$ concentration should be about 1% wt/v. Above this

Table 4.5 Comparison of detection limits (ng ml^{-1}) for hydride generation techniques[a]

Element	Hydride AAS	Nebuliser ICP–AES	Hydride ICP–AES	Nebulise ICP–MS	Hydride ICP–MS
As	1.0	110	0.8	1.0	0.005
Se	1.0	70	0.8	6.0	0.02
Hg	0.01	120		3.0	0.4
Sb	1.0	90	1.0	0.5	0.004
Bi	1.0	90	0.8	0.3	0.02
Te	1.0	70	1.0	1.0	0.1

[a] From Powell et al. (1986).

the concentration is not critical and has little effect on the generation of element hydrides. However, at higher levels the plasma becomes unstable (Powell et al., 1986), probably due to excessive amounts of hydrogen being generated and carried to the ICP.

Table 4.5 shows some detection limits that may be obtained with hydride generation ICP–MS, and other spectrometric techniques. These values are not likely to be obtained in real sample analysis, due to higher blanks, instrument drift and shorter practical integrations.

To increase the efficiency of hydride generation, an oxidising agent may be added before the reduction step. Hydrogen peroxide is effective and has low blank values. Peroxodisulphate is more efficient but was found to produce high blank levels (Wang et al., 1988).

Interference effects   Interference from other chemical species can only occur at the reduction stage, as hydride formation ensures virtually complete separation from other solution constituents. The most serious interferences come from certain transition elements such as Cu and Fe, although not all hydride forming elements are affected. Thompson et al. (1978b) studied the effects of various matrix species on the recovery of As, Bi, Sb, Se and Te for hydride generation ICP–AES and concluded that interference is negligible in the determination of As and Sb, but for the determination of Bi, Se and Te a separation of the analytes from interfering transition metals is needed. The most serious interferences to the generation of lead hydride are Cu and Fe. Sodium cyanide (0.02% m/v) and sulphosalicylic acid (0.4% m/v) dissolved in the NaBH$_4$ solution may be used to eliminate Cu and Fe interferences respectively (Wang et al., 1988).

The determination of As and Se by hydride generation ICP–MS can be complicated by the presence of $^{40}Ar^{35}Cl^+$ and $^{40}Ar^{37}Cl^+$ polyatomic ions caused by excess chloride vapour entering the plasma with the carrier gas. This may be reduced by replacing hydrochloric acid with nitric or sulphuric acid. However, the reduction process is not as efficient (Dean et al., 1990). A more elegant method for the elimination of chloride interference is to use a tubular membrane gas liquid separator. This prevents chloride vapour entering the carrier gas flow and allows unambiguous determination of [75]As and [77]Se (Branch et al., 1991).

4.3.2.4 *Applications of hydride generation ICP–MS* The few applications that have been reported, include the analysis of water samples, sediments, rocks and biological materials.

Lead in NIST SRM 1643a, trace elements in water, has been determined, using isotope dilution (see chapter 6) as a method of calibration (Wang et al., 1988). Isotope dilution used as the precision of isotope ratios was better than that of individual isotopes. Furthermore, chemical interferences from Cu and Fe can suppress the total signal, but do not affect the isotopic ratio. In this application NBS SRM 983, radiogenic lead, at a concentration of 3 ng ml^{-1} was used as the spike. In addition to NIST SRM 1643a, synthetic lead standards at 1, 2 and 3 ng ml^{-1} were analysed by the same procedure. The results are shown in Table 4.6. The precision was between 0.03% and 3.2%. The poorest reproducibility measured was for 1 ng ml^{-1} lead due to the relatively high blank value (0.38 ng ml^{-1}). The absolute error was in the range of -2.6% to $+1.6\%$. The limit of detection was restricted to 0.01–0.05 ng ml^{-1} by the reagent blanks.

The precision and accuracy of lead isotope ratios were also determined with a set of lead isotope standards including NBS SRM 981, common lead and isotopically characterised galena samples. The reported and measured (mean) isotope abundances and statistical parameters for five repetitive determinations of NBS SRM 981 are listed in Table 4.7. A variety of matrices were tested for their effect on the generation of lead hydride. The most serious interferences were caused by Fe and Cu. Sulphosalicylic acid and sodium cyanide, dissolved in $NaBH_4$ were used to eliminate these.

A comparative investigation between pneumatic nebulisation and continuous hydride generation has been carried out for isotopic analysis of Se in

Table 4.6 Lead concentration measured by isotope dilution, hydride generation, ICP–MS; 3 ng ml^{-1} of NBS SRM 983 were used as spike and the [206]Pb and [208]Pb isotopes were measured; each measurement was repeated three times[a]

Sample type	Expected (ng ml^{-1})	Measured (ng ml^{-1})	SD (ng ml^{-1})	RSD (%)	Error (ng ml^{-1})
Synthetic, 0 ng ml^{-1}	0.0	0.38	0.0058	1.53	$+0.38$
Synthetic, 1 ng ml^{-1}	1.0	0.98	0.03	3.2	-0.02
Synthetic, 2 ng ml^{-1}	2.0	1.97	0.0006	0.03	-0.03
Synthetic, 3 ng ml^{-1}	3.0	3.05	0.015	0.5	$+0.05$
NBS SRM 1643a[b]	27 ± 1	26.3	0.32	1.2	-0.6

[a] From Wang et al. (1988).
[b] Diluted 10-fold.

Table 4.7 Lead isotope abundances in NBS SRM 981
common lead measured by hydride generation, ICP–MS[a]

	Isotope abundance (%)			
	^{204}Pb	^{206}Pb	^{207}Pb	^{208}Pb
Reported	1.43	24.17	22.13	52.26
Measured	1.42	24.14	22.08	52.34
RSE[b] (%)	+0.90	+0.14	+0.22	−0.15
RSD (%)	0.76	0.47	0.38	0.25

[a] From Wang et al. (1988).
[b] RSE, relative standard error $= (A_{measured} - A_{reported})/A_{reported} \times 100(\%)$; the reported values were measured by thermal ionisation mass spectrometry.

biological samples, which is of interest in human metabolic studies (Janghorbani and Ting, 1989). Background count rates for the hydride system were found to be 3–5 times larger than those for nebulisation. However, the signal to background ratios for Se were 30–50 times greater. Absolute detection limits with nebulisation sample introduction were 60, 20 and 20 ng for ^{74}Se, ^{77}Se and ^{82}Se, respectively. Those for hydride generation were 2, 0.6 and 0.6 ng, respectively.

The memory effect of the hydride procedure for Se determination was found to be severe. When a 10 ng ml^{-1} solution of Se was replaced with a 10% HCl blank, the signal intensity decayed to 3.3% of its original value in 3.5 min and took about 10 min for this to decrease to 1.2%. For isotope ratio determinations it is essential that background levels are achieved between samples.

The ratios ^{74}Se:^{77}Se and ^{82}Se:^{77}Se have been determined in a number of biological matrices, including human blood plasma, bovine liver (NIST 1577a), and rat brains (Janghorbani and Ting, 1989). The precision of the ratios was between 0.4 and 1.3% for ^{74}Se:^{77}Se and between 0.2% and 1.0% for ^{82}Se:^{77}Se, depending on the sample matrix. Accuracy in the determined ratios of ± 1.3% could be achieved with this method of sample introduction.

Although mercury does not form a volatile hydride, it can be reduced to volatile elemental mercury using the same apparatus and reagents that are used for hydride generation. In determining mercury, sample preparation and reagent contamination can be major problems. The risk of contamination, loss of mercury by adsorption to the walls of the storage vessel and loss of volatile mercury are all problems associated with sample preservation. Mixing liquid samples with 2 mM potassium permanganate acidified with 1 mM hydrochloric acid can stabilise mercury in solution for at least 1 month. Analysing samples immediately after sampling has taken place may reduce storage and preservation problems. Mercury contamination in reagents can be reduced by the use of sub-boiling acids and by bubbling nitrogen through the reducing agent (5% $NaBH_4$ in 0.1% NaOH) overnight.

Table 4.8 Mercury vapour ICP–MS analysis of some environmental materials[a]

Sample	ICP–MS (ng l^{-1})	Reference (ng l^{-1})[b]
Lake Vänern, 1	1.51	1.73
Lake Vänern, 2	1.46	1.70
Lake Vänern, 3	1.15	1.17
Lidan River	2.90	3.08
Klarälven River	2.34	2.55
MESS-1	164 ± 9.18[c]	171 ± 14[c,d]

[a] From Haraldsson *et al.* (1989).
[b] Reference results obtained by double amalgamation technique.
[c] In ng g^{-1}.
[d] Certified value (National Research Council, Canada).

A new quartz phase separator can lead to severe memory problems. Separators that are stored filled with water or acid for 48 h cause high blank values for several hours. Well used separators give better blank level, as do those heated to 130°C for 2 h.

For the generation of volatile mercury, tin(II) chloride is preferred as a reducing agent to sodium borohydride. This does not produce hydrogen, which leads to a more stable plasma and therefore better precision. However, the ICP–MS instrument will become severely contaminated with tin and also with lead which is a contaminant of the tin(II) chloride. Therefore it is not recommended as a reducing agent. Despite the problems of hydrogen production, a precision of 2–3% can be obtained for 100 pg of mercury reduced with $NaBH_4$. The absolute limit of detection for mercury with a vapour system is about 8 pg (Haraldsson *et al.*, 1989).

Sample introduction by vapour generation has been applied to the determination of mercury in reference sediment materials and natural water samples. The results of this study by Haraldsson *et al.* (1989) are shown in Table 4.8. For water analysis this method is very attractive as it offers low limits of detection and minimises the problems of contamination from sample handling. Mercury may be determined in a solid material after a suitable closed vessel acid digestion, such as the one suggested by Haraldsson *et al.* for the decomposition of sediments.

4.3.3 *Osmium tetroxide vapour generation*

4.3.3.1 *Introduction* Osmium and rhenium are heavy transition metals. Osmium has seven naturally occurring isotopes and rhenium has two. The ^{187}Re isotope is radioactive and decays via β decay to ^{187}Os and has a half-life of 4.35×10^{10} years. Osmium is a platinum group element whilst

rhenium has chemistry similar to molybdenum or tungsten. The difference in chemical behaviour of Re and Os results in a wide range of Re:Os ratios in nature. This elemental pair is potentially useful as a geochronometer (Lindner *et al.*, 1986, 1989) or isotopic tracer that can be used for direct dating many types of ore deposits and for studies of meteorites, mantle evolution and the Cretaceous-Tertiary boundary (Bazan, 1987; Dickin *et al.*, 1988; Richardson *et al.*, 1989; Russ *et al.*, 1987).

Despite the potential of the Re:Os system as an isotopic tracer, it has not been widely used. The high ionisation energy of Os (8.73 eV) precludes its determination by thermal ionisation mass spectrometry, but the high temperature of the ICP results in a degree of ionisation of 78%. Techniques such as secondary ion mass spectrometry (SIMS), resonance ionisation mass spectrometry (RIMS) and laser ablation microprobe mass analysis (LAMMA) (Lindner *et al.*, 1986, 1989) have all been used, but generally yield poor precision. Osmium is one of the least abundant stable elements and the radiogenic isotope (^{187}Os) makes up only 1–2% of the total osmium. Thus, counting statistics are one of the chief limitations in the precise determination of ^{187}Os abundances in geological samples, despite a limit of detection of approximately 0.02 ng ml^{-1} for Os.

A few ng ml^{-1} Os in solution can be oxidised so that volatile osmium tetroxide may be distilled directly into the plasma. This avoids the use of the inefficient spray chamber system and allows nearly two orders of magnitude improvement in Os count rates compared to solution nebulisation. Volatile osmium tetroxide distillation has a feature which is important for Re:Os and Os isotope ratio studies. The most abundant isotope of rhenium (62.9%) has a direct isobaric overlap with ^{187}Os. As the ^{187}Os signal is so small, Re is potentially a very serious interferent. Distillation of Os directly into the plasma yields an almost perfect separation from Re, so that with a small interference correction, even Re-rich geological samples can be analysed directly from solution without prior chemical separation. Separation is essential for ICP–MS determination of Os using a nebuliser and spray chamber. Although two isotopes of W have direct isobaric overlap with ^{184}Os and ^{186}Os, the distillation also provides a complete separation of Os from W.

4.3.3.2 *Experimental*

Instrumentation The osmium tetroxide generator consists of a glass sample chamber in which the dissolved sample is combined with the reactants. The oxide is produced in the sample chamber by means of a reaction in which the sample and a strong oxidising agent react to form OsO_4. The gas is transported upwards by the Ar carrier gas flow and is cooled in a condenser, maintained at about 5°C by ice-water. The condenser conserves reactants and removes most of the water vapour from the OsO_4 carrier gas stream resulting in a dry plasma. The improvement in sensitivity may, in part,

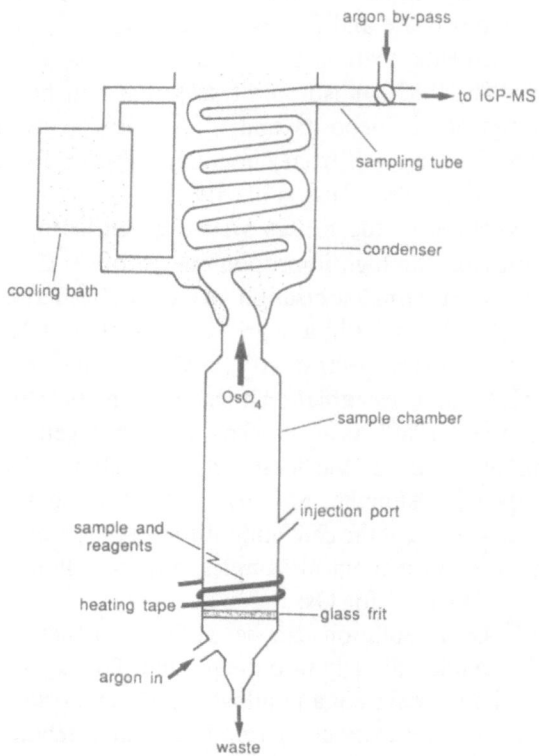

Figure 4.11 Schematic diagram of discrete batch osmium tetroxide vapour generator. After Richardson *et al.* (1989).

be related to the reduced solvent cooling of the plasma. The gas is then introduced into the plasma. A schematic diagram of a discrete batch OsO_4 generator is shown in Figure 4.11 and is based on the design described by Russ *et al.* (1987) and Bazan (1987). The application of a continuous flow apparatus with ICP–AES has also been described by Bazan, but its use with ICP–MS has not been reported.

Oxidising agent and other reagents   Periodic acids have been found to be the most suitable oxidising agents. Bazan (1987) found 10% *ortho*-periodic acid (H_5IO_6) to be the most appropriate after investigating eight oxidising agents: 2.5%, 5% and 10% H_5IO_6, 10% $KMnO_4$, 10% $K_2Cr_2O_7$, 30% H_2O_2, conc. HNO_3 and fuming HNO_3. The Os signal intensity and the ease of handling were used as criteria in determining the most suitable agent. Richardson *et al.* (1989) elected to use 5% H_5IO_6 following a comparison of the oxidising ability of H_5IO_6, HNO_3 and H_2O_2 at different concentrations. Russ *et al.* (1987) and Dickin *et al.* (1988) selected 10% periodic acid (HIO_4), although Russ *et al.* (1987) found that after a few minutes white crystals

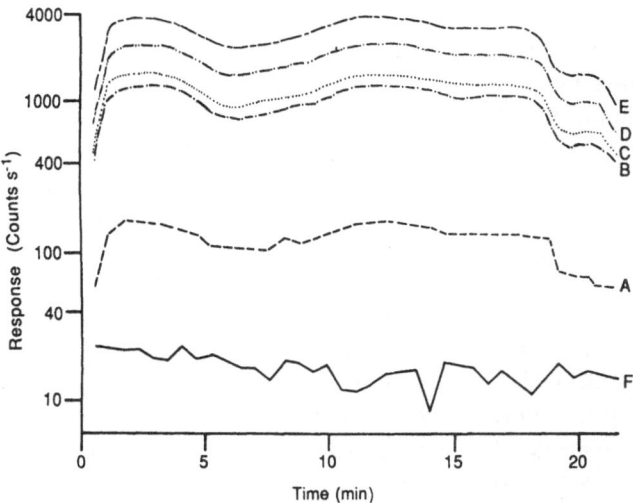

Figure 4.12 Example of count rates over a period for 10 ng Os. (A) ^{187}Os; (B) ^{188}Os; (C) ^{189}Os; (D) ^{190}Os and (E) ^{192}Os. The ^{197}Au signal (F) is typical of instrumental background. After Dickin *et al.* (1988).

sometimes formed on and clogged the frit at the bottom of the reaction flask. This problem was overcome without loss of Os sensitivity by reducing the HIO_4 concentration to 2.5%.

Bazan (1987) and Russ *et al.* (1987) suggested that more Os signal was obtained if, before samples were analysed, they were left overnight, after being made basic with a 0.1 N ammoniacal solution. However, Dickin *et al.* (1988) found that total Os counts were not dependent on the pH of the sample.

It is desirable to achieve an Os count rate which is constant for the duration of the sample analysis. An example of count rates over a period for five Os isotopes is given in Figure 4.12. Initially chilling the sample and reagents to about 0°C prior to injection into the reaction chamber can help to retard Os vapour generation at the start of the run. The Os signal plateau can then be sustained at the end of the run by gradually increasing the temperature of the heating tape by adjusting the applied voltage. However, when using this method, the boiling point of the reactant mixture may be reached before all the Os has been driven off, and increased heating is ineffective. This problem can be overcome by adding H_2SO_4 to the sample/oxidising agent mixture, which raises the boiling point by several degrees, thus allowing a higher reaction temperature to drive off the last of the Os from the sample.

Procedure With the batch system, the torch can be isolated from the sample carrier gas line. This allows the plasma to be maintained and hence the plasma can be left operating while sample change-over takes place. Between samples the generator including the condenser should be cleaned, using the procedure outlined in the next section. The reaction vessel can then be loaded with 1 ml of oxidising agent inserted through the injection port, followed by 1 ml of sample solution and 1 ml 50% H_2SO_4. To avoid problems of plasma instability when the generator is opened to the torch, the reaction chamber should first be flushed with argon to force air out of the system. Argon is then passed through the reactants, the heating coil is switched on raising the reaction chamber temperature to about 130°C, and data collection can begin. A steady Os signal can be obtained for about 20 min, sufficient time to make several replicate Os isotope measurements.

Memory effects   One of the most serious sources of systematic error in the analysis of Os is a pronounced sample memory effect, even after the generator is cleaned. Osmium appears to plate on to, and be retained by, the sampling assembly (torch, sampling tube and generator) more readily than other elements. The cleaning process outlined below should be followed between analyses, even of the same sample:

 (i) Take the generator off-line and flush the remaining reagents into the waste tube.
 (ii) Rinse the sample chamber in cold distilled or deionised water up to the level of the injection port. Flush the water out.
 (iii) Add cold 5 M HNO_3 up to the level of the condenser. Flush the HNO_3 out.
 (iv) Purge the sampling tube and torch of sample by injecting 3 ml of 5 M HNO_3 into the sample chamber, putting the generator on-line and passing Ar through the generator. Run this solution until the count rates for ^{190}Os and ^{192}Os are below about 200 counts s^{-1} for 100 s. Increasing the Ar carrier gas flow decreases the wash time.
 (v) Rinse with distilled or deionised water up to the condenser to remove the HNO_3 prior to the next run.
 (vi) If the Ar carrier gas flow was increased for flushing, return it to the original flow rate.

This technique physically flushes out the sample and oxidises or ionises any Os that remains in the sample chamber, sampling tube or torch. In addition to this process the memory problem can be alleviated by replacing the entire sampling assembly daily, i.e. between groups of 6–12 analyses of the same sample, or between different samples. The whole assembly should then be cleaned in an aqua regia bath overnight.

ICP–MS settings   The plasma operating conditions used with the OsO_4 generator are the same as those used with a conventional nebuliser/spray chamber sample introduction system, with the exception of the carrier gas flow. This in general needs to be higher than required for a nebuliser system. Richardson *et al.* (1989) found it necessary to operate with a carrier gas flow of 1.8 l min^{-1}. However, it should be noted that this work was carried out on a Sciex ELAN with a sampling cone to load coil separation of 19 mm. This distance requires a high flow-rate in order to allow the normal analytical zone to reach the sampling cone. The carrier gas flow rate that gives maximum Os signal is the optimum.

Data can be collected using either peak hopping or scanning modes of operation. The procedures for isotope ratio determination, outlined in chapter 11, should be used.

4.3.3.3 *Applications.* This technique has been used to determine Os isotope ratios in samples containing $\leqslant 5$ ng of common osmium (Russ *et al.*, 1987). Here the ratios ^{190}Os: ^{192}Os, ^{189}Os: ^{192}Os and ^{188}Os: ^{192}Os were determined to a precision of better than $\pm 0.5\%$ (1σ). For the minor isotopes, the ratios ^{187}Os: ^{192}Os and ^{186}Os: ^{192}Os were determined to $\pm 1\%$ and ^{184}Os: ^{192}Os (4×10^{-4}) was determined to $\sim 10\%$.

Osmium isotope ratios were determined on pure Os solutions and a natural sulphide ore standard by Dickin *et al.* (1988). The means of all runs on samples with Os concentrations of a few ng ml^{-1} gave isotope ratio results with precisions of 0.3% (2σ) and accuracies 0.1–0.2% against SIMS data.

The OsO$_4$ generator has been used for a number of other applications such as determining Re:Os ratios (Richardson *et al.*, 1989) and the half-life of ^{187}Re (Lindner *et al.*, 1986, 1989).

4.3.4   Reactive gases

4.3.4.1 *Introduction*   Silane (SiH$_4$) is the main gas used in the production of semiconductors. The direct analysis of silane by ICP–MS has been found to be a practical proposition for both the measurement and identification of elemental impurities below 10^{-9} g. Several modifications to the instrument must be made. In order that the amount of matrix material that deposits on the sampling cone orifice is minimised, a cone fabricated from an alloy that will run at a higher temperature than a conventional nickel cone must be used.

4.3.4.2 *Gases*   To facilitate the introduction of gas samples, the conventional nebuliser and spray chamber should be replaced with a suitable gas handling rig, such as the one shown in Figure 4.13. The exit line of the rig is coupled directly to the central injector of the plasma torch.

The gas handling rig consisted of a gas mixing and dilution system with flow rates controlled by mass flow controllers. Using this system high purity argon, SiH$_4$, hydrogen and calibration gases can be mixed in precise amounts prior to introduction into the plasma.

The system may be optimised with 2 ppb ^{127}I in the carrier gas flow and maximum signal obtained at a flow of 1.3 l min^{-1}. Hutton *et al.* (1990) found that when silane was present in the argon the optimum was obtained at a

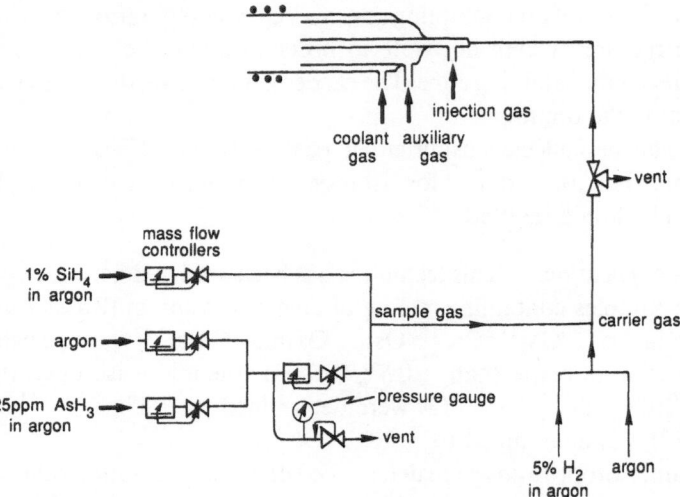

Figure 4.13 Layout of a gas handling rig, suitable for the introduction of pyrophoric gases. After Hutton *et al.* (1990).

lower flow of $0.7 \, l \, min^{-1}$. This was also found to give better signal stability. An addition of $7.5 \, ml \, min^{-1}$ hydrogen to the carrier gas was found to yield a 70% increase in analyte signal.

4.3.4.3 *Quantification* The gas dilution system that is available on the rig shown in Figure 4.13 allows variable independent additions of gaseous forms of elements which are under investigation in the silane. For example, to determine As in SiH_4, a calibration curve of over three decades of concentration can be obtained from ppt level additions of AsH_3 to the argon

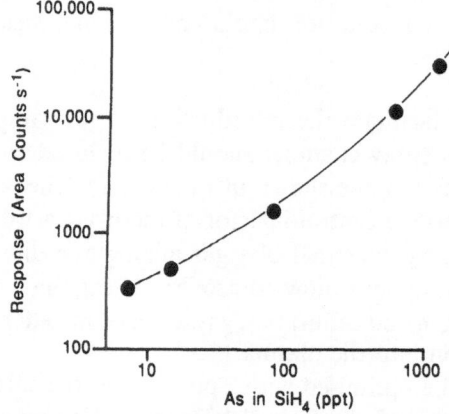

Figure 4.14 Calibration curve for ^{75}As in $600 \, cm^3 \, min^{-1}$ of SiH_4 obtained by variable additions of AsH_3 to silane. After Hutton *et al.* (1990).

(Figure 4.14). Limits of detection in the SiH_4 of about 0.55 ppb for ^{75}As and 0.65 ppb for ^{127}I are obtained.

Details of the procedure for silane analysis are given by Hutton *et al.* (1990).

4.4   Liquid chromatography

4.4.1   *Introduction*

The toxicological and biological importance of many metals and metalloids depends mainly on their chemical form. However, most measurement techniques for trace elemental analysis determine only the total amount of individual elements and do not provide information about the chemical states or elemental species present.

Liquid chromatography (LC) is a particularly attractive approach to speciation because it can be used to separate non-volatile and thermally labile compounds, unlike gas chromatography (GC) which is limited to volatile and thermally stable compounds. Conventional detectors for LC such as refractive index, ultraviolet (UV) or conductance detectors provide a response that is more or less non-specific. Therefore element-specific or multi-element detectors are also needed.

The element-specific detectors that have found most use are flame and electrothermal atomisation atomic absorption spectrometry (FAAS and ETA–AAS). The direct coupling of these techniques to LC has been reviewed by Ebdon *et al.* (1987). As a chromatographic detector, FAAS often fails to give sufficiently low detection limits, except where pre-concentration is possible. Although ETA–AAS does exhibit useful detection limits it is not able to monitor the LC eluent continuously. The time required to run an atomiser dry-ash-atomise-cool cycle results in infrequent aliquots being analysed, thus 'real time' chromatographic interpretation is not possible.

Techniques such as atomic fluorescence spectrometry and atomic emission spectrometry with a microwave, direct current or inductively coupled plasma offer multi-element detection and long linear dynamic ranges. A multi-element detector can also avoid chromatographic interference caused by co-eluting peaks of different metals. However, they generally exhibit inadequate limits of detection for many applications.

Inductively coupled plasma mass spectrometry offers an exciting alternative multi-element detection mode for coupled LC systems. Limits of detection are often significantly lower than those obtained from ICP–AES, and similar to those of ETA–AAS. The elution rates from LC systems are similar to the liquid flows required by ICP–MS to operate in a real-time mode. The ability to measure isotope ratios on eluting peaks is potentially of use in biological speciation tracer studies as well as in measuring concentrations by isotope dilution. This saves time incurred in running several chromatograms to obtain a calibration curve and compensates for matrix effects.

Two of the main limitations of solution analysis by ICP–MS are spectral and physical interferences caused by the matrix. Liquid chromatography offers the possibility of separating matrix elements from the analyte.

4.4.2 *Principles*

The coupling of an LC system to an ICP–MS instrument is straightforward. The eluent that leaves the analytical column is conducted to the nebuliser of the ICP by a length of microbore tubing, made either of a plastic material or stainless steel. The schematic in Figure 4.15 shows a typical HPLC–ICP–MS configuration. A high pressure pump (usually dual piston) drives the liquid mobile phase through a valve (where the sample is injected) to the analytical column, where species separation takes place. The liquid then flows to the nebuliser and the aerosol is introduced to the plasma. Various other columns, valves and detectors can be inserted along the path of the liquid as required. A scavenger column, after the pump but before the sample injection valve, cleans up the liquid phase by extracting any trace elements that are present. Between the sample injection valve and analytical column, a guard column can be inserted to protect the analytical column from the effects of species such as lipids, which otherwise degrade the separation. An ultraviolet or other detector can be placed after the analytical column, so that the retention time of compounds of interest can be determined. A post-column flow injection system can be inserted to allow the quantification of samples and standards using the same flow conditions as the HPLC column.

For multi-element determinations, analysis can be carried out in either peak hopping or scanning mode, depending on the number of isotopes under investigation, their mass separation and a number of other factors outlined

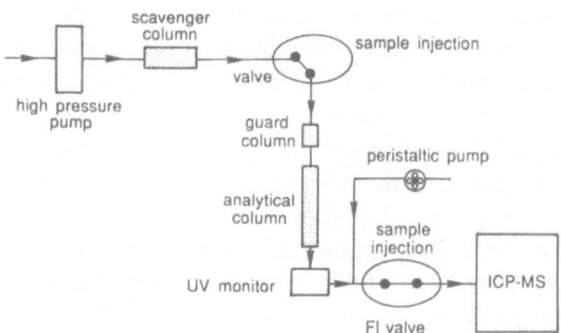

Figure 4.15   Schematic of a typical HPLC–ICP–MS configuration. After Dean *et al.* (1987a).

Table 4.9 Summary of LC–ICP–MS studies[a]

Element	Interface HPLC to ICP–MS	Chromatography	Species and sample type	Comments	Reference
As	2-ft × 0.01 inch i.d. Flexon HP tubing	Anion-exchange column, 5 μm Adsorbosphere-NH_2 packing, slurry packed. Pellicular amino guard column. Eluent: 30% methanol/15 mM $NH_4H_2PO_4$/1.5 mM CH_3COONH_4; pH 5.75 with glacial acetic acid	As^{3+}, As^{5+}, DMA and MMA in freeze dried urine standards	Calibration by standard addition. Absolute detection limit of 36–96 pg in urine. $^{40}Ar^{35}Cl$ interference on As^{3+} by co-elution of Cl species. Less than 2% O_2 added to nebuliser gas flow reduce carbon build up on sampling cone	Heitkemper et al. (1989)
As	1.2 m × 0.24 mm i.d. Teflon tubing	Anion-pairing: 1. RP-18 Spheri-10; 2. Radial compression system C_{18}, 10 μm; 3. PRP-1, 10 μm. Anion-exchange: 4. Radial-PAK SAX; 5. PRP-X100. Cation-pairing: 6. PRP-1; 7.5 μm C_{18}. Eluents: 1. 0.005 M TBAH-5% methanol, pH 7.0. 2. as for 1. 3. 0.005 M TBAH-4% methanol, pH 6.0. 4. 0.025 M Na_2HPO_4/0.025 M NaH_2PO_4/5% methanol, pH 0.8. 5. 0.008 M Na_2HPO_4/0.008 M NaH_2PO_4/5% methanol, pH 7.0. 6. 0.05 M sodium dodecylsulphate/5% methanol/2.5% acetic acid, pH 2.5. 7. 0.01 M sodium dodecylsulphate/5% methanol/2.5% acetic acid, pH 2.5	As^{3+}, As^{5+}, DMA, MMA and As-betaine in DORM-1	Calibration by standard addition. Absolute detection limit of 50–100 pg. Anion pairing less tolerant to matrix than anion exchange. Cation pairing suitable for the determination of DMA and As-betaine	Beauchemin et al. (1989)
As	Plastic capillary tubing	1. 5 μm ODS column in series with Partisil 5 μm SAX column. 2. Benson 7–10 μm anion exchange resin. Eluent: Column equilibration with 1 mM K_2SO_4, pH 10.5, sample injection with 50 mM K_2SO_4, pH 10.5	As^{3+}, As^{5+}, DMA, MMA and As-betaine in standard solutions	Limits of detection 5–10 ng ml^{-1}	Branch et al. (1989a)

Element	Tubing	Column and eluent	Compounds	Detection notes	Reference
As	0.25 mm i.d. Teflon tubing	1. Asahipak GS220. 2. Intersil ODS-2 column. Eluents: 1. 25 mM tetramethylammonium hydroxide/25 mM malonic acid, pH 6.8. 2. 10 mM tetramethylammonium hydroxide-water/methanol, pH 6.8	15 As compounds including As^{3+}, As^{5+}, As-betaine, arsenosugars and dimethyl arsinoylethanol in human urine	Absolute detection limit of 20–150 pg of As	Shibata and Morita (1989)
As	0.25 mm i.d. Teflon tubing	5 μm C_{18} column. Eluent: 0.01 M sodium dodecylsulphate/5% methanol/2.5% acetic acid	As^{3+}, As^{5+}, DMA, MMA, As-betaine and As-choline in DORM-1	84% of total As found to be As-betaine. As-betaine absolute limit of detection of 300 pg	Beauchemin et al. (1988a)
As	2 ft 4 in × 0.02 in i.d. Polyplex tubing	Anion/R ion chromatography column. Eluent: 5–50 mM phthalic, pH 2.55	As^{3+}, As^{5+} and DMA in urine reference materials	Calibration by standard addition. $^{40}Ar^{35}Cl$ interference reduced by resolving chloride from As compounds. 20-fold dilution of urine to avoid column overloading from chloride. Limits of detection: As^{3+}, 340 pg ml^{-1}; As^{5+} 420 pg ml^{-1}, DMA, 700 pg ml^{-1}	Sheppard et al. (1990)
Sn	Silicone rubber tubing	Partisil 10 μm SCX column. Eluent: 80 + 20 methanol/water, 0.1 M with respect to ammonium acetate	Tributyltin in organotin standards and water samples	Calibration by standard addition. Absolute limit of detection of 200 ng ml^{-1} for TBT (as Sn). Low flow torch on ICP–MS gives poorer (500 ng ml^{-1}) limit of detection. Low flow torch avoids carbon build up on sampling cone	Branch et al. (1989b)
Sn	0.4 m × 0.25 mm i.d. Teflon capillary tubing	Spherisorb ODS-2 (C_{18}), 5 μm column. Silica precolumn for analytical column saturation with silica. Eluent: 0.1 M SDS and 0.02 M SDS micellar mobile phase	Trimethyltin chloride, triethyltin bromide, tripropyltin chloride, monomethyltin trichloride, dimethyltin dichloride and trimethyltin chloride standards	Absolute limits of detection for compounds of 27–126 pg Sn. Calibration curves of 3.5 orders of magnitude. External calibration	Suyani et al. (1989a)

Table 4.9 (Contd.)

Element	Interface HPLC to ICP-MS	Chromatography	Species and sample type	Comments	Reference
Sn	0.4 m × 0.25 mm i.d. Teflon capillary tubing	**Ion pair chromatography:** Spherisorb ODS-2 (C_{18}), 5 μm. Eluent: 0.004 M sodium pentane sulphonate in 80:19:1 methanol/water/acetic acid, pH 3. **Ion exchange chromatography:** Adsorbosphere SCX, 5 μm. Eluent: 0.1 M ammonium acetate in 85% v/v methanol/water	Trimethyltin chloride, tributyltin chloride and triphenyltin acetate standards	Limits of detection for compounds of 200 pg ml^{-1} Sn. Calibration curves of 3 order of magnitude for ion exchange and 2 orders for ion pair. Ion pair suffered from poor resolution. External calibration	Suyani et al. (1989b)
Hg	Mercury cold vapour generator between HPLC and ICP-MS	Pico-Tag C-18 column. Eluents: 0.06 mol l^{-1} ammonium acetate, 3% acetonitrile and 0.005% v/v 2-mercaptoethanol; pH 6.8 for methylmercury and pH 5.3 for thimerosal studies	Methylmercury acetate, ethylmercury chloride and mercury (II) chloride in tuna fish reference material and thimerosal in contact lens cleaning solution	Limits of detection of 7–20 ng ml^{-1} Hg, improved to 0.6–1.2 ng ml^{-1} with post column mercury cold vapour generation. Linearity over 3–4 orders of magnitude	Bushee (1988)
Hg	0.06 m FEP tubing	Pico-Tag C-18 column. Eluents: 0.06 mol l^{-1} ammonium acetate, 3% acetonitrile and 0.005% v/v 2-mercaptoethanol, pH 5.3	Thimerosal in vaccines, toxoids and saline diluents	Thimerosal content determined quantitatively and decomposition products determined qualitatively	Bushee et al. (1989)
Cd		Superose-12 size exclusion column. Chelex-100 scavenger column. Eluent: 0.12 M Tris HCl, pH 7.5	Cd species in pig kidney following cooking and in vitro gastro-intestinal digestion	The majority of soluble Cd in retail pig kidney is associated with a metallothionen-like protein that survives both cooking and in vitro gastro-intestinal digestion	Crews et al. (1989)
Au, Cd, Cu, Zn	1 m × 0.25 mm i.d. PTFE capillary tubing	**Anion exchange chromatography:** Alltech WAX 300 column. Eluent: 20 mM–200 mM Tris buffer, 15 min linear gradient, pH 6.5. **Size exclusion chromatography:** Bio-Sil TSK 250 column. Eluent 25 mM Tris buffer, pH 7.7	Elemental binding to immunoglobins and serum albumin of Cu, Au and Zn. Blood samples from rheumatoid arthritis patients on gold drug therapy	Significant matrix effects. Limit of detection, 0.2–0.7 ng ml^{-1}	Matz et al. (1989)

P, S	0.025 m × 0.25 mm i.d. stainless steel tubing. Ultrasonic nebuliser	**Inorganic phosphates:** PRP-1 Styrene-divinyl-benzene copolymer, 10 μm particles. Eluent: 0.005 M (TEA) NO_3/2% methanol, pH 6 **Adenosine phosphates:** PRP-1. Eluent: 0.01 M (TEA) Br/5% methanol, pH 10 **Amino acids:** PRP-1. Eluent: 0.01 M (TEA) Br/1% acetonitrile, pH 7.5 **Sulphates:** Vydac 201 TP Silica-based C18 stationary phase 10 μm particles. Eluent: 0.005 M $(TBA)_3$ PO_4/5% methanol, pH 7.1	P and S in separated inorganic phosphates and sulphates, organic phosphates and amino acids	Limit of detection for P in phosphates of 0.4–4 ng and for S of 7 ng in sulphates	Jiang and Houk (1988)
As, Se (30 other elements)	0.025 m × 0.25 mm i.d. stainless steel tubing. Ultrasonic nebuliser	Econosphere C_{18}, 5 μm particles. Eluents: PIC-B5 and PIC-A, pH 3.0 and 7.1, respectively	Separation of As and Se species in standards	Survey of over 30 elements for anionic and cationic response. As and Se limits of detection of 1 ng	Thompson and Houk (1986)
Cd	–	Superose-12 size exclusion column. Eluent: 0.12 M Tris HCl, pH 7.5	Cd in ferritin and equine metallothionein	Several theoretical criteria were examined for optimising HPLC–ICP–MS coupling	Dean et al. (1987a)
Cd, Cu, Zn	–	PRP-1 column, 10 μm particles. Eluents: For U separation, 1% methanol, 0.001 M N-MFHA and 0.01 M pyridine, pH 5.0. For Ti and Mo separation, 2% methanol, 0.003 M N-MFHA, pH 1.0 (Mo) and 2.2 (Ti)	Mo(VI), Ti(IV) and U(VI) matrix species retained on column. Cd, Cu and Zn analytes washed off to ICP–MS. Procedure carried out on sediments MESS-1 and BCSS-1 and nickel base alloy BAS-346	Separation procedure removes Mo(VI), Ti(IV) which have oxides that interfere with Cd, Cu and Zn. Removal of U prevents matrix suppression effects. Limits of detection for Cd, Cu and Zn of 1–2 ng ml^{-1} and recoveries of 100%	Jiang et al. (1987)

[a] DMA, dimethylarsinic acid; MMA, monomethylarsonic acid; TEA, tetraethylammonium; TBA, tetrabutylammonium; PIC-B5, sodium pentanesulphonate; PIC-A, tetrabutylammonium phosphate; SDS, sodium dodecyl sulphate; N-MFHA, N-methylfurohydroxamic acid.

in chapter 6. If only one element is under investigation, a single isotope could be monitored continually.

4.4.3   *Instrumentation, reagents and operating parameters*

The use of an LC–ICP–MS system as a single isotope or element detector requires little comprehension of ICP–MS and a great deal of understanding of LC. To carry out multi-element LC–ICP–MS, the user should have a much greater appreciation of the operation of an ICP–MS instrument. Table 4.9 is a summary of a number of LC–ICP–MS applications. The greater part are single element speciation studies using HPLC, and many of those are studies of arsenic species in biological materials. However, each of the As speciation studies was carried out using a different method, i.e. with different column packings, mobile phase reagents and pH. The same is true of other elements studied. The user intending to carry out any LC–ICP–MS is advised to pay a great deal of attention to the chromatography aspects of the apparatus and seek the appropriate advice. Table 4.9 can be used to direct the interested reader to the appropriate publications on the subject.

4.4.4   *Applications*

A summary of the applications of LC–ICP–MS is given in Table 4.9. The majority are concerned with speciation of a particular element or small suite of elements and the remainder deal with on-line matrix removal.

Different species of arsenic have a variety of chemical and toxicological properties. The most toxic species are As(III) and As(V). Monomethyl-arsonic acid (MMA) and dimethylarsinic acid (DMA) are moderately toxic whilst arsenobetaine (AB) and arsenocholine (AC) are relatively non-toxic (Beauchemin *et al.*, 1988a). In animals, biomethylation of inorganic As is regarded as a detoxification mechanism. The products are either excreted or stored. A variety of HPLC–ICP–MS methods have been reported for the determination of As(III), As(V), MMA, DMA, AB and AC in biological materials. Limits of detection are generally at picogram levels. The various species of other elements, such as Cd, Hg and Sn, have also been separated using HPLC–ICP–MS methods.

Rheumatoid arthritis (RA) is a disease of the immune system that leads to inflammation at bone joints and ultimately to the destruction of the cartilage and bone. Gold based drugs have been shown to be effective in inducing remission in RA, but little is known of their metabolism or mode of action. Characterisation of Au metabolites is important, to help in understanding the action of, and reasons for toxic reactions to gold drug therapy. Interfacing an HPLC to an ICP–MS provides an on-line system that can separate Au metabolites and is sensitive enough to detect these metabolites at physiological concentrations ($50\,\mathrm{ng\,ml^{-1}}$). The system described by Matz *et al.* (1989) could separate a number of Au metabolites, with picogram limits of detection.

Liquid chromatography can be used to separate matrix species from the analyte in order to prevent effects like refractory oxide interferences or analyte signal suppression. The oxides of molybdenum coincide with all the most abundant isotopes of Cd, whilst those of titanium interfere with Cu and Zn isotopes. An on-line LC–ICP–MS method for separating molybdenum and titanium matrices from sediment and alloy reference materials, in order that Cd, Cu and Zn could be accurately determined was described by Jiang *et al.* (1987).

4.5 Flow injection

4.5.1 *Introduction*

The term flow injection (FI) analysis was first used by Růžička and Hansen (1975) to describe an analytical technique in which a discrete sample volume is injected into a continuously flowing carrier stream. During passage to the downstream detector the sample zone is dispersed in the carrier, which may contain a reagent, and may be merged with other reagent-containing streams. The reaction product is measured as a transient signal, by a flow through detector such as an ICP–MS instrument.

Flow injection methodology offers advantages over continuous nebulisation where (i) sample pre-treatment is necessary involving separation and pre-concentration, (ii) large dilution factors are required, (iii) there is limited sample volume, (iv) samples have a high dissolved solids content, (v) a range of calibration standards is required, (vi) standard additions are needed, or (vii) where variations in solution properties (e.g. viscosity) may affect continuous nebulisation. An FI–ICP–MS system offers the unique potential for on-line isotope dilution.

4.5.2 *Apparatus*

The FI apparatus is fairly simple. It requires one or more precision peristaltic pumps, lengths of micro-bore tubing (e.g. PTFE or Tygon), and a sample injection valve. Depending on the application, additional equipment may be required such as T-junction connectors for merging different sample streams, mixing coils to aid merging of two or more reagents, and sample pre-treatment columns for on-line sample pre-concentration or matrix separation. The simplest FI configuration is shown in Figure 4.16. The total volume of the system is small, and therefore, relatively small volumes of reagents and samples can be used. A typical sample size is 100 μl. The complexity of the system can be altered to suit the application. A review of flow injection

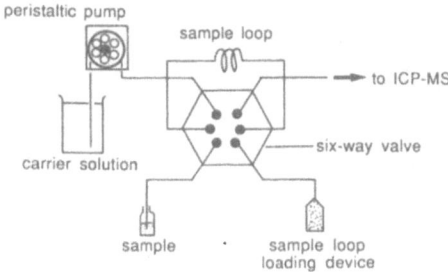

Figure 4.16 Schematic of a simple flow injection system.

techniques for atomic absorption spectrometry has been published (Tyson, 1985), and contains details of a variety of FI configurations and apparatus that can be readily used with ICP–MS, and some theoretical considerations for the FI technique.

4.5.3 *Sample introduction*

Although it is recommended to use a suitable injection valve and loop for reproducible sample injection into the carrier stream, it is not always necessary. Discrete microlitre volumes can be introduced using a small pipette joined with plastic tubing to the upstream tip of the uptake tubing (Houk and Thompson, 1983). The pipette is manually placed into the test solution until it is filled to the line. It is then withdrawn quickly and immersed in a rinse vessel containing reference blank solution. The resulting plug of test solution is segmented by air bubbles on both ends. This method is cheap and straightforward, but is likely to be subject to poor reproducibility.

Flow injection valves with a sample loop offer precise dispersion of a fixed volume into a carrier stream. The valve is operated so that in the 'load' position the sample loop is isolated from the carrier stream and can be filled with test solution (Figure 4.17a). A simple syringe or a peristaltic pump connected to valve can be used for loading. In the 'injection' position the carrier stream is directed through the loop, washing the test solution off in a discrete 'plug' of sample (Figure 4.17b). The valve can be switched manually or by remote control. Additional loops and multi-position valves allow complex switching arrangements for merging different test solutions or automated loading and injection.

4.5.4 *Operating parameters*

4.5.4.1 *Flow rate* Measured transient peak height and peak area varies in the manner shown in Figure 4.18. These effects have been well characterised in flow injection atomic absorption spectrometry (Tyson, 1985). The optimum

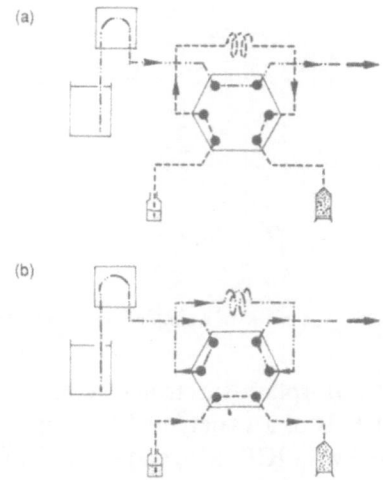

Figure 4.17 Direction of liquid for flow injection sample introduction. (a) Injection valve position to load sample loop; (b) valve position to inject sample into carrier stream.

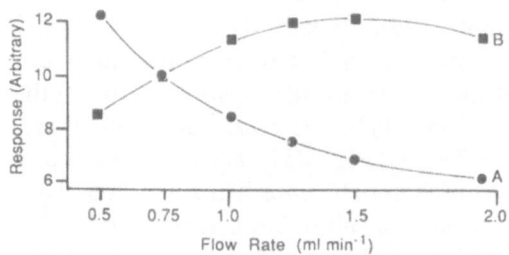

Figure 4.18 Effect of flow rate on transient (A) peak area and (B) peak height using 100 μl injection loop. After Dean *et al.* (1987a).

flow rate can readily be determined if only one measurement method is to be used. If both methods are to be used, a compromise flow rate should be chosen, as the two measurement methods have conflicting maxima. Generally the optimum flow rate for peak height measurement is close to that at which solutions would be nebulised conventionally.

4.5.4.2 *Dispersion* This can be quantified using the peak height response according to the relationship

$$D = H_m/H_p$$

where D is the dispersion, H_p is the instrument response when a discrete solution flows through to the detector, and H_m is the response obtained when an undiluted solution flows through. The injected volume and the length of tubing between the FI valve and nebuliser, control the dispersion of the

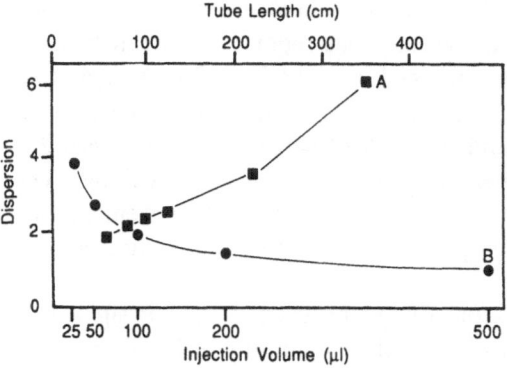

Figure 4.19   Effect of (A) tube length and (B) injection volume on dispersion. After Dean *et al.* (1987a).

discrete sample in the carrier flow. For a fixed length of tubing increasing the injected volume decreases sample dispersion, but increasing the length of the tube causes greater dispersion (Figure 4.19). The use of these parameters to increase or decrease dispersion is important for the analysis of samples with high dissolved solids or high acid concentrations. Conversely, where sensitivity is of prime importance, dispersion could be minimised by using large injection volumes.

4.5.5   *Applications*

The first report of the use of FI–ICP–MS described a manual discrete sample injection system using pipettes (Houk and Thompson, 1983). With this technique isotope ratios were determined for Mg and Ni, at levels of 2–$40\,\mu g\,ml^{-1}$ in discrete volumes of 50–$200\,\mu l$. The precision of these isotope ratios was ± 1–3%. Thompson and Houk (1986) reported on the use of FI to estimate the dispersion in a liquid chromatography introduction system. Over 30 elements were studied and limits of detection were in the range 0.01–$0.1\,ng$ for most elements using $10\,\mu l$ injections.

Some characteristics of FI–ICP–MS were reported by Dean *et al.* (1988). The effect of carrier solution flow rate on peak height and peak area was studied, together with the effect of injection volume and tube length on dispersion. It was found that the properties of this technique were very similar to those of FI–AAS.

Solutions used for analysis by ICP–MS are often restricted to those containing $< 0.2\% \, m/v$ dissolved solids. This is primarily due to problems associated with matrix deposition on the sampling cone aperture. Flow injection of limited sample volumes can be effective in minimising this effect.

Extraction of organomercury as the chloride from two marine biological reference materials (dogfish muscle tissue, DORM-1 and lobster hepatopancreas, TORT-1) with toluene, followed by back extraction into an aqueous medium of cysteine acetate was reported by Beauchemin et al. (1988e). As the final extracts contained $> 4\%$ sodium, FI was used to avoid clogging of the interface. The reproducibility for five $100\,\mu l$ injections of $100\,ng\,ml^{-1}$ Hg was about 7%, and the limit of detection was between 2.2 and $4.8\,ng\,ml^{-1}$ depending on mercury isotope and injected volume. The Hg signal was suppressed by a factor of 8 by concomitant elements in the sample, but calibration by isotope dilution gave results that agreed well with the certified values.

Hutton and Eaton (1988) reported that analyte signal fell by about 90% in 1 h if a $2\%\,m/v\,Al_2O_3$ matrix was continuously nebulised. However, the long term signal stability was found to dramatically improve when FI was used to introduce $500\,\mu l$ aliquots of solutions with this matrix. Over a 3 h period signal reproducibility with FI–ICP–MS was found to be $< 5.5\%$. Aluminium samples were analysed at $1.6\%\,m/v\,Al_2O_3$ after digestion with HNO_3/HCl and the results were in good agreement with those obtained by glow discharge mass spectrometry (GDMS). The use of FI also permitted rare earth elements (REE) to be determined in brine samples diluted to only $1.75\,m/v$ NaCl. Limits of detection for all REEs were $< 0.05\,ng\,ml^{-1}$ and results were in good agreement with those obtained by radio neutron activation analysis.

Flow injection systems can be used to carry out on-line matrix separation. In the determination of platinum in airborne particulate matter, Mukai et al. (1990) successfully used a cation-exchange resin column for tapping major matrix elements and hafnium, as these caused platinum signal suppression and spectral interference from HfO, respectively. Aliquots of prepared solution were drawn into the $800\,\mu l$ sample loop with a polyethylene syringe. The limit of detection was $0.1\,ng\,ml^{-1}$.

Flow injection can be used for on-line dilution of samples. This can be achieved by adjusting the dispersion of the injected sample. An extension of this principle was reported by Israel et al. (1989). Tandem injections were made, such that successive aliquots merged with the tail of the antecedent. With this approach a steady state could be achieved. Signal stability was poor, but could be significantly improved by merging the stream with a diluent. On-line dilution of up to 900-fold could be achieved with this technique.

The feasibility of using FI for on-line isotope dilution was reported by Viczian et al. (1990a). This was carried out by mixing the sample and spike solution with the on-line flow injection system. The spike was loaded continuously in the valve loop, and the samples were loaded individually through the sample valve. Only the sample volume consumed during the measurement was mixed with the spike and the remainder of the sample was intact. After optimisation of the system, Pb was measured in NIST 981 (common lead) to within 0.5% of the true value.

4.6 Direct sample insertion

4.6.1 *Principles*

As with the ETV device, the direct sample insertion (DSI) device was first applied in ICP–AES studies. The sample solution (approx. 10 µl) is pipetted on to a wire loop or cup that can be made from materials such as graphite, Mo, Ta or W. The probe is then moved along the axis of the torch to a position about 10 mm from the ICP where drying takes place. Upon completion of this stage, the device is propelled rapidly into the core of the plasma, and measurement takes place. The duration of the transient peak is about 0.5 s, about an order of magnitude less than the signal obtained from an ETV.

4.6.2 *Applications*

After optimising all operating conditions, including insertion distance from the top of the load coil, Boomer *et al.* (1986), using a tungsten double loop DSI device, found improvements of up to 40-fold in limits of detection for Ag, As, Cd, Cu, Li, Mn and Pb compared to nebulisation. In addition, the polyatomic peak $^{40}Ar^{16}O$ was reported to be reduced 30-fold.

In a comparison of DSI with ETV, Hall *et al.* (1988a) reported that although better precision and limits of detection could be anticipated using DSI for the analysis of relatively 'clean' and dilute solutions, the ETV technique has greater flexibility in providing accurate analysis of more complex solutions containing high concentrations of total dissolved salts. The DSI device introduces the whole of the sample into the plasma making matrix removal prior to analysis impossible.

A mechanical, stepper-motor driven computer controlled DSI device that can easily be attached to a commercial ICP–MS system, together with its software support, has been designed and developed (Karanassios and Horlick, 1989a). Spectra of a dry plasma and background spectra were obtained for DSI cup probes made of graphite, Mo and Ta as well as a W wire loop. These spectra were much simpler compared to those obtained using a nebuliser (Karanassios and Horlick, 1989b). It was also shown that DSI techniques can be used to eliminate some spectral interferences caused by the sample matrix. The absence of water in a DSI system reduces oxide species to about 0.1% of the parent. In addition, by relying on differential thermal volatilisation and chemical modification with NaF, the suppressive matrix effect of U can be eliminated (Karanassios and Horlick, 1989c).

5 Interferences

5.1 Introduction

With the introduction of any new analytical technique come claims for improved analytical capabilities and performance. When the first commerical ICP–AES systems were launched in the 1970s great hopes existed for 'interference free analysis'. It was not long, however, before a number of serious interference effects were identified making 'real' sample analysis quite complex. Some of the interference effects could be overcome by matrix matching of standards and samples while others necessitated a separation of the analyte, such as the REE, from the matrix. With the introduction of ICP–MS in the early 1980s there was great hope that this would finally be an 'interference free' technique. However, even in the first papers to be published on the plasma source technique (Gray, 1975; Houk *et al.*, 1980) a number of interference effects were identified. Since this time, some extensive studies have been carried out to identify, quantify, understand and even eliminate the main problem areas.

The interferences which occur in ICP–MS fall broadly into two groups, 'spectroscopic' and 'non-spectroscopic' or 'matrix effects'. The first type may be subdivided into four areas, those due to (a) isobaric overlap, (b) polyatomic or adduct ions, (c) refractory oxide ions and (d) doubly charged ions. Polyatomic and refractory oxide ions have been separated here for ease of discussion. In addition, polyatomic ions are thought to result mainly from ion molecule reactions during the expansion process. Refractory oxide species, as their name suggests, may in fact be present in the plasma as the strong oxide bonds are not always broken down. The second type of interference effects are more complex, and possibly less well understood, but may be broadly divided into (a) suppression and enhancement effects and (b) physical effects caused by high total dissolved solids. The extent of the interference problems in most cases, is related to the nature of the sample matrix and much can be done to minimise or even eliminate potential problems by careful sample preparation (see chapter 7).

5.2 Spectroscopic interferences

5.2.1 *Isobaric overlap*

An isobaric overlap exists where two elements have isotopes of essentially the same mass. In reality, the masses may differ by a small amount, perhaps

0.005 m/z, which cannot be resolved by the quadrupole mass analyser used in commerical ICP–MS systems. To discriminate between such small differences in mass, a high resolution system such as a double focusing mass spectrometer must be used.

Most elements in the periodic table have at least one (e.g. Co), two (e.g. Sm) or even three (e.g. Sn) isotopes free from isobaric overlap. However, the exception to this is In which has an overlap at 115 m/z with ^{115}Sn and 113 m/z with ^{113}Cd. As a general rule, isotopes with odd masses are free from overlap, while many with even masses are not. It is notable that there are no isobaric peak interferences below 36 m/z. For some elements, the most abundant, and therefore most sensitive isotope, may be subject to isobaric overlap, e.g. ^{48}Ti (73.7% abundant) and ^{48}Ca. The severity of this type of interference is dependent to some extent on the sample matrix and relative proportions of the elements concerned. For example Ba, La and Ce all have an isotope at 138 m/z, with ^{138}Ba being the most abundant. If ultratrace levels of barium are to be determined in a sample, then this would be the preferred one since it provides the most sensitive measurement. In most sample types, the concentration of Ba is many times greater than that of La or Ce while ^{138}La and ^{138}Ce are of low abundance. Therefore, in practice, ^{138}Ba can be used for measurement without the need for any correction to the data. In contrast, however, a more serious problem exists between the overlap of ^{204}Hg and ^{204}Pb. The precise measurement of lead isotope ratios in geological samples has been carried out for many years by thermal ionisation mass spectrometry but measurement times are very long and extensive sample preparation is required. Lead isotope ratios measured by ICP–MS, are of poorer precision (typically 0.2%) but this may be adequate in some applications (see chapter 11). Since ^{204}Pb is the only non-radiogenic isotope (^{206}Pb, ^{207}Pb and ^{208}Pb are all radioactive decay products) its accurate measurement is critical, since all the other Pb isotopes are ratioed back to ^{204}Pb. Mercury, although not present at very high levels in geological samples (typically $< 100 \, \mathrm{ng \, g^{-1}}$), is a common contaminant in the acids used for dissolution, particularly HNO_3. A correction may therefore be necessary to 204 m/z to account for the contribution of ^{204}Hg. This is most readily done by measurement of another Hg isotope, e.g. ^{201}Hg (12.2% abundant) and from that calculating the relative contribution of Hg to the peak of 204 m/z.

$$^{204}\mathrm{Pb} = 204_{\mathrm{integral}} - (201_{\mathrm{integral}}/13.2) \times 6.8$$

In principle, any isobaric overlap can be corrected for in this way, providing that another isotope of the interfering element is itself free from interference. Indeed, some commercial ICP–MS instruments have software facilities to carry out such corrections. However, it must be emphasised that in practice, a certain amount of error is always introduced into the measurement when corrections are applied. The magnitude of this error must be assessed in the particular matrix of interest. Trace Fe determination in Ni based alloys is a case in point. Although there are two Fe isotopes free from isobaric overlap [56]Fe and [57]Fe, both of these are subject to significant interferences from polyatomic ions (see below). [58]Fe could be used for analysis in samples containing a high ratio of Fe:Ni (a common occurrence) where the correction for the contribution to 58 m/z from Ni would be relatively small. However, in samples such as Ni based alloys (McLeod *et al.*, 1986), this ratio is reversed and the signal due to the interference from [58]Ni would be greater than that from [58]Fe itself.

The effectiveness of a correction for the isobaric overlap from [48]Ca on [48]Ti was demonstrated by Date *et al.* (1987a) for the determination of Ti in two calcium rich geological samples NBS-1b (argillaceous limestone) and NBS-120b (phosphate rock) (Table 5.1). The spectrum for 500 μg ml^{-1} solution of calcium (Figure 5.1) shows five of the six calcium isotopes. [48]Ti is directly coincident with the 0.19% abundant isotope of Ca. Therefore, in the presence of such high Ca concentrations significant error would occur during the determination of Ti unless a correction was made for the contribution of Ca to this peak. In order to avoid using [48]Ti for the determination, the next most abundant free isotope, [47]Ti or [49]Ti, would normally be selected. For samples high in P, [47]Ti also suffers from a

Table 5.1 ICP–MS data for titanium in limestone NIST-1b and phosphate rock NIST-120b[a]

Ion	NIST-1b[b]	NIST-120b[b]
[47]Ti	296	1600[c]
[48]Ti	1530	1990
[49]Ti	309	800
[48]Ti corrected	280	858
Reference value	296	850 $\pm$ 230

[a] From Date *et al.* (1987a).
[b] Concentrations are in μg g^{-1}.
[c] PO interference.

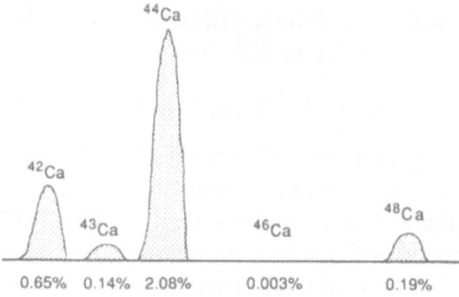

Figure 5.1 Spectrum obtained in the range 42–48 m/z for a Ca reference solution at 500 μg ml^{-1} showing five of the Ca isotopes. [44]Ca = 2.86M counts s^{-1} (from Date *et al.*, 1987a).

polyatomic ion interference (from $^{31}P^{16}O$). The only titanium isotope free from any interference in this context is ^{49}Ti. By measuring ^{43}Ca and taking the ^{43}Ca to ^{48}Ca ratio from measurements on a synthetic solution, a correction may be made for ^{48}Ca on ^{48}Ti. The success of this correction is illustrated in Table 5.1 and this method can be usefully applied in a number of applications.

In addition to isobaric overlap between elements present in a sample matrix or in the dissolution acids, a number of overlaps exist with Ar, the plasma gas, and with Kr and Xe which can occur as impurities in liquid argon supplies. By far the greatest population of ions in the plasma is formed from argon, the plasma support gas and successful correction for the overlap of ^{40}Ar (99.6%) with ^{40}K (0.01%) and ^{40}Ca (96.9%) cannot normally be made since the peak at $40\,m/z$ is always saturated. Alternative isotopes should in this case be used. Although the levels of Kr and Xe are usually low (equivalent ^{59}Co concentration of $< 10\,ng\,ml^{-1}$), the level of Kr in particular can vary considerably depending on the purity of the argon used and the liquid level in the argon tank. If Sr isotope ratio determinations are to be made, the level of Kr present may be a limiting factor for accurate analysis. Two of the Sr isotopes, ^{84}Sr and ^{88}Sr, have an isobaric overlap with Kr. Bottled high purity argon gas should be used in these circumstances since it typically contains the lowest levels of rare gas impurities. In addition, Kr has an isobaric overlap with the three most abundant isotopes of Se (Figure 5.2a). Xenon also occurs

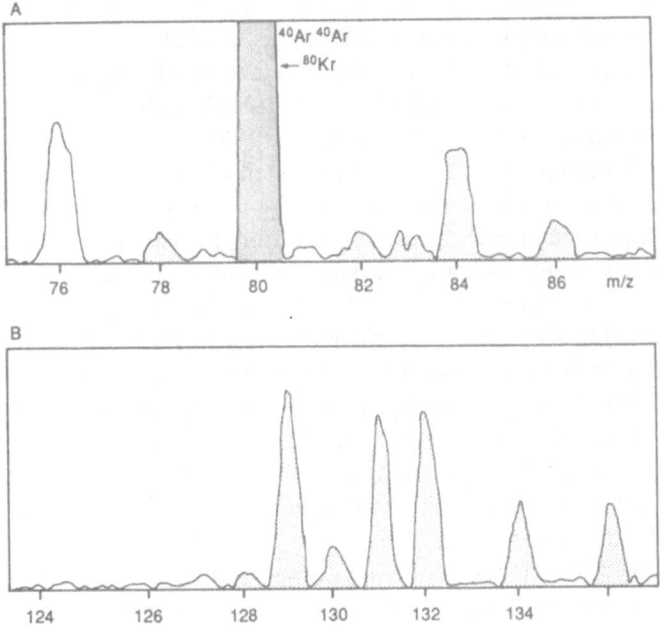

Figure 5.2 Mass spectra for 2% HNO_3 showing (A) Kr 78–86 m/z and (B) Xe 126–136 m/z isotopes. Full scale deflection for (A) 156 counts s^{-1} and (B) 78 counts s^{-1}.

in argon gas supplies and has nine isotopes (124–$136\,m/z$), seven of which have an isobaric overlap with Ba or Te (Figure 5.2b). Xenon is not usually of major concern for the determination of these elements since both have other isotopes free from interference available. If these isotopes are used for determination, then appropriate corrections may be required to account for interference effects.

5.2.2 *Polyatomic ions*

More serious practical interference problems than those caused by elemental isobaric overlap, occur due to the formation of 'polyatomic' or 'adduct' ions. The ions, as their name suggests, result from the short lived combination of two or more atomic species, e.g. ArO^+. Argon, hydrogen and oxygen are the dominant species present in the plasma and these may combine with each other or with elements from the analyte matrix. The major elements present in the solvents or acids used during sample preparation (e.g. N, S and Cl) also participate in these reactions. Although the composition of the gas extracted from the plasma at the interface is effectively frozen within about $1\,\mu s$ of leaving the plasma, fast ion molecule reactions can occur between species present in the gas. A very large number of polyatomic ion peaks can therefore occur but these are only significant up to about $82\,m/z$. The extent of polyatomic ion formation, and thus the effective interference problems, depends on many factors including extraction geometry, operating parameters for plasma and nebuliser systems and most importantly on the nature of the acid and sample matrix (Gray and Williams, 1987b).

A number of authors have examined the factors affecting the formation of polyatomic ion interferences (Horlick *et al.*, 1985; Vaughan and Horlick, 1986; Tan and Horlick, 1986; Gray and Williams, 1987b; Vaughan *et al.*, 1987). In particular, Tan and Horlick list in some detail all of the possible polyatomic interfering ions. The listing is comprehensive but gives little indication of the relative magnitude of each species. In addition, the extent of polyatomic ion formation can be dependent on the specific instrument design (e.g. Gray and Williams, 1987b; Horlick *et al.*, 1985; Zhu and Browner, 1987).

In practical analysis, however, there are relatively few serious interference effects from such species providing that sample preparation receives due attention. The mass spectrum from 34 to $80\,m/z$ is shown in Figure 5.3 for deionised water, the ideal matrix. The largest peak seen in this part of the mass range is from $^{40}Ar^{40}Ar$ at $80\,m/z$ (40–$41\,m/z$ is skipped) with the small $^{40}Ar^{36}Ar$ dimer at $76\,m/z$. Of about equal size are $^{40}Ar^{16}O$ and $^{40}Ar^1H_2$ (or $^{40}Ar^2H$) at 56 and $42\,m/z$, respectively. A small peak is also observed at $57\,m/z$ from $^{40}Ar^{16}O^1H$. The only other significant peak results from $^{12}C^{16}O_2$ at $44\,m/z$. The gas peaks from O^+, N^+ and Ar^+ are not shown since these masses were skipped during the aquisition of this spectra to prolong the life of the ion detector.

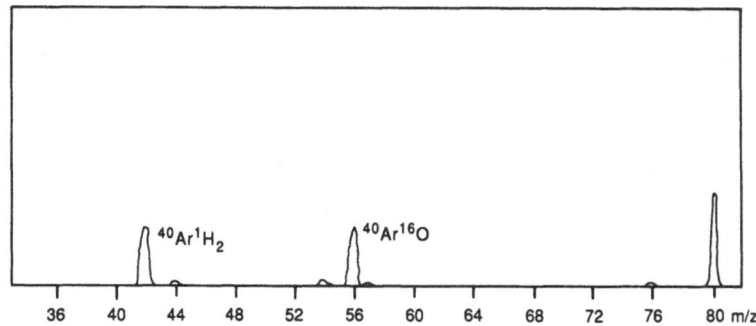

Figure 5.3 Mass spectrum for deionised H_2O from 32 to $80 \, m/z$. The main polyatomic peaks are visible at $56 \, m/z$ (ArO), $42 \, m/z$ (ArH$_2$) and $80 \, m/z$ (Ar$_2$). Full scale deflection 2000 counts s^{-1}. The mass region from 30 to $40 \, m/z$ is omitted.

The number of interfering polyatomic ion peaks seen in Figure 5.3 is small, but it is rare to have such a simple matrix for analysis. Samples are usually acidified (e.g. natural waters) or digested (e.g. biological, soils) with a variety of inorganic acids prior to analysis. The choice of acid at this stage can have far reaching effects with respect to the generation of polyatomic ion interferences as seen in Figure 5.4. The polyatomic ion peaks in both H_2O_2 and HNO_3 are identical to those identified in deionised water and these media are therefore considered ideal matrices. However, the spectra in an HCl or H_2SO_4 matrix are more complex. In addition to the peaks seen in deionised water, a number of additional polyatomic peaks occur. In an HCl matrix, the largest occur at 51 and $53 \, m/z$ (from $^{35}Cl^{16}O$ and $^{37}Cl^{16}O$) with two smaller peaks at 52 and $54 \, m/z$ resulting from a combination of $^{36}Ar^{16}O$ with $^{35}Cl^{16}O^1H$ and $^{38}Ar^{16}O$ with $^{37}Cl^{16}O^1H$, respectively. In addition two further peaks are seen at 75 and $77 \, m/z$ due to the formation of ArCl causing significant interferences on As and Se in a HCl matrix. In a 1% H_2SO_4 matrix the most significant polyatomic peaks occur at 48 (SO), 64 ($^{32}S^{16}O_2$, $^{32}S_2$), 49 ($^{32}S^{16}O^1H$ with $^{33}S^{16}O$) and $50 \, m/z$ ($^{34}S^{16}O$ with $^{33}S^{16}O^1H$) which all result from combinations of S and O from the acid medium. An additional peak is also seen at $65 \, m/z$ from $^{33}S^{16}O_2 + {}^{32}S^{16}O_2{}^1H$. The equivalent concentrations (referenced to Co$^+$) of some polyatomic ions from each of the three mineral acids is shown in Table 5.2 for a PlasmaQuad PQ2. It should be remembered that S and Cl in any form, i.e. from the sample itself, from reagents, etc. will cause these interferences.

A correction may be made for the overlap of a polyatomic ion peak with an elemental peak. For example, it may be possible to successfully use a reagent blank to correct successfully for gas peaks such as $^{40}Ar^{16}O$ and $^{14}N^{16}O$. However, these peaks may be relatively large compared with the analyte contribution at any given mass and significant errors in the data may result if a correction is applied. In addition, experience in our laboratory suggests that the polyatomic gas peaks are less stable than analyte ion peaks,

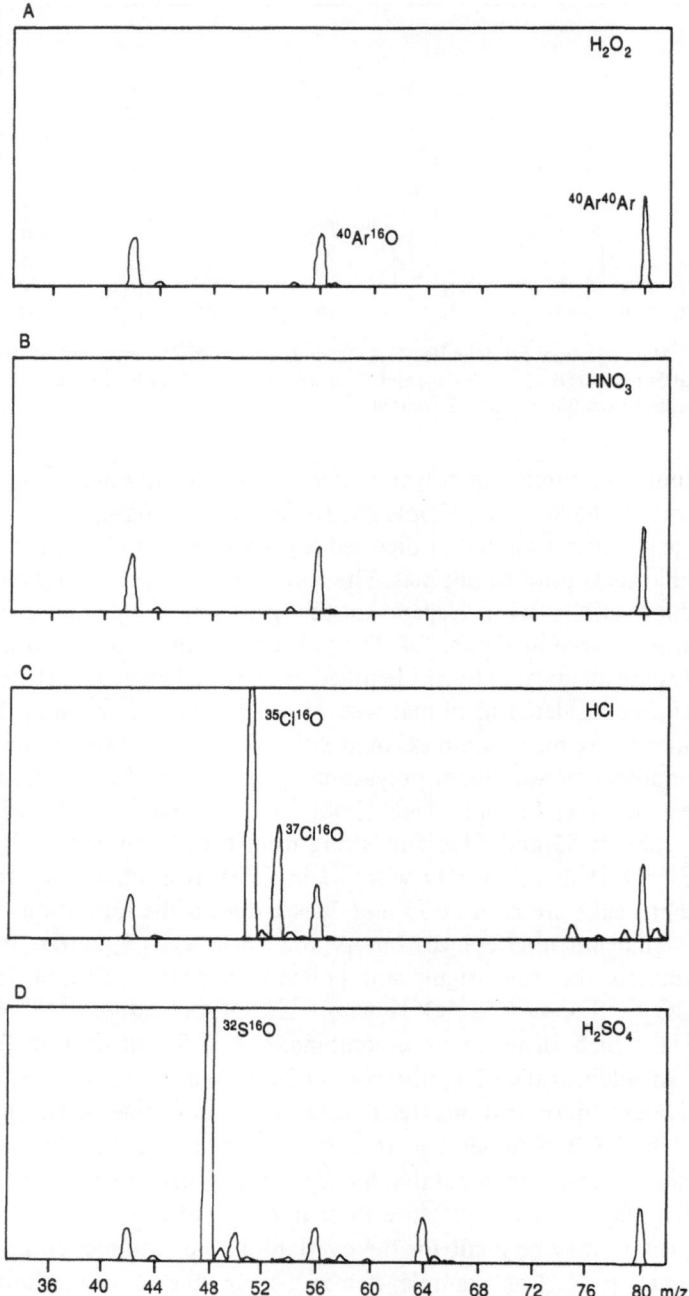

Figure 5.4 Mass spectrum from 32 to 80 m/z for (A) H_2O_2, (B) HNO_3, (C) HCl and (D) H_2SO_4. Full scale deflection 2000 counts s^{-1}. The mass region from 30 to 40 m/z is omitted.

Table 5.2 Levels of polyatomic ion interference using a water cooled spray chamber

Mass (m/z)	Element	Abundance (%)	1% HNO₃ (eq. conc.)[a]	1% HCl (eq. conc.)	1% H₂SO₄ (eq. conc.)
31	P	100	6.03	6.22	7.29
44	Ca	0.14	8.50	10.3	23.4
48	Ti	73.7	0.62	1.24	906
49	Ti	5.5	0.13	2.93	12.4
50	Ti, V, Cr	5.3, 0.25, 4.35	0.16	0.49	45.7
51	V	99.7	0.66	230	1.57
52	Cr	83.8	0.61	6.59	0.88
53	Cr	9.5	0.17	75.0	0.58
54	Cr, Fe	2.36, 5.8	11.9	16.5	9.12
55	Mn	100	0.70	0.77	0.68
56	Fe	91.7	81.9	84.2	52.7
57	Fe	2.14	1.68	2.08	1.24
64	Ni, Zn	0.95	0.28	7.14	108
65	Cu	48.9	0.22	0.84	9.85
66	Zn	27.8	0.33	3.59	6.49
67	Zn	4.1	0.09	0.94	0.63
75	As	100	0.20	43.1	0.35
77	Se	7.5	0.24	14.8	0.28
80	Se	50.0	678	716	424

[a] eq. conc. = equivalent Co concentration in $ng\,ml^{-1}$.

introducing a further source of error in making corrections. In the case of the determination of trace levels of Fe, for example, it is better to use ^{57}Fe which, although less abundant than ^{56}Fe, has only a small interference from $^{56}Ar^{16}O^{1}H$ (equivalent to $2\,ng\,ml^{-1}$). In general, therefore, the most serious interferences due to the formation of polyatomic ions are formed between the most abundant isotopes of H, C, N, O, S, Cl and Ar for example, as in $^{40}Ar^{16}O$. The smaller isotopes of these elements, e.g. $^{38}Ar^{17}O$, will also combine, but the interference will be less severe.

If an assessment is to be made of polyatomic ion interference in real samples, i.e. those with a complex matrix, it is useful to examine how their level is modified under different ICP operating conditions and thus to minimise their impact. The effect of RF forward power and nebuliser gas flow rate on the formation of ArO⁺, ClO⁺ and ArAr⁺ has been studied on the VG Elemental PlasmaQuad. It is useful for comparative purposes to quote the size of the interfering polyatomic peaks in terms of their equivalent concentration to a known elemental response; e.g. that of $1\,\mu g\,ml^{-1}$ Co. The magnitude of the interfering ion can than be assessed in terms of practical analysis and allows comparison to be made among different instruments.

The variation in analyte response with nebuliser flow rate for eight elements covering from 4 to 238 m/z is shown in Figure 5.5. Under the conditions used here (1300 W RF forward power) all elements studied develop a maximum response between 0.7 and $0.9\,l\,min^{-1}$ with differences in actual

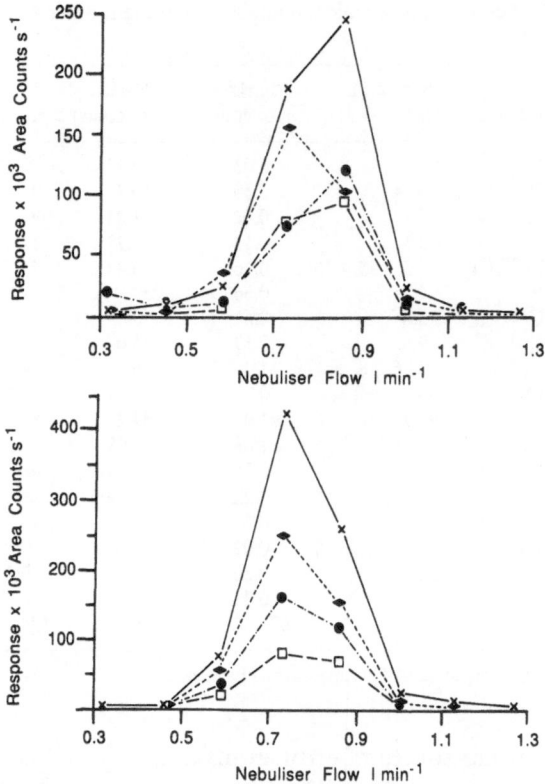

Figure 5.5 System response versus nebuliser gas flow rate at 1300 W forward power; response normalised to 100% abundance. Key: upper plot, ×, Co; ♦, Ce; ●, Al; □, Li. Lower plot, ×, W; ♦, U; ●, In; □, Pb (after Gray and Williams, 1987b).

response being mainly related to the first ionisation energy of the element. Providing that all other conditions such as sampling distance remain constant, it is possible to choose an optimum nebuliser gas flow rate for any fixed power, which is applicable to all elements, a very desirable feature of a technique to be used for multi-element analysis. The data shown in Figures 5.6 and 5.7 indicate that at the three power settings examined, the level of ArO^+, ClO^+ and the ratio of $ArO^+:Co^+$ are all at, or close to, a minimum at a similar nebuliser flow rate, that also gives maximum analyte sensitivity. From these figures it is clear that it should be possible to optimise the instrument to give maximum elemental sensitivity while keeping ArO^+, ClO^+ and $ArAr^+$ close to a minimum.

Many of these interferences are directly caused by the formation of polyatomic ions which include O and H. These elements are derived from the high ion populations produced by dissociation of water vapour from solutions. A considerable reduction in these can be obtained if the amount of water vapour reaching the plasma is reduced. This is readily achieved

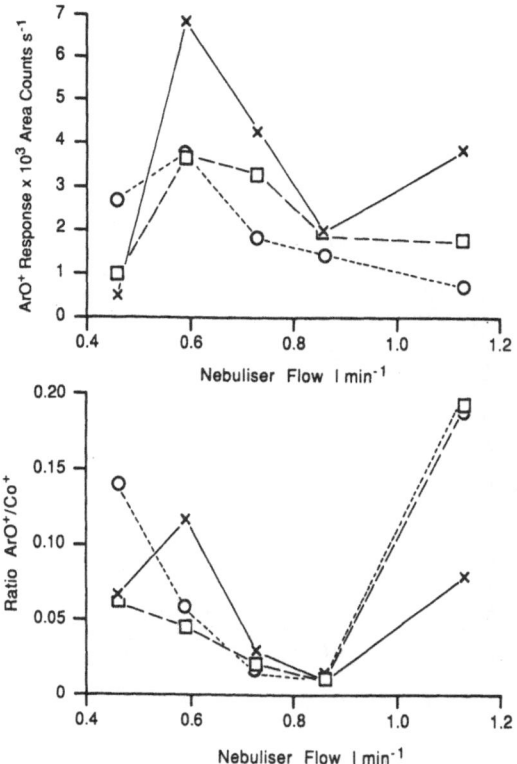

Figure 5.6 Variation of ArO response with nebuliser gas flow rate at three forward power levels. ×, 1500 W; □, 1300 W; ○, 1100 W (from Gray and Williams, 1987b).

using a temperature regulated spray chamber and is discussed in some detail in chapter 3. A number of third party computer programs are currently available, which allow the user to readily assertain which polyatomic ion interferences are likely to occur in a given matrix (Vaughan and Horlick, 1987; Williams, 1989). In addition, some instrument manufacturers provide software which allows automatic interference corrections to be implemented.

5.2.3 Refractory oxides

Although polyatomic ions probably present the most serious interference problem, in some matrices refractory oxide ions must also be considered. These species occur either as a result of incomplete dissociation of the sample matrix or from recombination in the plasma tail. Whatever the origin of these ions, the result is an interference 16 (MO^+), 32 (MO_2^+) or 48 (MO_3^+) mass units above the M^+ peak as shown in Figure 5.8. In general, the relative level of oxides expected can be predicted from the monoxide bond strength of the element concerned. Those elements with the highest oxide bond strength

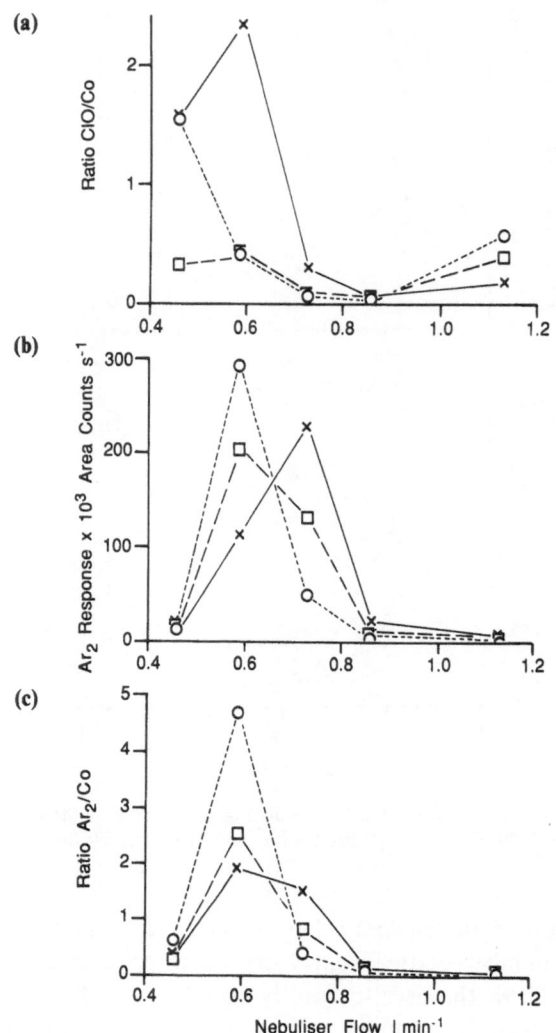

Figure 5.7 Ratio of (a) ClO:Co, (b) Ar response ratioed to Co and (c) Ar_2:Co, with nebuliser gas flow rate at three power levels ×, 1500 W; □, 1300 W; ○, 1100 W (from Gray and Williams, 1987b).

usually give the greatest yield of MO^+ ions, e.g. Ce (Table 5.3). The levels of oxide species are usually quoted with respect to the elemental peak, i.e. MO^+/M^+, typically as a percentage, although strictly the ratio should be expressed as $MO^+/(MO^+ + M^+)$. In general, the level of oxide formation rarely exceeds about 1.5% for most elements, although it must be stressed that plasma operating conditions can influence the formation of oxide ions (e.g. Gray and Williams, 1987a; Horlick *et al.*, 1985). As with polyatomic ion formation, the RF forward power and nebuliser gas flow rate have a significant effect on the level of MO^+ ions generated. Oxygen is introduced into the

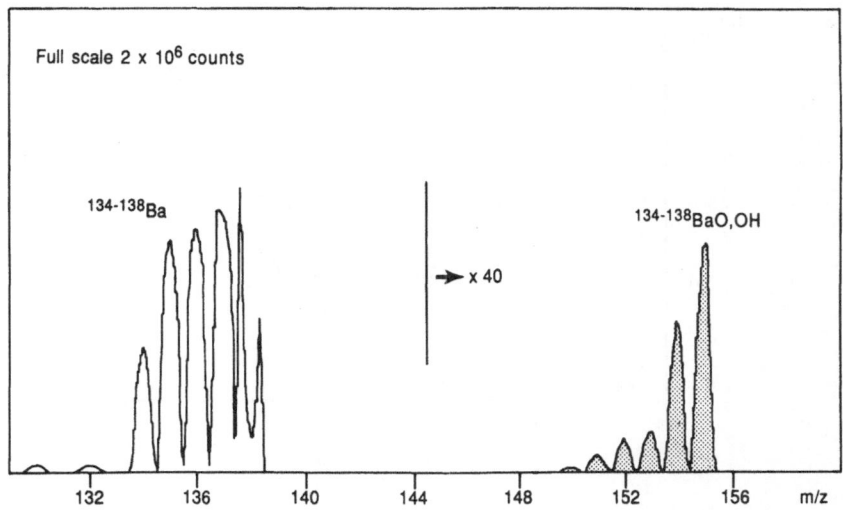

Figure 5.8 Mass spectrum from 130 to 156 m/z showing Ba, BaO and BaOH species for 100 μg ml^{-1} Ba (from Jarvis *et al.*, 1989).

Table 5.3 Ratio of oxide to element response for analytes with a range of bond strengths[a]

Element	Bond strength (kJ mol^{-1})	MO$^+$/M$^+$
Rb	255	5.5×10^{-7}
Cs	297	2.8×10^{-8}
Co	368	1.7×10^{-5}
Pb	409	1.2×10^{-5}
Fe	427	1.1×10^{-5}
Cr	512	3.6×10^{-5}
Al	563	1.1×10^{-5}
Ba	597	8.3×10^{-5}
P	607	3.7×10^{-3}
Mo	619	9.5×10^{-4}
Sm	662	2.3×10^{-3}
Ti	760	1.8×10^{-3}
Zr	795	4.7×10^{-3}
Ce	795	1.3×10^{-2}
Si	799	1.5×10^{-3}

[a] From Gray (1989a).

plasma in water as vapour and in each aerosol droplet produced by the nebuliser. The water causes a reduction of the plasma temperature, because energy is used to dissociate the water molecules. This results in a significant change in plasma equilibrium. The relationship between MO$^+$/M$^+$, nebuliser gas flow rate and RF forward power is broadly similar for the main two

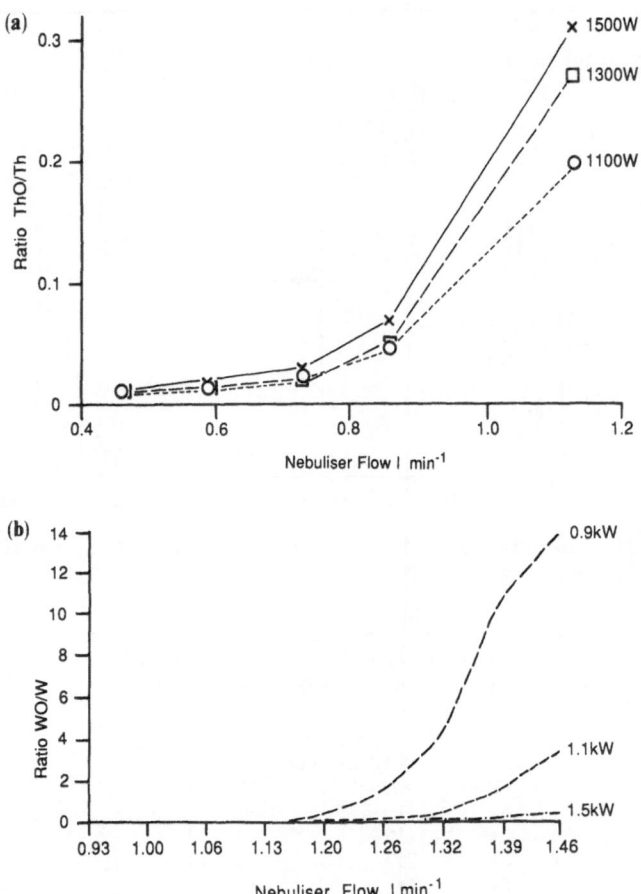

Figure 5.9 Variation in oxide ratio with nebuliser gas flow rate at three power settings: (a) ThO:Th and (b) WO:W (from Gray and Williams, 1987b; Horlick *et al.*, 1985).

commercial ICP–MS systems currently available (Plasma Quad and Elan; Figures 5.9a and b). The highest MO^+/M^+ ratios are seen at the highest nebuliser gas flow rates although the effect of low power is rather different on the two systems used here, with a significant increase in MO^+/M^+ ratio at very low power (0.9 kW) on one system (Horlick *et al.*, 1985) (Figure 5.9b).

The nebuliser gas flow rate determines the position along the axis by which dissociation of molecules is virtually complete (Gray, 1989a). At the flow rates normally used for multi-element work, this point is kept well in advance of the position at which ions are extracted. However, if the flow rate is increased, dissociation, which requires a finite residence time in the hottest part of the plasma (i.e. close to the load coil), occurs further along the axis. The proportion of oxide detected is seen to rise as undissociated oxide ions are extracted into the expansion stage (e.g. McLaren *et al.*, 1987a). Increasing

Table 5.4 A table of barium oxide and hydroxide levels for two generations of ICP–MS instruments, expressed as percentages of parent ion[a]

| | Ba (μg ml^{-1}) | | | | | |
| | 1 | | 10 | | 100 | |
Ion	PQ1[b]	PQ2[b]	PQ1	PQ2	PQ1	PQ2
$^{132}Ba^{16}O$	0.40	–[c]	0.09	–	0.08	–
$^{134}Ba^{16}O$	0.07	–	0.07	–	0.07	0.01
$^{135}Ba^{16}O + OH$	0.09	0.03	0.10	0.04	0.10	0.04
$^{136}Ba^{16}O + OH$	0.13	0.03	0.16	0.06	0.15	0.06
$^{137}Ba^{16}O + OH$	0.12	0.04	0.16	0.05	0.13	0.06
$^{138}Ba^{16}O + OH$	0.07	0.03	0.08	0.03	0.08	0.03
$^{138}Ba^{16}O^{1}H$	0.09	0.05	0.11	0.05	0.10	0.06
$^{132}Ba^{16}O_2$	–	–	–	–	–	–
$^{134}Ba^{16}O_2$	–	–	–	–	–	–
$^{135}Ba^{16}O_2$	–	–	–	–	0.0002	–
$^{136}Ba^{16}O_2$	–	–	0.002	–	0.0003	–
$^{137}Ba^{16}O_2$	–	–	0.001	–	0.0003	–
$^{138}Ba^{16}O_2$	–	–	0.0004	–	–	–

[a] From Jarvis et al. (1989).
[b] PQ1 = VG Elemental PlasmaQuad 1; PQ2 = VG Elemental Plasma-Quad 2.
[c] Less than 0.0001%.

the residence time of the sample in the plasma, should therefore lead to better dissociation and lower oxide levels.

Recent changes in instrumentation by manufacturers have led to a number of improvements. The relative differences in the levels of oxide produced on two generations of ICP–MS instruments the VG Elemental PlasmaQuad 1 and the PQ2 are illustrated in Table 5.4 for Ba (see Jarvis et al., 1989 for discussion). These data were obtained under 'normal' multi-element operating conditions appropriate for each instrument. The integrals for some of the more abundant isotopes, e.g. ^{138}Ba at $10\,\mu g\,ml^{-1}$ and $^{134-138}Ba$ at $100\,\mu g\,ml^{-1}$ are beyond the range of the detector system and their response has been calculated from a smaller isotope. The $[BaO]^+$ and $[BaOH]^+$ peaks, i.e. 150–155 m/z are between 0.07 and 0.15% of the parent ion for the PlasmaQuad 1 but are significantly lower at 0.01–0.06% for the PQ2 (calculated as $[BaO]^+ + [BaOH]^+/Ba^+$). These proportions are typical of values reported for most of the commercially available ICP–MS instruments of these vintages.

In addition to monoxide species, dioxide and trioxide ions have been recorded (e.g. Jarvis et al., 1989). The MO_2^+ levels for barium are less than 0.002% (Table 5.4) but may be significant in a barium matrix for example where $^{137}Ba^{16}O_2$ would cause an interference on mono-isotopic ^{169}Tm (Figure 5.10). Although trioxide species are rarely recorded, their occurrence may be critical in some matrices. Figure 5.11 shows a spectrum obtained for

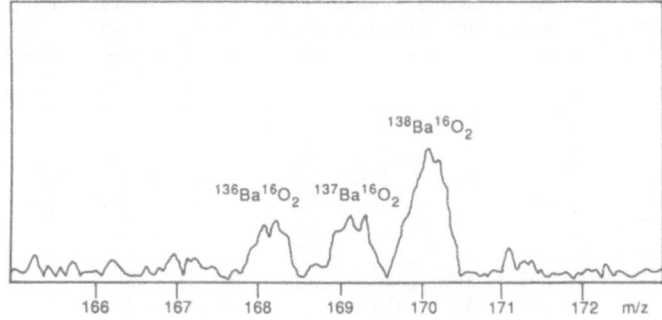

Figure 5.10 Mass spectrum from 165 to 172 m/z showing BaO_2 species for 100 $\mu g\,ml^{-1}$ Ba. Area counts s^{-1} for 168 $m/z = 133$, 169 $m/z = 167$ and 170 $m/z = 378$ (after Jarvis et al., 1989).

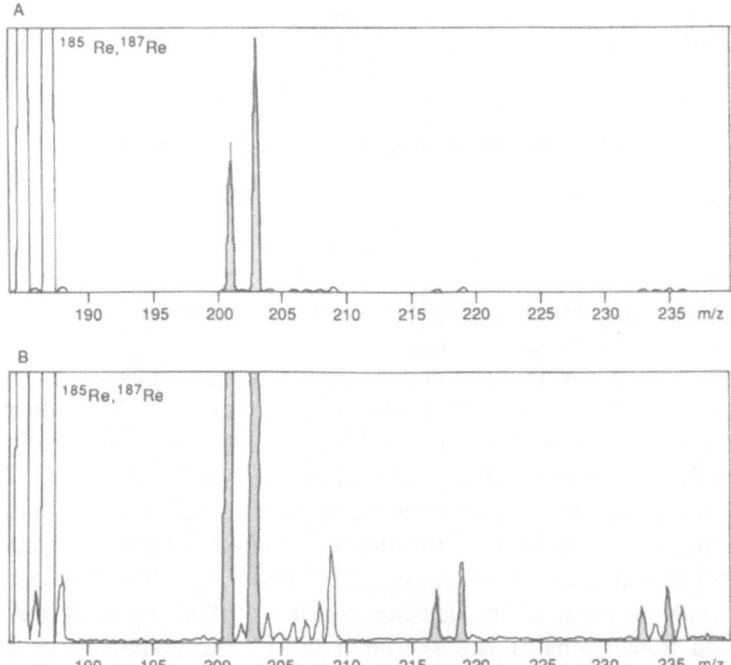

Figure 5.11 Mass spectrum for 150 $\mu g\,ml^{-1}$ rhenium solution in 1% HCl showing ReO, ReO_2 and ReO_3 (shaded peaks) and associated OH species, (A) 5000 and (B) 300 counts s^{-1} full scale deflection.

150 $\mu g\,ml^{-1}$ solution of Re in 1% HCl. The peaks for the two isotopes of Re^+ (185 and 187 m/z) are saturated and their integral cannot be measured at this concentration. ReO^+ (at 201 and 203 m/z) occurs at about 1% of the parent ion. In addition, ReO_2^+ and ReO_3^+ peaks are clearly seen 32 and 48 m/z above the parent peaks along with their associated OH^+ peaks. It is

Table 5.5 Matrix element composition of Ni-base certified reference materials BAS 345 and 346

Element	Concentration (wt%)	Element	Concentration (wt%)
Al	5.58	Mo	3.01
B	0.19	Ni	~60
C	0.153	Ti	4.74
Co	14.7	V	1.00
Cr	9.93	Zr	0.44

interesting to note that, in this case, the trioxide species are more abundant than the dioxide species. It is most likely that these oxide peaks are due to a combination of dissociation and recombination processes.

As with most interference effects in ICP-MS, the extent of the problem is dependent on the sample matrix and the analyte level of interest. The analysis of Ni based alloys, in solution, is an example of the severity of the oxide interference problem. In a study by McLeod *et al.* (1986) two Ni-based alloys (SRM BAS 345, 346) were brought into solution using a mixture of HCl, HNO_3 and HF. The major element composition of the SRMs is given in Table 5.5. Although nickel is the main constituent (60% w/w) Co, Al, Mo, Cr and Ti are also present at a few wt%. The results obtained in solutions containing different levels of total dissolved solids are shown in Table 5.6. At 0.01% TDS, matrix effects were minimal (see below), but considerable error on measured values for a number of trace elements was apparent (Table 5.6). The principal interferences recorded on the elements of interest in this work are listed in Table 5.7. Perhaps the most severe interference effects are due to the formation of MoO^+ and MoO_2^+ species. Although Mo is present in the solid at a concentration of 3.01% w/w ($3 \mu g\,ml^{-1}$ in solution), it has seven naturally occurring isotopes. Interferences from both mono- and dioxide species can therefore occur across a wide range of masses from 108 to 132 m/z. The only Cd isotope free from isobaric overlap (111 m/z) therefore suffers from an oxide interference in this matrix.

One of the most serious refractory oxide interference problems results from the formation of light REE oxides which can potentially cause serious interferences on the heavy REE. Since the REE occur in a group from ^{139}La to ^{175}Lu, any element above 155 m/z could be affected by light REE oxide species. Indeed, some authors have reported levels of CeO^+ of about 30% (Vaughan and Horlick, 1989). This level of formation is unusually high, particularly for modern instruments and may reflect the difficulty of obtaining maximum elemental sensitivity whilst minimising interferences on the particular instrument concerned. On other instruments, however, a correction for LREE oxide formation has been shown not to be necessary in matrices ranging from geological samples and soils to plants, biological samples and metals.

Table 5.6 Trace element determination in Ni-based alloys by ICP–MS at various dissolved solids concentrations[a]

| Element | Certified value ($\mu g\,g^{-1}$) | BAS 346 | | |
		0.01% TDS	0.1% TDS	1% TDS
^{69}Ga	52	63	58	56
^{71}Ga		51	50	31
^{75}As	50 ± 2	ND[b]	140	61
^{78}Se	9 ± 1	ND	ND	6.0
^{82}Se		ND	ND	17
^{107}Ag	35 ± 2	38	33	15
^{109}Ag		41	36	16
^{111}Cd	0.40 ± 0.05	36	27	14
^{114}Cd		37	27	13
^{118}Sn	91 ± 7	110	94	42
^{120}Sn		100	95	40
^{121}Sb	47 ± 3	58	43	19
^{123}Sb		54	45	19
^{128}Te	12 ± 1	69	44	18
^{130}Te		67	44	20
^{203}Tl	2	ND	2.2	1.1
^{205}Tl		ND	2.0	1.1
^{206}Pb	21 ± 2	ND	18	8.3
^{208}Pb		ND	19	8.6
^{209}Bi	10 ± 1	11	10	4.7

[a] From McLeod *et al.* (1986).
[b] ND = not detected.

Table 5.7 Matrix metal oxide interferences in the analysis of Ni-based alloys[a].

Element	Mass (m/z)	Interference
Ga	69	^{53}CrO
Se	78	^{62}NiO
Se	82	^{50}TiO$_2$
As	75	^{59}CoO
Ag	107	^{91}ZrO
Cd	110	^{92}MoO
Cd	111	^{93}MoO
Cd	112	^{94}MoO
Cd	113	^{95}MoO
Cd	114	^{96}MoO
Cd	116	^{98}MoO
Te	128	^{96}MoO$_2$
Te	130	^{98}MoO$_2$

[a] From McLeod *et al.* (1986).

Table 5.8 A comparison of europium data obtained by ICP–MS[a]

Sample (ng ml^{-1})	Ba (μg ml^{-1})	Eu[b] (ng ml^{-1})	Eu[c] (ng ml^{-1})	Ba:Eu	^{135}Ba:^{151}Eu
10	100	8.61	10.3	10000	1195
20	100	19.7	19.2	5000	515
20	200	16.5	19.0	10000	1224
50	100	52.7	49.6	2000	198
100	100	104	95.8	1000	96

	Reference (μg g^{-1})	Eu (μg g^{-1})			
SCo-1[d]	1.16	1.14			

[a] From Jarvis *et al.* (1989).
[b] Calculated after BaO interference correction (using ^{151}Eu peak).
[c] Calculated using the 2$^+$ ion mode (see text).
[d] SCo-1 values from Govindaraju (1989).

This is not the case however, for the analysis of ultrapure LREE compounds which may contain > 99.99% of a single REE. The refractory oxide formation from these matrices can be predicted and where possible alternative isotopes can be used. In a few cases this is not possible and an alternative method of approach is needed. A correction for an oxide overlap can be made by analysing a solution containing only the element forming the oxide and calculating the percentage contribution of the MO$^+$ ion to the analyte peak. The isotope ratios of the interference should be examined to eliminate any contribution from contamination present in the interfering element solution.

As with all interference correction procedures this can lead to considerable error. A practical example of the results of such a correction are given in Table 5.8. Barium, a relatively abundant element in many matrices, has a monoxide bond strength of 563 kJ mol^{-1} (BaO). A number of potential interferences exist from the overlap of each of the BaO$^+$ (and BaOH$^+$) species on some of the light and middle mass REE from 146 to 155 m/z. For all but one element an alternative isotope is available free from interference, however both of the Eu isotopes are affected, ^{151}Eu from ^{135}Ba ^{16}O (and ^{134}Ba ^{16}O^{1}H) and ^{153}Eu from ^{137}Ba^{16}O (and ^{136}Ba^{16}O^{1}H). For a correction to be made it is not necessary to know the relative proportions of oxide and hydroxide, but merely the total contribution of signal from a known concentration of barium to the 151 peak. However, correction may introduce significant error since high concentrations of barium must be measured. For example, unless system sensitivity is reduced, 100 μg ml^{-1} Ba can only be directly measured using either ^{130}Ba, ^{132}Ba or ^{134}Ba since detector saturation occurs on the more abundant isotopes at this concentration. From the integrals obtained for these isotopes the integral for the 135 peak can be calculated. From this number the integral contribution and therefore a correction factor can be deduced. Thus there is scope for significant error at several stages in the correction procedure. The results obtained after such correction procedures have been applied, are shown in Table 5.8. In general, the accuracy is rather poor and the data indicate that in practice, oxide or more importantly in this application, hydroxide formation, is probably not very stable and the correction may not be successful. The most effective way to reduce or eliminate an interference problem of this severity is by separation of the analyte from the matrix by ion exchange chromatography, solvent extraction or co-precipatation for example. In the case of barium and europium, the instrument can be operated in such a way as to allow the determination of europium using its doubly charged species (see below).

5.2.4 *Doubly charged ions*

In the ICP, most ions are produced as singly charged ions although some multiply charged species also occur. The extent of doubly charged ion formation in the plasma is controlled by the second ionisation energy of the element and the condition of plasma equilibrium (see Jarvis *et al.*, 1989 for discussion) and an estimate of the relative abundance of these species can therefore be made from these figures (Figure 5.12). Only those elements with a second ionisation energy lower than the first ionisation energy of argon will undergo any significant degree of 2^+ formation. The elements concerned are typically alkaline earths, some transition metals and the REE. Nebuliser gas flow rates can affect the level of doubly charged ions produced. At very low flow rates, plasma temperature is increased and equilibrium is shifted towards higher yields of 2^+ ions. In instruments using an asymmetrically grounded load coil, high nebuliser gas flow rates tend to produce higher plasma potentials which can generate some additional 2^+ ions. At normal operating conditions, the production of 2^+ ions is generally small ($<1\%$) (e.g. Figure 5.13).

In practice, the effect of a small proportion of 2^+ ions is two-fold. Firstly, it results in a small loss of signal and therefore sensitivity for the singly charged species but, more importantly, generates a number of isotopic overlaps at one half of the mass of the parent element. Fortunately, the number of elements affected is few but interference corrections may be needed if alternative isotopes cannot be used. An example is cited by Date *et al.* (1987a) for the determination of trace levels of Ga in two geological reference materials, manganese nodules NOD A-1 and NOD P-1. The matrices of these two samples are dominated by Fe and Mn at levels approaching $500\,\mu g\,ml^{-1}$ in solution. There are a number of potential interferences from the formation of polyatomic ions of Fe and Mn. Although the elements affected are few in number, Ga and Ge suffer from polyatomic overlap by MnO^+ (^{71}Ga), $MnOH^+$ (^{72}Ge) and FeO^+ (^{72}Ge) in this matrix (Figure 5.14). Since the magnitude of the correction to ^{71}Ga would be very large in the presence of such high levels of Mn, ^{69}Ga is preferred for analysis. However,

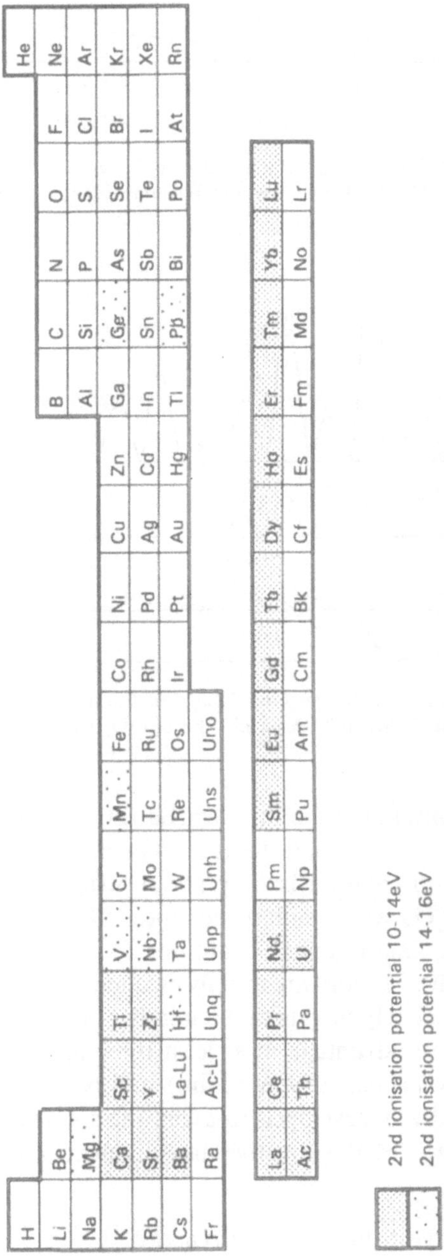

H																	He
Li	Be											B	C	N	O	F	Ne
Na	Mg											Al	Si	P	S	Cl	Ar
K	Ca	Sc	Ti	V	Cr	Mn	Fe	Co	Ni	Cu	Zn	Ga	Ge	As	Se	Br	Kr
Rb	Sr	Y	Zr	Nb	Mo	Tc	Ru	Rh	Pd	Ag	Cd	In	Sn	Sb	Te	I	Xe
Cs	Ba	La-Lu	Hf	Ta	W	Re	Os	Ir	Pt	Au	Hg	Tl	Pb	Bi	Po	At	Rn
Fr	Ra	Ac-Lr	Unq	Unp	Unh	Uns	Uno										

La	Ce	Pr	Nd	Pm	Sm	Eu	Gd	Tb	Dy	Ho	Er	Tm	Yb	Lu
Ac	Th	Pa	U	Np	Pu	Am	Cm	Bk	Cf	Es	Fm	Md	No	Lr

2nd ionisation potential 10-14eV

2nd ionisation potential 14-16eV

Figure 5.12 Distribution of elements with second ionisation energies lower than the first ionisation energy of argon at 16eV.

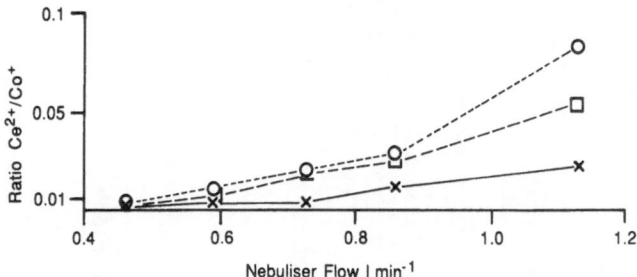

Figure 5.13 Variation of doubly charged ion ratio Ce^{2+}:Ce with nebuliser gas flow rate. x, 1500 W; □, 1300 W, ○, 1100 W (from Gray and Williams, 1987b).

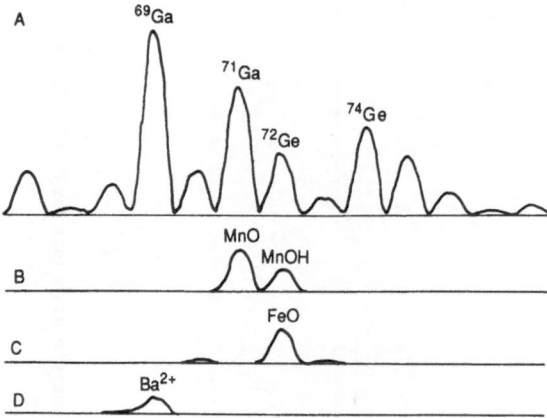

Figure 5.14 Spectra obtained in the range $66-78\,m/z$ for solutions containing (A) $1\,\mu g\,ml^{-1}$ trace elements, (B) $500\,\mu g\,ml^{-1}$ Mn, (C) $500\,\mu g\,ml^{-1}$ Fe and (D) $5\,\mu g\,ml^{-1}$ Ba (from Date et al., 1987a).

these samples also contain barium at a Mn:Ba ratio of 100:1 and therefore a significant contribution is made to the ^{69}Ga peak from $^{138}Ba^{2+}$. The concentration of Ga in solution is only about $30\,ng\,ml^{-1}$. Unfortunately, no reference data for Ga were available for these SRMs and an assessment of the accuracy of the data obtained could not be made.

Although high levels of 2^{+} ions are usually an undesirable feature in routine analysis since they are likely to generate a number of interference problems, they have been used to advantage in at least three applications (Date and Hutchison, 1987; Jarvis et al., 1989; Meddings and Ng, 1989). In one (Jarvis et al., 1989), a method was developed to avoid the spectral interferences on europium from $[BaO]^{+}$ by the sensitive measurement of Eu^{2+} ions.

5.2.5 Alleviation of spectroscopic interferences

It is clear that there are a number of serious spectroscopic interferences which, even at the low levels generated on most modern instruments, may still be too high to permit ultratrace level determination of some elements. Of the main groups described, polyatomic ions (particularly the interference from from $ArCl^+$, ArO^+ and ClO^+) are the most serious in practical analysis. There are several ways in which some of the interference effects described can be reduced for particular applications. In general, these 'remedies' change the relative proportions of interferences and therefore are restricted in their application to specific elements. Such methods include (a) instrument optimisation, (b) the use of mixed gases or solvents, (c) sample introduction, (d) fundamental instrument design and (e) alternative plasma sources.

5.2.5.1 *Instrument optimisation*   Section 5.2.2 highlights the importance of correct RF power and nebuliser gas flow rates in particular, and the significant effect these parameters can have on the signal response and the level of polyatomic, oxide and doubly charged ion species produced. On modern commercial ICP–MS systems, it is possible to optimise the instrument to give low relative levels of many spectroscopic interferences. In some instances it is, however, possible to adjust the system so that one or two interfering species are produced. A practical example of selective optimisation is given by Jiang *et al.* (1988) for the measurement of K isotope ratios. Changes in the position of the sampling cone orifice relative to the load coil, combined with the use of low forward power and high nebuliser gas flow rates, resulted in the background spectrum becoming dominated by NO^+ species. Under these operating conditions nearly all of the Ar^+ and ArH^+ ions were suppressed. These operating conditions would be quite unsuitable for multi-element analysis but are used to advantage in this particular application (see section 11.3.9).

5.2.5.2 *Mixed gases and solvents*   A number of authors have reported that the presence of a molecular gas in the argon changes the fundamental properties of the plasma (e.g. Choot and Horlick, 1986). For example the addition of a few percent of hydrogen or nitrogen enhances the ionisation process. This feature has been attributed to the higher thermal conductivity of these diatomic gases which lead to a more efficient transfer of energy within the plasma. The use of mixed gases has recently been investigated by Beauchemin and Craig (1990) to reduce interference effects in ICP–MS. A small addition of H_2 or N_2 can lead to a reduction in the size of the argon dimer in some instances. Evans and Ebdon (1989, 1990) investigated the use of oxygen and nitrogen additions to suppress the formation of ArCl and ArAr and thus reduce the interference on As and Se from these species. The size of both the ArCl and ArAr peaks was reduced relative to a peak integral for In. However, with the introduction of N_2, sensitivity was also much reduced from 23 198 to only 4333 counts s^{-1} for 100 ng ml^{-1} of In (Table 5.9).

Table 5.9 Sensitivity for In using different modification methods and intensity ratios for some polyatomic ion interferences[a]

Method modification	In response (area counts s^{-1})	In:ArCl	In:ArAr
None	18 213	0.19	2.2
Propan-2-ol spike	3 646	29	60
O$_2$ introduction	23 198	2.3	11
N$_2$ introduction	4 333	144	58

[a] From Evans and Ebdon (1990).

5.2.5.3 *Sample introduction* The method of sample introduction used can critically affect the level of some spectroscopic interferences. In general, a technique which can be used to introduce a 'dry' sample should lead to a reduction in the level of polyatomic and oxide species which involve O and H in their formation. The two methods which have been used to the greatest extent in ICP–MS applications are laser ablation and electrothermal vaporisation. The former is discussed in some detail in section 10.3 and can be used for a wide range of sample types. The latter has been successfully applied for the determination of Fe isotopic ratios in blood samples (section 11.3.3) and has been applied as a sample introduction technique in a number of applications (see chapter 4).

5.2.5.4 *Fundamental instrument design* The total elimination of spectroscopic interference effects does not seem possible with the instrument configuration used in most ICP–MS instruments. Although some interferences can be reduced by careful optimisation, argon, oxygen and hydrogen ions dominate the ion population in the plasma and so elimination of polyatomic ions formed from these species is unlikely to be possible. High resolution ICP–MS instruments have been developed which use a conventional ICP ion source (e.g. Bradshaw *et al.*, 1989). These are discussed in chapter 2. Due to its high cost this instrument is unlikely to become widely available but may be useful in some specialist applications.

5.2.5.5 *Alternative plasma sources* All currently available plasma source mass spectrometry instruments use an argon ICP as the ion source. However, a number of workers have investigated the use of alternative systems. Polyatomic ion interferences from Ar can be substantially reduced using a He ICP (Koppenaal and Quinton, 1988). However, a number of alternative interferences are reported principally those involving recombinations with He and the lower gas temperature of the He ICP has restricted its application. As an alternative to the use of an inductively coupled plasma, Brown *et al.* (1988) have investigated the use of a microwave induced He plasma (He MIP) for the determination of F, Cl and Br while Wilson *et al.* (1987b) used a

N_2 MIP. In the latter, the background species were composed mostly of NO^+.

5.3 Non-spectroscopic interferences

This second group of interferences can be broadly divided into two categories: (a) physical effects resulting from the dissolved or undissolved solids present in a solution; (b) analyte suppression and enhancement effects. A number of studies have been carried out to explore, understand and minimise some of these effects. Although the former are now reasonably well understood, a very confusing and inconsistent picture is brought to light with respect to analyte signal suppression. These inconsistences probably result from complex mechanisms which are not fully controlled during experiments and it becomes very difficult to compare one set of results with another. The results reported so far appear to be very instrument specific and a practical explanation of the effect which takes into account the observed behaviour has not yet been devised.

5.3.1 *High dissolved solids*

From some of the earliest work on ICP–MS, it became clear that the system was not tolerant to solutions containing significant amounts of dissolved solids (Houk *et al.*, 1980). A consequence of attempting to analyse solutions with greater than about $500\,\mu\mathrm{g\,ml^{-1}}$ total dissolved solids (TDS) was considerable signal drift over very short periods of time. Early instruments were fitted with smaller diameter sampling apertures (typically 0.5 mm) than those used on modern instruments, and even these modest levels of TDS caused progressive blocking of the sampling aperture (e.g. Houk *et al.*, 1980; McLeod *et al.*, 1986). It is now usual practice to use a 0.7–0.8 mm skimmer and 1.0–1.3 mm sampling aperture which are more tolerant to higher levels of dissolved solids. However, depending on the specific matrix, it is usual to limit analyte solutions to a total dissolved solids content of nominally $< 2000\,\mu\mathrm{g\,ml^{-1}}$. The actual level of TDS may be significantly less than this figure. For example, geological samples such as limestones, are composed largely of $CaCO_3$ and during sample preparation approximately 40% of the sample mass is lost as CO_2. At this level, signal drift with time is minimal and can easily be corrected for using either an external drift correction procedure or internal standards (see chapter 6). The effects of aspirating a limestone sample at a dilution factor of 500 are shown in Figure 5.15. Even after 20 min of continuous aspiration, the analyte signal precision is between 5 and 10% where counting statistics are not a limiting factor. By comparison (Figure 5.16) the same sample, but at a dilution factor of 100 (nominal $10\,000\,\mu\mathrm{g\,ml^{-1}}$ or $5545\,\mu\mathrm{g\,ml^{-1}}$ TDS), causes a severe loss in signal after only 10 min. This is reflected in the poor precision recorded, 17–20% RSD.

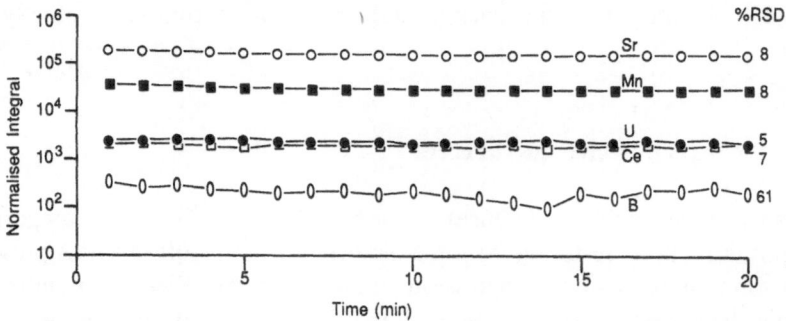

Figure 5.15   Precision data for limestone CCH-1 analysed at a concentration of 0.2 g per 100 ml (× 500 dilution).

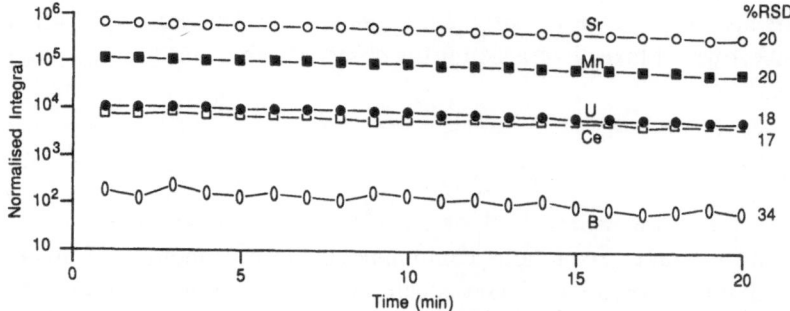

Figure 5.16   Precision data for limestone CCH-1 analysed at a concentration of 1 g per 100 ml (× 100 dilution).

The practical effects of aspirating solutions containing a high level of a refractory element were investigated by Williams and Gray (1988). The effects on analyte signal with time were monitored whilst continuously aspirating $1000\,\mu g\,ml^{-1}$ Al. Within the first 20 min a rapid decrease in signal occurs, before the analyte response stabilises, as material is deposited on the sampling cone orifice (Figure 5.17). The authors conclude that signal loss is not simply a result of a reduction in the number of ions entering the ICP–MS systems, but more likely a modification of the ion extraction process as a result of the reduction in orifice diameter. To help alleviate the problem of rapid signal loss at the beginning of an analytical run, the system can be primed. A solution of similar composition to that of the unknown samples should be aspirated for about 20 min prior to analysis, signal loss during the early stages of analysis is then reduced and quantitative determination can be made once a steady state is reached. This procedure has been successfully used in our laboratory to reduce signal loss for a range of sample types containing a heavy or refractory matrix.

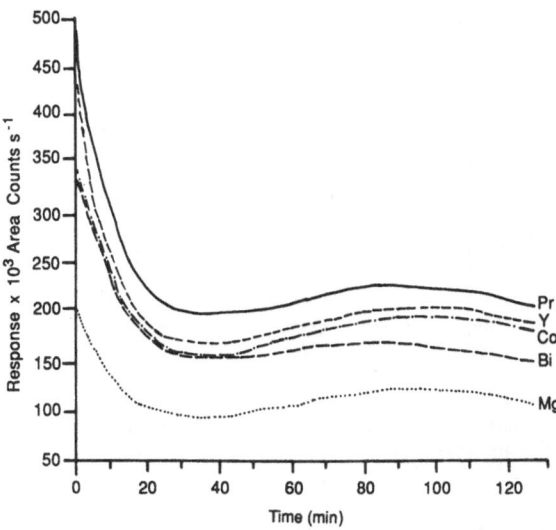

Figure 5.17 Effect of continuously aspirating $1000\,\mu g\,ml^{-1}$ Al on some analyte elements (after Williams and Gray, 1988).

5.3.2 *Suppression and enhancement effects*

The ICP as an atomisation-excitation source for atomic emission spectrometry is relatively free from ionisation interferences caused by easily ionisable elements in solution (Houk *et al.*, 1980). However, ionisation effects seem in general to be more severe in ICP–MS than in ICP–AES since there are no equivalents of atom lines in emission spectrometry. During the development of the technique, Olivares and Houk (1986) demonstrated the effects of a Na matrix at concentrations from 10^{-5} to $10^{0}\,mol\,l^{-1}$ on the response for Co at $1.2\,\mu g\,ml^{-1}$ (Figure 5.18). At a concentration of only about $200\,\mu g\,ml^{-1}$ the analyte count rate is suppressed although at this level it is not serious. Similar effects are recorded on more modern systems but the magnitude of these effects is variable (Tan and Horlick, 1987). In addition to a useful review of documented matrix effects, a comprehensive study was carried out by Beauchemin *et al.* (1987a) to examine the effect of 0.01 M concentration of a range of matrix elements (Li, B, Na, Mg, Al, K, Ca, Cs and U) on the signal from $100\,\mu g\,ml^{-1}$ of some analyte elements (V, Cr, Mn, Ni, Co, Cu, Zn, Cd, Pb). Some matrix elements, notably Na, Mg, K, Ca and Cs apparently caused an enhancement in analyte signal, while Li had little effect and B, Al and U caused analyte suppression. The authors concluded that since enhancements were observed with easily ionisable elements, a simple shift in ion–atom equilibrium in the plasma was not a likely mechanism for the observed effects since then only suppression should have been recorded.

Thompson and Houk (1987) performed a number of experiments under different operating conditions but concluded that in general analyte signals

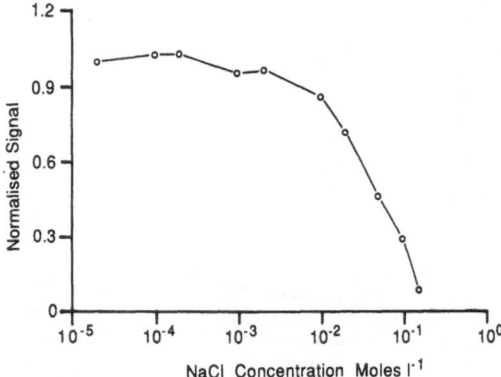

Figure 5.18 Experimentally determined suppression of Co ion in the presence of a NaCl matrix (after Olivares and Houk, 1986).

were all suppressed by concomitant or matrix elements (including Na and U) under all the conditions tested. These conditions included gas flow rates (34–40 psi), forward power (1.0–1.4 kW), ion lens voltages and two different nebulisers. Enhancement of analyte signal was observed in only a few isolated cases in contrast to data from Beauchemin *et al.* (1987a). The mechanism proposed to account for these matrix effects have focused on the ICP or the supersonic expansion, although several workers have suggested that some of the observed effects may arise in the ion optics. Gillson *et al.* (1988) attribute the apparent matrix effects to changes in the flux and composition of the ion beam. These changes are thought to arise due to space charge effects within the skimmer. A number of alternative mechanisms have however been suggested. In any given system several effects operate and it is a combination of these which are observed and reported. The work by Tan and Horlick (1987) suggests that matrix effects are strongly dependent on nebuliser gas flow rate while plasma power and sampling depth do not seriously affect their magnitude. Instrument optimisation for reduced levels of matrix effects should therefore be possible.

In practice, matrix effects such as those described above, can be difficult to measure and quantify. High concentrations of matrix elements lead to blocking of the sampling cone orifice and therefore erratic loss of signal (see above). Rigorous analytical protocol is required to clearly distinguish between analyte signal loss and analyte signal suppression. In addition, high matrix concentrations tend to lead to poorer precision (e.g. Wilson *et al.*, 1987a) thus some apparent enhancement and suppression effects may simply be a result of poor RSD. Memory effects can also be severe over extended periods of time presenting a more complex picture (Beauchemin *et al.*, 1987a). The use of flow injection may help to alleviate some of these problems particularly for experimental investigations.

In general, the lower the atomic mass of the analyte and the lower the degree of ionisation of the analyte in the plasma, the greater will be the effect of a given added concomitant or matrix element on the ion count rate of the analyte. For a given analyte, the greater the atomic mass and degree of ionisation of the added element, the greater will be the effect of the matrix element on the analyte count rate (Gregoire, 1987a). In practical analysis, many of these interference effects are not severe and may not always need to be corrected for. For example, in the work of Date *et al.* (1987a) experiments were carried out to investigate the interference effects from a pure Ca, Fe or Mn matrix. At 500 μg ml^{-1}, analyte suppression was minimal (typically less than 5%) and a correction was not considered necessary in either of these matrices.

If necessary, a number of methods can be used to overcome some of the non-spectroscopic interference effects. Dilution of samples to bring the concentration of matrix elements to typically less than 500–1000 μg ml^{-1} can be successfully employed (e.g. Ridout *et al.*, 1988; Gregoire, 1987a) although determination limits are then compromised. Internal standards have also been used (e.g. Thompson and Houk, 1987) but to be successful, analyte and internal standard must be closely matched in both mass and ionisation energy if they are to behave alike in the presence of a matrix element (see section 6.9.2). Instrumental optimisation in the presence of a matrix element has also proved successful for reducing the level of observed suppression (Wang *et al.*, 1990). Matrix matching of standards and samples may be a practical solution if high purity reagents are available. Care should be taken to ensure that both acid concentrations and matrix element levels are identical in all samples and standards to eliminate transport effects due to different viscosities. Standard addition calibration techniques can also be used (see section 6.9.3). The most satisfactory method for some matrices, however, may be to separate the analytes from the matrix using techniques such as ion exchange separation or co-precipitation. In this case the matrix can be totally removed with the additional benefit of analyte pre-concentration. Finally, all samples should be scanned prior to quantitative determination to establish the nature of the matrix and the possible interference effects which are likely to occur. The most effective remedy can then be implemented.

6 Calibration and data handling

6.1 Introduction

As with most instrumental methods of analysis, ICP–MS cannot give an absolute value for the concentration of an element or isotope but is always a comparative technique. The quantification of measured data from an ICP–MS instrument is achieved by comparison of the measured counts from an unknown sample with those from a substance containing a known amount of the element or isotope of interest. Calibration strategies are usually similar, regardless of the form in which the sample is presented for analysis. In addition, the physical form of the sample and standard are usually the same, i.e. solid samples are calibrated with solid standards, although there may be some exceptions to this rule. For elemental determination, calibration may take place prior to the actual analysis of the unknown samples. The measured counts for each sample are then compared with the standard calibration and concentration data are calculated on-line during acquisition. Alternatively, all of the raw data for both standards and samples may be collected and processed off-line, on completion of an analytical run. However, calibration curves are not usually stored from day to day and a new calibration is generated with each analytical run.

Before considering the various methods of calibration, it is useful to examine the alternative methods available for data collection, which vary to some extent from one instrument to another, and to define some of the general concepts and terms used to describe data quality.

6.2 General concepts

6.2.1 Mass scale calibration

Mass scale calibration is necessary to determine the digital to analogue conversion setting of the quadrupole control system required to set it at any given mass. The mass scale is normally calibrated across the entire mass range and typically about six points are used. However, if data are to be collected only from a narrow mass range, then the mass calibration should be checked and redefined if necessary.

6.2.2 Accuracy, precision and reproducibility

The term accuracy is used here as an estimate of the closeness of the *measured* concentration or ratio, to the 'true' or agreed upon value. It may be defined as

$$(Actual - measured/actual) \times 100$$

The term precision is used here to describe the short term reproducibility of a signal, concentration or ratio. It is expressed as one standard deviation of the mean or as the standard deviation calculated as a percentage of the mean, hence % relative standard deviation (% RSD). The term reproducibility is used to describe the 'long term precision' of a measurement and is useful for evaluating the general performance of an instrument over several hours.

6.3 Instrumental modes of data collection

There are two principal modes which can be used for data collection, (a) peak hopping or jumping and (b) scanning.

6.3.1 Peak hopping

In this mode, the mass spectrometer is used to collect data at a number of fixed mass positions (typically 1–3) for each isotope of interest. The location of the central position of the peak is particularly important in this mode of operation since it is used to locate the starting point for the measurement of each peak. If three points per peak are used then one measurement is taken either side of the central point. Thus at each single point of measurement, peak height is measured. There are a number of advantages and disadvantages to this mode of data acquisition. On most instruments, the amount of time spent collecting data at each isotope can be varied, so that the response from a small isotope can be collected for longer to improve counting statistics. However, so far this feature does not appear to have been widely utilised. An additional advantage is that time is not wasted collecting data for isotopes which are not of interest. This can, however, also be a major disadvantage of using the peak hopping mode since no record is available should data for additional isotopes subsequently be required and, more importantly, interference and matrix effects are often missed because the spectra cannot be examined. In theory, the peak hopping mode of measurement should provide a number of advantages where:

  (a) only a small number of isotopes are required (< 20)
  (b) the isotopes of interest are spread across the mass range
  (c) isotope ratio measurements are made.

The dwell time on each isotope can be varied according to the isotopic abundance, thus improving the counting statistics on the smaller isotope.

6.3.2 Scanning

An alternative mode of operation is to collect data for a relatively large number of points (approx. 15–20 per peak) so that the peak shape is defined for each isotope and the area under the curve is integrated. Providing that sufficient storage channels or memory locations are available (typically 4096 for 4–240 m/z), a complete spectrum, containing information for all isotopes within the mass range 4–240 m/z, can be collected and stored. The maximum scan speed is ultimately controlled by the scan speed of the mass spectrometer and the rate at which data storage/transfer takes place. A lower limit to the rate of scan may be set by the data storage. In order to take advantage of the scanning rate permissible with the quadrupole, a fast multi-scaler facility is usually used to precede the data handling computer as a buffer store. The usual mode of operation is to scan the quadrupole across the mass range of interest, and to successively build up a picture of the mass spectrum. The advantages of the scanning mode of operation lie principally in the fact that data are available not only for the isotopes of immediate interest but also over a wider mass range for archival purposes. In addition, interfering peaks are more easily indentified if complete spectral information is available. Although initially the peak hopping and scanning mode of operation appear to be rather different, the two in fact become essentially the same when all isotopes are measured in a given mass range with a large number of points for each peak.

Procedures which combine the relative merits of peak hopping and scanning modes of operation are sometimes useful. For example, some instruments allow narrow parts of the mass range to be scanned, e.g. over only 5–10 isotopes, which are separated by large areas (sometimes termed skipped scan regions) across which the quadrupole is rapidly scanned and no data is collected.

Two methods of defining the peak for integration are used in the scanning mode of operation. These are constant width and valley integration.

6.3.2.1 *Constant mass width*   In this integration mode a fixed peak width is integrated regardless of the peak height, width or resolution. For example, if a peak width of 0.7 m/z is used, 0.35 m/z is integrated on either side of the mass marker (Figure 6.1). Due to the near gaussian shape of the peaks, a large proportion of the peak area is included in a 0.7 m/z window excluding only the 'wings'. The advantage of this integration method is that the often 'noiser' margins of the peak are excluded. Tailing of one peak into another is less critical when only the central part of the peak is integrated in this way. The reproducibility of the integration does, however, depend on the stability of the mass calibration and it is therefore essential to check this on a regular basis.

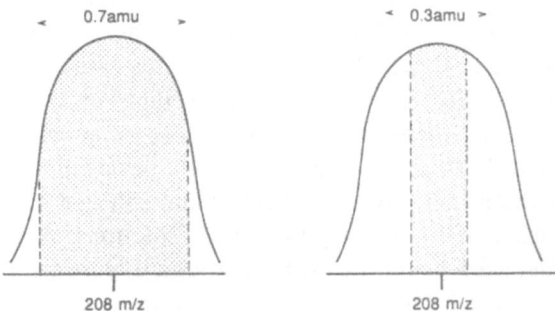

Figure 6.1 Constant mass width integration.

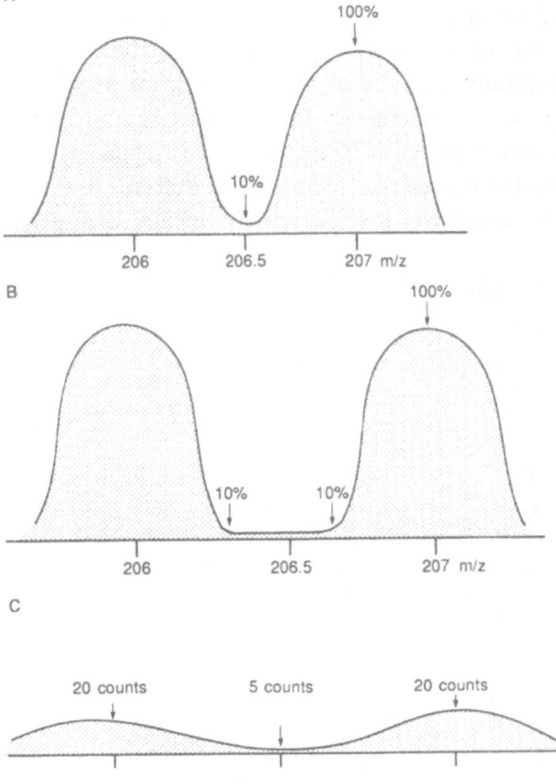

Figure 6.2 Valley integration. (A) Peaks resolved such that a signal ≤ 10% of adjacent peak height is reached between peaks; (B) peaks over resolved, valley between peaks has signal ≤ 10% of adjacent peak height; (C) small peaks, 10% peak height may not be obtained under these circumstances.

6.3.2.2 *Valley integration*   In this second approach, the peak is integrated on both the high and low mass sides until a predefined single channel value is reached. This value is usually set to be 10% of the maximum peak height (Figure 6.2). Since a peak search routine is used to locate the maximum peak height the mass calibration is less critical than in other methods of integration. However, there are a number of occasions when a 10% value will not be reached. This may occur when one peak tails into another, in which case the resolution should be checked and adjusted if necessary. Very small peaks, such as those measured in 'blank' solutions, may never reach a height more than a few counts above background (Figure 6.2c). Under either of these conditions, a default value of 1 m/z (i.e. one mass width) is usually integrated.

6.4 Linearity of response

In some instrumental analytical techniques (e.g. AAS) the linear dynamic range is limited to only about two orders of magnitude. In ICP–MS, the response of signal versus concentration is typically linear over 6–8 orders of magnitude depending upon the mode of instrument operation. Instruments which operate using an ion detector in the pulse counting mode are generally restricted to about 6 orders (0.01 ng ml^{-1} to 10 μg ml^{-1}) providing that a dead time correction is applied. Operating the detector in the analogue mode or reducing the sensitivity by alternative means can generate a linear

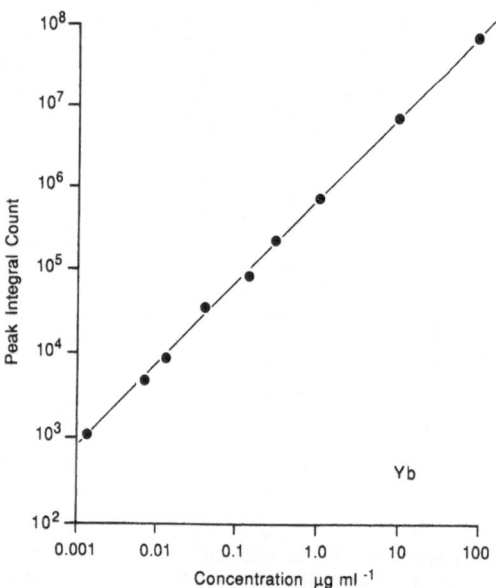

Figure 6.3   Calibration curve for Yb from 0.001 to 100 ng ml^{-1} in 1% HNO$_3$ (from Jarvis, 1989a).

calibration up to about $500\,\mu g\,ml^{-1}$, typically two orders above that of the pulse counting limit. An example of the linear response for Yb is shown in Figure 6.3. The data shown here were obtained using simple aqueous solutions but are equally applicable to more complex matrices. For the analysis of solids, for example by laser ablation, a similar dynamic range is available. It is therefore possible to calibrate the instrument using only a single standard data point for each element whether samples are in a solid or liquid form.

6.5 Blanks

Prior to analysis, most materials undergo some form of sample preparation. In some cases this may be limited to simple grinding and pressing into a pellet, acidification (e.g. water samples) or more extensive digestion with a mixture of acids. In each case there is scope for some degree of contamination to be introduced into the sample at a number of stages. The contribution from this contamination should be taken into account and this can, in some cases, be done by preparation of a 'procedural' or 'reagent' blank. This blank is taken through the same preparation procedure as the actual sample but contains no sample material. For samples which have been brought into solution, this is a relatively straightforward process. However, in the case of solids, a blank may be difficult to prepare. In some cases, high purity solid reagents can be mixed and prepared along with the actual samples and contamination can be monitored. It should be emphasised that a standard blank and procedural blank fulfil separate requirements. If a mixed element standard solution is prepared for external calibration in 0.5 M HCl, then a standard blank containing just 0.5 M HCl should also be prepared. This blank can be used to correct for gas peaks and polyatomic ion interferences which occur in the standard as a result of the acid used and to correct for contamination. The blank is also used as the zero point for the calibration. A procedural blank in contrast, can be used to correct for interferences which occur as a result of the reagents used for digestion, for contamination and in addition to correct for polyatomic ion interferences.

6.6 Factors affecting signal stability

A number of potential sources of signal instability can be identified. These include fluctuations in power supplies, electronics noise, plasma noise, sample introduction method and sample matrix effects. In general, on modern instruments, fluctuations in power input supplies are minimal and the latter areas are more likely to make a significant contribution to signal stability. Of particular importance are the effects of high dissolved solids or of refractory matrices, such as aluminium or zirconium.

The long term effects of aspirating solutions containing high dissolved solids are that the sampling cone becomes progressively blocked eventually resulting in a rapid loss of sensitivity. However, on a short time scale, such solutions can cause considerable instability in the plasma due to the high sample and solvent loading, and undissociated material is able to pass through the plasma. Volatilisation noise also occurs due to local release of analyte from large micro-particulates. The ion extraction process may then be affected and signal stability impaired. A similar effect may occur during laser ablation if the mass of ablated material is large. The particulate material is not fully dissociated in the plasma and an unstable signal may result.

The introduction of solutions containing high concentrations of refractory elements may also affect signal stability. The spectra (180–200 m/z) illustrated in Figure 6.4 were collected for two solutions both containing 5000 μg ml^{-1} total dissolved solids (TDS). The upper spectrum (A) is from a silicate rock and shows only a small peak at 181 m/z from Ta and an otherwise blank spectrum. The lower illustration (B) is from a 5000 μg ml^{-1} Al solution. No peaks are clearly defined but there are several spikes of noise which occur as relatively large signals in single channels. These spikes are thought to result from the extraction of unvaporised and undissociated particulate material from the plasma.

The sample introduction system may itself be a source of signal instability (see chapter 3) particularly if the signal produced is a transient one. In practice, short term signal stability is most seriously affected by the plasma itself, by

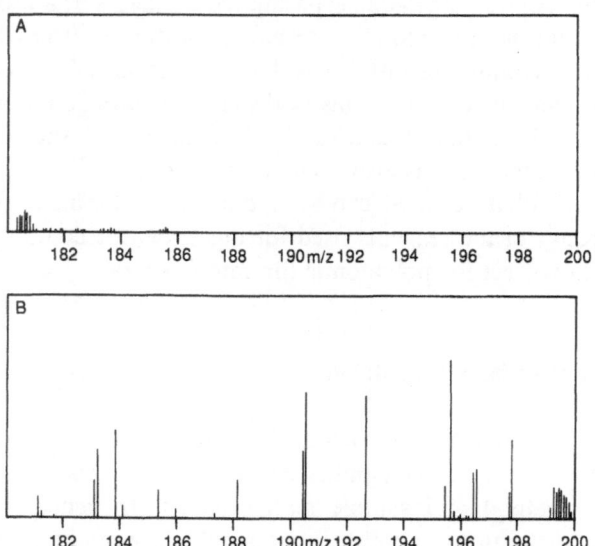

Figure 6.4   Mass spectra (180–200 m/z) for two solutions containing 5000 μg ml^{-1} total dissolved solids: (A) silicate rock and (B) Al. A number of single channel noise spikes are seen in the lower spectrum.

incorrect operation of some types of nebuliser and from the pulsing effects of a peristaltic pump used with some types of nebuliser (see chapter 3).

6.7 Qualitative analysis

The facility to collect signals for all masses during a series of scans or peak jumps can be used for a purely qualitative examination of a sample. Regardless of which mode of data acquisition is used, it is possible to collect information over the complete mass range (4–240 m/z) in less than 60 s. The spectra can then be visually examined for the presence or absence of an analyte, and to identify possible sources of interference. Sample matrices which are unfamiliar to the analyst should always be subject to qualitative analysis prior to full quantification. Some of the software provided with commercial instruments, allows the user to display several spectra simultaneously and to subtract one from another to remove background species. The vertical scale can usually be expanded or selected parts of the mass range plotted to allow a detailed examination of each spectrum.

6.8 Semi-quantitative calibration

On many ICP–MS instruments, a plot of mass against sensitivity yields a relatively smooth curve if degree of ionisation and isotopic abundance are taken into account (Figure 6.5). This response curve can be used to calibrate the instrument to provide semi-quantitative data. In practice, this calibration method is ideal as a survey tool, particularly if an unfamiliar matrix or sample type is to be analysed or if only the approximate levels of the components are required. The profile of the response curve is usually defined using 6–8 elements suitably spread across the mass range. The response for each element is corrected for isotopic abundance, concentration and degree of ionisation (Appendix 1) and a second order curve fitted to the data.

The actual shape of the curve is dependent on a number of factors including tuning of the ion optics and the operating conditions of the quadrupole. Although on some instruments the response curve may be stored, it should be redefined prior to analysis (on a daily basis) since its shape is highly dependent on the way in which the instrument is optimised. In addition to the general curve shape, the offset of the curve (i.e. sensitivity) may vary each time the instrument is set up. The amount of offset can be determined by measurement of the sensitivity of an element positioned in the centre of the mass range, e.g. [115]In or [103]Rh. This procedure may need to be carried out several times in an 8-h period. Once the response curve is established, the concentration of all elements can be determined in an unknown sample by reference back to the response curve. The accuracy of the data produced by

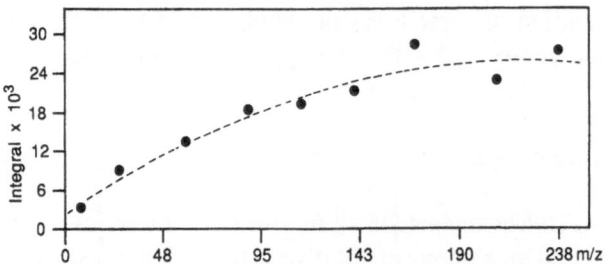

Figure 6.5 Instrument response curve from 4 to 240 m/z.

Table 6.1 Semi-quantitative analysis of NIST reference water 1643b[a]

Element	Mean ($n = 16$) (ng ml^{-1})	Standard deviation (ng ml^{-1})	Reference value (ng ml^{-1})
Be	0.78	0.19	1.9
B	4.6	0.70	9.4
Na	680	240	800
Mg	1500	250	1500
V	5.40	0.47	4.5
Cr	2.4	0.4	1.9
Fe	164	8.1	9.9
Mn	3.5	0.33	2.8
Ni	6.8	0.79	4.9
Co	3.5	0.28	2.6
Zn	3.8	1.4	6.6
As	7.0	0.65	4.9
Sr	25	1.4	23
Mo	16	1.2	8.5
Cd	3.9	0.45	2.0
Ag	2.0	0.41	0.98
Ba	3.3	0.28	4.4
Tl	0.85	0.15	0.80
Pb	1.5	0.20	2.4
Bi	1.1	0.14	1.1

[a] From Ekimoff *et al.* (1989).

this method is variable and highly dependent on the element sought and the sample matrix. The data shown in Table 6.1 for a standard reference water sample were acquired on a Sciex Elan 250 using the software package supplied with the instrument (Ekimoff *et al.*, 1989). The accuracy of a range of elements across the mass range is from -59% to $+122\%$ calculated as the difference between the measured and reference value as a percentage of the reference. The precision of the data is from 5 to 50% RSD.

6.9   Quantitative analysis

6.9.1   *External calibration techniques*

The most widely used calibration method is that using a set of external calibration standards. For solution analysis these may have a simple acid or aqueous matrix containing the analytes of interest. Several standard solutions are prepared which cover the range of expected concentrations. For direct solid sample analysis, e.g. laser ablation, standards should be matrix matched to the unknown samples.

Simple aqueous standards are usually adequate for the calibration of liquid samples providing that the unknowns are sufficiently dilute, $< 2000\,\mu g\,ml^{-1}$ TDS, and samples do not contain a matrix of one element only. Above this level viscosity and matrix effects can be significant but can, to some extent, be corrected for by using matrix matched samples and standards. The use of standard reference materials is not recommended for primary calibration for a number of reasons. Natural reference materials do not contain *known* amounts of an element, rather they have a set of *agreed* values. While some agencies do not release materials until after completion of the certification process, some rely on the subsequent accumulation of data after release for refinement of reference values. Some elements will have been determined by many different laboratories and methods, while some may be poorly characterised. In addition, dissolution techniques are destructive and most reference materials are expensive and in limited supply while the solutions produced may degrade with time. Standard reference materials are therefore better used as quality control samples to assess both accuracy and precision.

The fitting of a calibration line to measured standard data is usually done by using least squares regression analysis. Under ideal conditions, measured data would form an exact linear function of concentration. However, errors are always superimposed on real data and therefore a statistical procedure such as a regression analysis is used to calculate the best fit calibration line (see Potts, 1987 for further discussion). The goodness of fit of the line to the measured data may be calculated and is termed the correlation coefficient. Although the linear regression fit is widely used in many analytical techniques, it is prone to some problems. For example, if the standard concentrations are not evenly spread so that the highest concentration is much larger than the others, the calibration line will be biased towards this outlying point which may result in a poorer fit for the other standard points.

6.9.2   *Raw data correction procedures*

During an analytical run the stability of the analyte signal may not remain constant for the reasons outlined above. Several approaches have been used to attempt to correct for changes in sensitivity, all of which rely on certain assumptions with respect to both the nature or the sensitivity change (i.e. sudden or gradual) and the change in relative sensitivity between elements.

6.9.2.1 *External correction* If analyte signal change is a linear function with either time or run order, an external drift correction can be applied. In practice, the same solution is analysed intermittently throughout an analytical run. The recorded change in signal for each element in that solution is assumed to be a linear function with either time or run order. The relative change in signal between two runs of such a solution is recorded and a correction applied to each of the intervening unknown samples.

The limitations of this method of data correction lie in the assumption that the signal change is linear. However, in practice this has been shown in many instances to be the case. The effect of running a solution containing a high concentration ($1000\,\mu g\,ml^{-1}$) of Al is shown in Figure 6.6. The gradual loss of each analyte signal with time is a function of sampling cone blockage. However, it should be noted that although the general pattern is one of a drop in sensitivity, the rate of change is not the same for all elements. These changes in analyte signal could, however, readily be monitored and a linear drift correction applied. In the author's experience, external drift correction can significantly improve both accuracy and precision in a range of sample matrices. The main advantage of this type of data correction is that nothing is added to the sample solutions themselves, whether in solution or solid form. In addition, the individual behaviour of each element is monitored independently and therefore specific corrections can be applied.

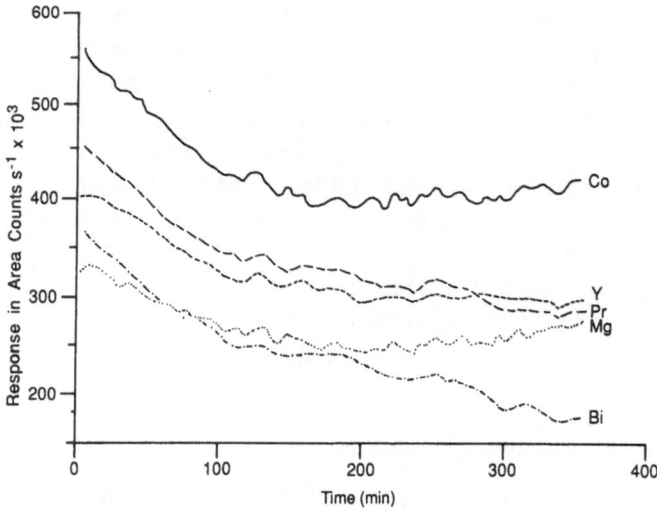

Figure 6.6 The effect of running $1000\,\mu g\,ml^{-1}$ matrix with time for five analyte elements. Although all elements lose signal with time, the rate of loss is different in each case (after Williams and Gray, 1988).

6.9.2.2 *Internal correction* The calibration or correction of one element using a second as a reference point is used in many forms of analytical atomic spectrometry, e.g. ICP–AES and is termed 'internal standardisation'. An internal standard can be used for several purposes:

(a) to monitor and correct for short term fluctuations in signal
(b) to monitor and correct for long term fluctuations in signal
(c) to calibrate for a second element
(d) to correct for unspecified matrix effects

The effectiveness of an internal standard requires that its behaviour accurately reflects that of other elements. For example, if it is used to calibrate for a second element, either their sensitivities should be the same or the sensitivity relationship between the two should be known and remain constant. If an internal standard is employed to correct for signal fluctuation, for whatever reason, then the relative signal change between the internal standard and the unknown element should be constant. The use of internal standards in ICP–MS is to some extent a historical inheritance from ICP–AES. Many users of ICP–MS find the implementation of a single internal standard for data correction very attractive, particularly when the data processing is carried out automatically by the system software. However, careful monitoring of analytical runs can reveal significant differences in relative elemental response throughout a run and it is rarely the case that one single element can reflect the behaviour of all the others.

Choice of internal standard For the analysis of samples in solution form, the addition of internal standards to samples is relatively straightforward. The choice of elements is limited by those which are naturally present since a known or constant amount of an internal standard is added to each blank, standard and sample. The internal standard should not suffer from an isobaric overlap or polyatomic ion interference or indeed generate them on isotopes of interest. As an alternative, an element which occurs naturally in the sample, and which has already been determined prior to ICP–MS analysis, can be chosen. In this case, the concentration will probably vary from sample to sample and this must be accounted for during the data manipulation stage. The use of a natural internal standard is the most effective way of using internal standards in solid sample analysis.

Whether an internal standard is 'added' or 'natural' a number of other factors should be considered. The element should be present at a concentration which will not yield signals limited by counting statistics. In addition, it has been suggested by some authors that it should be close in mass and ionisation energy to that of the element to be determined. Two elements which are frequently used as internal standards, are In and Rh. Both lie in the central part of the mass range (^{115}In, ^{113}In and ^{103}Rh), occur at very low concentrations in many sample types, are almost 100% ionised (In = ~98.5%

and Rh = $\sim 93.8\%$), do not suffer from isobaric overlap and are either monoisotopic (Rh = 100%) or have one dominant isotope (^{115}In = 95.7%). Although both In and Rh would appear to be ideal candidates for internal standards, in practice this is not always so (see below).

Some authors (e.g. Beauchemin *et al.*, 1988b) have suggested using a gas or polyatomic ion peak such as the argon dimer as an internal reference point. However, polyatomic ions result from different formation mechanisms and may therefore change during a run in a different way to the analytes. The results of monitoring the change in signal at $80\,m/z$ and some analyte elements are shown in Figure 6.7a. At the start of the analytical run there is a rapid decrease in the ^{40}Ar^{40}Ar signal which clearly does not reflect the response pattern exhibited by the analyte elements. Indeed the result of using ArAr as an internal standard in, this case, is highly detrimental (Figure 6.7b). The use of such peaks as internal standards is therefore not recommended.

Internal standards for short term signal fluctuation The correction of data for non-linear changes over short time scales, e.g. a few minutes, can be made only by using a component of the solution of interest. The suitability of any internal standard (including In and Rh) should be tested in each matrix of interest for all elements which require correction. In general, providing that the signal change between the internal standard and element of interest is in the same direction, i.e. both showing an increase, then short term precision will be improved over data calculated without any form of correction. However, if signal fluctuations are large over short time periods, this may indicate an instrument fault, incorrect optimisation or poor suitability of the sample matrix chosen. Attention to the latter may prove to be more successful than data correction using internal standards.

Internal standards for long term signal fluctuation In the case of longer term signal changes, e.g. several hours, the use of internal standards may in some cases improve precision. The data shown in Figure 6.6 were collected over a 6-h period. It is clear that although some elements show a decrease in sensitivity with time, e.g. Bi (although different elements decrease at different rates) some elements show an increase. Yttrium and Co are similar in mass and first ionisation energy (^{89}Y = 6.53 eV, ^{59}Co = 7.87 eV) but show contrasting behaviour while ^{141}Pr is much higher in mass than ^{89}Y and has a lower ionisation energy (5.42 eV), it exhibits a similar pattern to Y of signal decrease with time.

It must be stressed that the suitability of any element as a potential internal standard should be established in the matrix of interest prior to its use for long term drift correction.

6.9.2.6 *Internal standards to correct for matrix effects* Non-spectroscopic matrix effects have been widely documented in ICP–MS analysis and the

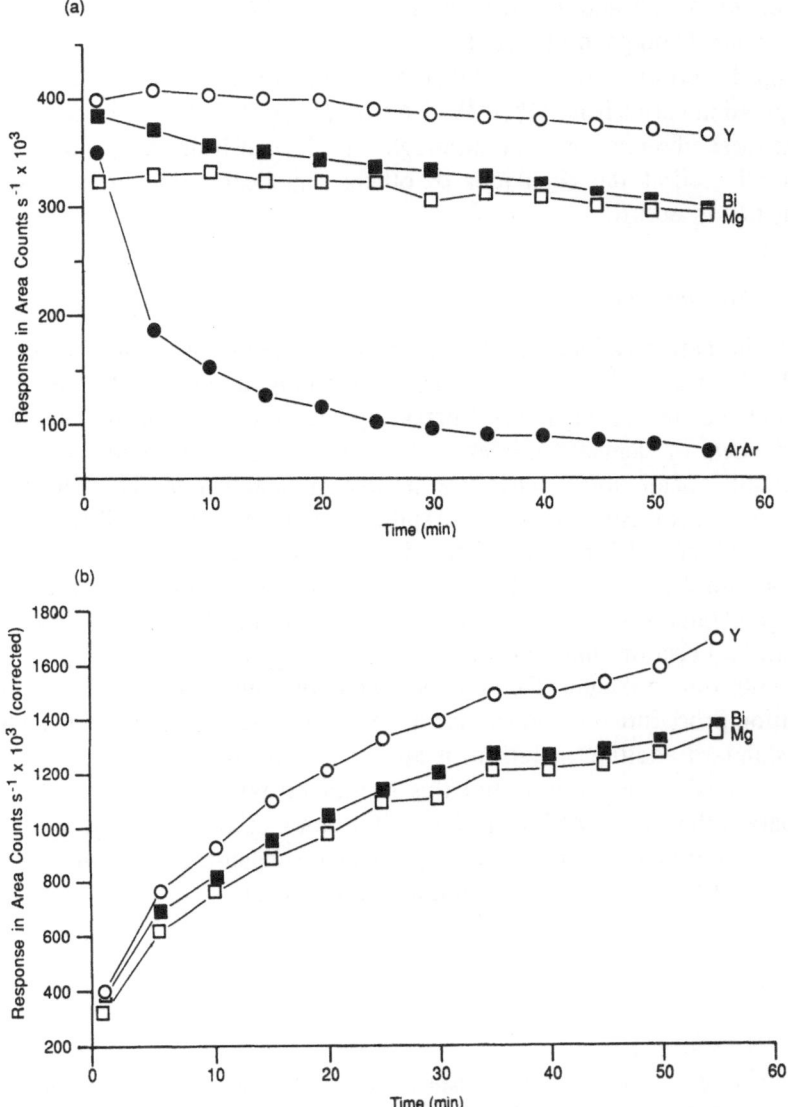

Figure 6.7 Response for $1 \mu g \, ml^{-1}$ Mg, Y and Bi in a $1000 \mu g \, ml^{-1}$ Al matrix: (a) without correction also showing the response of the Ar dimer and (b) internal correction using the Ar dimer.

use of internal standards to correct for analyte suppression or enhancement has been investigated (Thompson and Houk, 1987). Studies of the effects of forward power variation, sampling position and nebuliser gas flow rate indicate that under any fixed set of operating conditions, certain elements can be grouped together with respect to their behaviour in a matrix. For example, in the presence of 0.022 M NaCl ($\sim 500 \mu g \, ml^{-1}$) matrix Nb, Mo, Hf,

Sr, In, Zr, Y, Yb and Tm can be grouped together as they exhibit similar behaviour (Thompson and Houk, 1987). The spread of first ionisation energies within this group is from 5.7 eV (Sr) to 7.0 eV (Mo) and the amount of signal suppression varied from 41% ([89]Y, [93]Nb) to 58% ([88]Sr). It must be concluded from these observations that, although one element from this group would generally reflect the behaviour of others, significant differences between elements do occur.

6.9.3 *Standard additions*

This alternative calibration strategy has been used with some success in ICP–MS. The calibration is performed by taking aliquots of the sample to be analysed, and adding to each increasing quantities of a reagent containing the element or elements of interest. The increments are normally equal and the number of aliquots taken is at least three, preferably more. The calibration set therefore consists of a number of spiked samples plus the unspiked original sample, all of which have an almost identical matrix.

The samples are analysed and a graph constructed of the integral of the isotope of interest versus the concentration of the element added. The intercept of the calibration line on the X axis (a negative number) gives the concentration in the unspiked sample. Theoretical modelling of the standard addition procedure has shown that the optimum precision is obtained when the standard addition increment is approximately equal to or greater than the expected concentration and this should be taken into account when preparing the spikes. Although this calibration method can produce highly accurate and precise data (Table 6.2) it is time-consuming to perform and is best suited to a relatively small number of elements.

Table 6.2 Comparison of standard addition and isotope dilution data for coastal seawater reference material CASS-1[a]

Element	Standard addition $(ng\,ml^{-1})$	Isotope dilution $(ng\,ml^{-1})$	Reference value $(ng\,ml^{-1})$
Mn	1.80 ± 0.06		2.27 ± 0.17
Co	0.039 ± 0.003		0.023 ± 0.004
Ni	0.302 ± 0.005	0.319 ± 0.004	0.290 ± 0.031
Cu	0.263 ± 0.007	0.311 ± 0.018	0.291 ± 0.027
Zn	1.05 ± 0.004	1.00 ± 0.04	0.98 ± 0.10
Cd	0.027 ± 0.001	0.026 ± 0.002	0.026 ± 0.005
Pb	0.22 ± 0.02	0.24 ± 0.02	0.25 ± 0.03
Cr		0.13 ± 0.01	0.12 ± 0.02

[a] From McLaren *et al.* (1985).

6.9.4 *Isotope dilution*

Stable isotope dilution is a powerful strategy for elemental analysis which can in principle be applied with any mass spectrometric technique (Heumann, 1988). The basis of the method is the measurement of the change in the ratio of signal intensities for two selected isotopes of an element of interest after the addition of a known quantity of a spike which is enriched in one of these isotopes. This measurement permits the calculation of the concentration of the element in the sample. The method can be applied for any element with at least two stable isotopes, or even for mono-isotopic elements with a suitable long-lived radioisotope (e.g. Mn).

The isotope dilution determination of cadmium is illustrated by the two ICP mass spectra in Figure 6.8. For cadmium, as for most of the elements, the relative abundances of the isotopes are invariant in nature. The spectrum on the left shows the characteristic isotopic signature for the eight stable isotopes of cadmium. The spectrum on the right has been altered by the addition of a ^{111}Cd enriched 'spike'. The signal intensity for ^{111}Cd is now approximately equal to that for ^{114}Cd, the most abundant isotope of Cd. By measuring the 'altered' ^{114}Cd/^{111}Cd, ratio, it is possible to calculate the concentration C of Cd in the sample, according to the following formula:

$$C = \frac{M_s K(A_s - B_s R)}{W(BR - A)}$$

where

M_s is the mass of the spike (usually expressed in ng or μg)

W is the weight of the sample

K is the ratio of the natural atomic weight and the atomic weight of the ^{111}Cd-enriched material

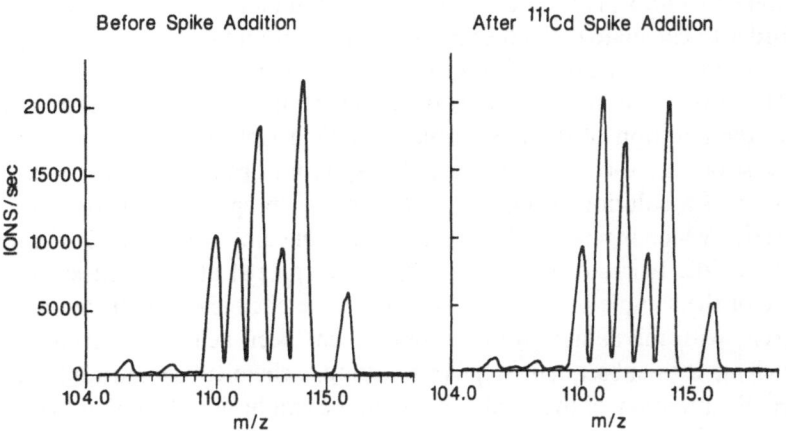

Figure 6.8 Determination of Cd by isotope dilution.

A is the natural abundance of ^{114}Cd (the 'reference' isotope)
B is the natural abundance of ^{111}Cd (the 'spike' isotope)
A_s is the abundance of ^{114}Cd in the ^{111}Cd-enriched spike
B_s is the abundance of ^{111}Cd in the ^{111}Cd-enriched spike
R is the measured ^{114}Cd/^{111}Cd ratio

As suggested by Figure 6.8, the amount of the spike is usually selected so that the measured ratio is near unity, to maximise the precision. (An exception to this practice arises when the natural abundances of the two chosen isotopes are nearly equal.) Also, it is normal to use the most abundant isotope (in this case ^{114}Cd) as the reference in order to have the best possible sensitivity. The steps in the isotope dilution analysis are discussed in detail in the next section.

6.9.4.1 *Steps in an isotope dilution analysis*   The first step in an isotope dilution analysis is to prepare an unspiked solution of the sample. This solution is required for two purposes. First, it can be used to obtain at least a rough estimate of the concentration(s) of the analyte(s) for establishing appropriate values for M_s, the spike weight(s). Secondly, it can be used to measure the ratio for the proposed pair(s) of isotopes. These data provide a very sensitive diagnostic of isobaric interferences. A significant difference between the measured value and the expected value based on natural abundances indicates an isobaric interference on at least one of the isotopes; since current ICP–MS instrumentation usually corrects automatically for natural isobaric interferences (e.g. overlap of ^{114}Sn on ^{114}Cd), interference by a polyatomic species is normally responsible for the discrepancy. The existence of polyatomic isobaric interferences can seriously degrade the accuracy of isotope dilution analysis, and the ideal is always to find two isotopes which are completely free of such interferences. For elements such as lead for which the isotopic abundances are not invariant in nature, the unspiked solution is used to measure the abundances of all of the isotopes in order to calculate the values of the 'natural' abundances A and B required for the subsequent calculation of isotope dilution results.

The next step in the analysis is the preparation of a solution of the sample after the addition of the appropriate enriched isotopic spike (or spikes, in the case of a multi-element analysis). The spike is normally added as a weighed aliquot of a solution of known concentration prepared from the enriched material, which is received from the supplier in solid form, often as the metal or its oxide. It is advantageous to add the spike(s) at the earliest possible stage of the sample preparation because, once chemical equilibration of the spike(s) with the analyte(s) in the sample has been achieved, partial loss of the element in subsequent steps (e.g. during a chemical separation) will not affect the accuracy of the results. It should be emphasised here that, with very few exceptions, chemical equilibration of the added spike(s) with the analyte(s) is an essential requirement of isotope dilution analysis. Thus, solid samples such as rocks and minerals, sediments and biological tissues must normally be subjected to a rigorous digestion procedure to ensure equilibration.

The third step of the analysis is the actual measurement of the 'altered' ratio(s). The details of this procedure will depend on the type of instrumentation, but of course it is desirable to measure the ratio(s) with the best possible precision. The normal procedure is to perform repetitive scans with a relatively short dwell time (not more than a few milliseconds) on each of the isotopes. At the same time, the ratio(s) should be measured for solutions of known isotopic abundances in order to determine the magnitude of instrumental mass discrimination effects. If such effects are significant, it is necessary to apply a correction to the ratio(s) measured for the sample before calculating the results.

The final step of the analysis is the calculation of results according to the formula shown in the previous section. Values for A_s and B_s are normally obtained from isotopic abundance data provided by the producer of the enriched material. These data must also be used to determine the atomic weight of the enriched material in order to calculate the value of K.

6.9.4.2 *Advantages and disadvantages of isotope dilution*   Isotope dilution methodology offers a number of distinct advantages over other calibration strategies for ICP–MS. As has been noted above, it will compensate for partial loss of the analyte(s) during the sample preparation, provided that any such loss occurs after chemical equilibration of the spike(s) with the analyte(s) has been achieved. Secondly, it is immune to a wide variety of physical and chemical interferences which would be expected to have an identical effect on two isotopes of the same element, since effects will cancel one another in the isotope ratio. Thirdly, the method encompasses what could be considered the ideal form of internal standardisation; the internal standard for each analyte element is one of its own isotopes. The net result is that neither the precision nor the accuracy of properly performed isotope dilution ICP–MS analyses is often matched by other ICP–MS calibration strategies.

Against the many advantages of isotope dilution must be weighed several limitations and disadvantages. A fundamental limitation which has already been mentioned is that the method is generally not applicable to mono-isotopic elements. A disadvantage is that enriched stable isotopes are available from only a very limited number of sources worldwide. Delays in procurement can be considerable, and costs can be substantial because minimum quantities must be ordered. It should be noted, however, that these minimum quantities often represent several years' supply for trace analyses, since only microgram, or even nanogram, quantities of the enriched isotope will be added to each sample. In most cases, the cost of the spike will be an insignificant fraction of the total cost of the analysis. A second disadvantage is that, ideally, the element(s) to be determined must have at least two isotopes free of isobaric interferences, whereas a single interference-free isotope suffices for the other calibration strategies. For example, the determination of Cd in marine sediments by external calibration poses no particular problem because the most abundant isotope, ^{114}Cd, is free of significant interferences (McLaren et al., 1987a), but determination by isotope dilution can be hindered by the isobaric interference of ^{94}Zr^{16}OH on ^{111}Cd (McLaren et al., 1987b, 1988).

Perhaps the most important practical limitation of isotope dilution ICP–MS is that it is too time consuming for routine use in many laboratories. It will continue to be used primarily by agencies involved in the production of certified reference materials, but it should also see some application to the production of in-house standards for large laboratories for which the cost of using certified reference materials in all quality control protocols may be prohibitive, or in critical analyses for which no suitable certified reference material exists. There are probably very few, if any, laboratories which could not benefit from at least occasional use of isotope dilution techniques.

The development of ICP–MS should lead to an increase in the use of isotope dilution methodology, because it offers some significant advantages over the traditionally used technique, thermal ionisation mass spectrometry (TIMS). The most important of these is the speed with which multi-element analyses can be performed. The versatility of the ICP as a universal and relatively interference-free ionisation source, combined with the long linear dynamic range of the detection systems of current generation ICP–MS instrumentation, permits the simultaneous determination of several elements with no sacrifice of precision or accuracy (Garbarino and Taylor, 1987; Fassett and Paulsen, 1989; McLaren et al., 1990). The fact that nebulisation of solutions is the normal sample introduction technique for ICP–MS also meshes well with isotope dilution methodology, which always involves the preparation of solutions, as explained in section 6.9.4.1. Thus, the time-consuming chemical separations and evaporations which are characteristic of TIMS (Heumann, 1988) can usually be avoided, with enormous savings of time. While the precision of isotope ratios determined by TIMS can be better by an order of magnitude or more than ICP–MS data (with relative standard deviations typically not less than 0.1%), the precision of isotope ratios determined by ICP–MS is usually more than adequate for trace element determinations, for which the limiting factor tends to be the inhomogeneity of the sample.

7 Sample preparation for ICP–MS

I. JARVIS

7.1 Introduction

Despite considerable new research on sample introduction techniques, solution nebulisation remains the preferred method for most ICP–MS applications. For the majority of sample types, therefore, dissolution is a prerequisite for routine analysis. Indeed sample dissolution is still one of the most common operations in analytical chemistry laboratories. Strong mineral acids are generally used for sample digestion, and appropriate combinations of acids have been used successfully to decompose most types of materials, from clinical, biological and botanical, through environmental and geological matrices such as soils, minerals, rocks and ores, to steels and alloys. The effectiveness of such attacks may be enhanced by the use of closed-vessel systems which allow increased temperatures and pressures during digestion. Alternatively, fusion of samples with alkali fluxes (e.g. Na_2CO_3, Na_2O_2, $LiBO_2$) followed by dissolution in a dilute acid, provides a different approach which is particularly effective for the digestion of refractory minerals and ores.

Most sample preparation procedures were developed for very specific applications (commonly for the determination of only one or a very small range of elements), and extreme caution must be exercised when applying them to multi-element techniques such as ICP–MS. The solution chemistry in many digestion methods is poorly understood, and our knowledge lags far behind developments in analytical instrumentation. Indeed, very few preparation procedures have been adequately tested by ICP–MS analysis, and the combinations of reagents, samples and methods that are safe and reliable are not well compiled.

This chapter provides an overview of sample preparation procedures which are judged to be particularly applicable to ICP–MS analysis. Only dissolution techniques will be considered in any detail, although some specific applications of separation and pre-concentration procedures to ICP–MS will also be presented. The later stages of sample presentation, such as methods for the conversion of samples to gases or vapours via hydride generation or electrothermal vaporisation are not considered. Aspects relating to the analysis of solid samples are described elsewhere (chapter 10). Finally, no attempt has been made to cover all dissolution methods, for which the reader is referred to the extensive reviews of Bock (1979) and Sulcek and Povondra (1989), but rather a limited number of tested methods are presented, many of which are in routine use in our laboratory.

7.2 General considerations

Some general aspects of sample preparation, such as the types of equipment and combinations of reagents which are best suited to ICP–MS analysis, will be considered before any specific digestion procedures are described.

7.2.1 *Laboratory equipment and practices*

The high sensitivity of ICP–MS and the wide range of trace elements which are incorporated into most analytical protocols, require that great care must be taken to avoid contamination during sample preparation. Contamination may arise from three main sources: (1) equipment used to grind, sieve and homogenise samples, (2) the laboratory environment and digestion apparatus and (3) analytical reagents used during sample preparation. A detailed discussion of the procedures necessary to obtain representative samples is outside the scope of this work (see e.g. Johnson and Maxwell, 1981; Potts, 1987; Cresser, 1990). However, many materials must be ground and homogenised prior to digestion, to ensure that a representative portion of the bulk sample is analysed. It is commonly accepted that a maximum grain size of 200 mesh (75 μm) is required to assure sample homogeneity. To minimise contamination during sample grinding and homogenisation, nylon sieves (unless labile organic compounds are also to be determined) and agate grinding apparatus provide the best media, other materials potentially leading to significant contamination by a range of trace elements.

For many applications cleanroom facilities (cf. Moody, 1982) are unnecessary if care is taken with overall laboratory cleanliness. In general, Teflon PTFE (polytetrafluoroethylene) and PFA (perfluoroalkoxy; a tetrafluoroethylene with a fully fluorinated alkoxy side-chain) digestion vessels and other apparatus are preferred to conventional borosilicate glass. Teflonware is very effectively cleaned by boiling for several hours in 8 M HNO_3, followed by rinsing with liberal amounts of deionised water. Oven drying at 105°C will expel traces of acid absorbed by the Teflon. Where required, glassware may be cleaned by soaking with 5% Decon 90 solution in an ultrasonic tank (5–30 min), prior to rinsing with deionised water, soaking overnight in 1 M HNO_3 and rinsing three times with deionised water prior to use. High purity graphite crucibles provide the best medium for fusion procedures, avoiding the metal contamination with is inherent in the use of alternative materials. Vitreous graphite degrades more slowly (but is more expensive) than other varieties and causes minimal contamination from particulate carbon.

Ultrapure reagents (e.g. Aristar grade) should be used in conjunction with 18–25 Mohm cm deionised water for sample digestion. Procedural blanks should always be prepared for each batch of samples. Sample solutions should be stored in clean new polypropylene bottles and analysed as soon as possible after preparation.

Table 7.1 Physical properties of common mineral acids

Acid	Formula	Concentration (%)	Molarity (M)	Specific gravity	Boiling point (°C)	Comments
Nitric	HNO_3	68	16	1.42	84	HNO_3
					122	68% HNO_3, azeotrope
Hydrofluoric	HF	48	29	1.16	112	38.3% HF, azeotrope
Perchloric	$HClO_4$	70	12	1.67	203	72.4% $HClO_4$, azeotrope
Hydrochloric	HCl	36	12	1.18	110	20.4% HCl, azeotrope
Sulphuric	H_2SO_4	98	18	1.84	338	98.3% H_2SO_4
Phosphoric	H_3PO_4	85	15	1.70	213	Decomposes to HPO_3

7.2.2 Choice of mineral acids

A relatively restricted number of mineral acids and other reagents are in common use for sample decomposition. Detailed discussion of the properties and applications of these reagents may be found elsewhere (Bock, 1979; Cotton and Wilkinson, 1988; Sulcek and Povondra, 1989). All chemicals are not, however, equally suited to ICP–MS applications, so some of the advantages and disadvantages of the main reagents will be discussed. The physical properties of the common mineral acids used in sample preparation are summarised in Table 7.1. Additionally, although it is assumed that readers will be familiar with general safety practices operable in chemical laboratories and with procedures necessary for the safe handling of reagents, the safety requirements of some of the more dangerous acids are discussed, in order to emphasise the special hazards of these materials.

7.2.2.1 *Nitric acid* Nitric acid is one of the most popular reagents used for sample decomposition. Concentrated HNO_3 (16 M, 68%) is a strong oxidising agent which will liberate trace elements from many materials as highly soluble nitrate salts. It is commonly used in the dissolution of metals, alloys and biological materials, although the addition of a complex-forming species such as HCl or HF is often necessary to effect the complete dissolution of many metals and ores. More concentrated 'fuming nitric acid' (20–23 M) may be used to rapidly destroy organic matter. Nitric acid forms an azeotrope (constant boiling point mixture) with water of 68% HNO_3, which has a boiling point of 122°C. This is the concentrated reagent strength generally used in the laboratory. The low boiling point of HNO_3, however, necessitates

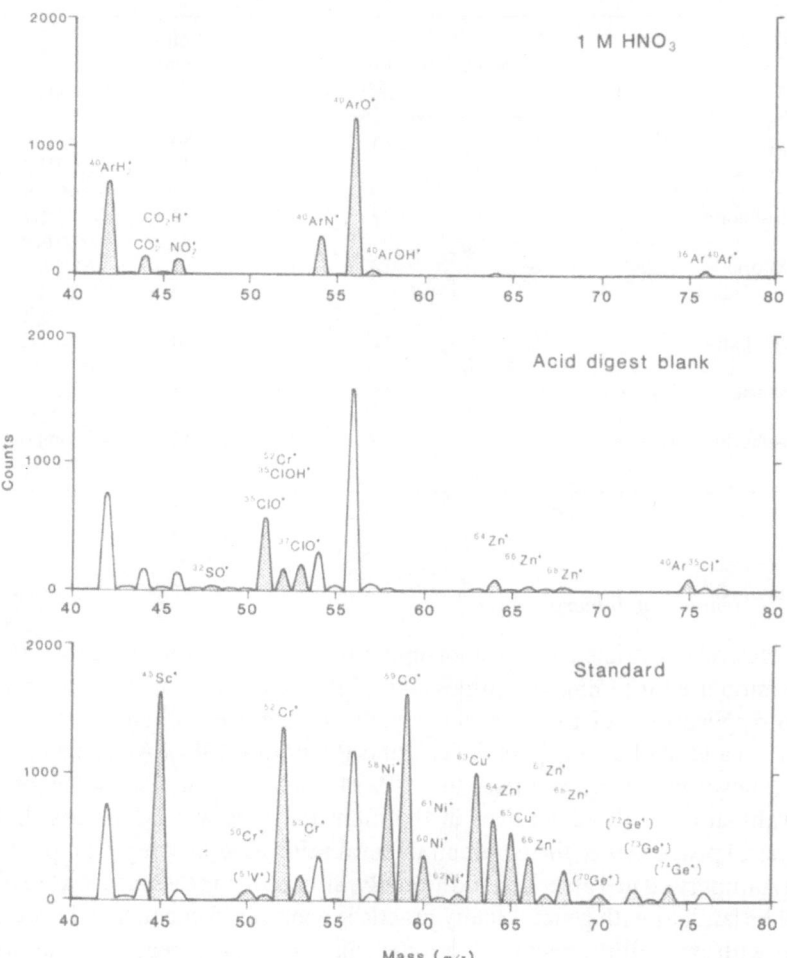

Figure 7.1   Spectra of the mass range 41–79 m/z for BDH Aristar 1 M HNO_3, a blank solution from a $HF/HClO_4$ digest (500-fold dilution) prepared in 1 M HNO_3, and a standard containing 100 ng ml^{-1} Co, Cr, Cu, Ni, Sc, Zn and 19 other trace elements, also in a 1 M HNO_3 matrix. Peaks in the HNO_3 blank spectrum are predominantly gas polyatomic species. In addition to gas peaks, the acid digest spectrum displays Cl polyatomic ions from residual $HClO_4$, plus small amounts of contaminant SO_4, Cr and Zn from the digest acids. In the spectrum of the multi-element standard the characteristic mass distributions of the trace elements are clearly displayed. The equal intensities of the mono-isotopic elements (^{45}Sc, ^{59}Co) demonstrate that the mass response is flat. Vanadium and Ge (in parentheses) are identifiable contaminants. Note the coincidence of Cl polyatomic species in the acid digest with the main isotopes of V and Cr displayed in the standard spectrum.

relatively long digestion times or the addition of other strong oxidising agents such as H_2O_2 or $HClO_4$, to completely destroy organic and many other matrices. Nevertheless, samples containing high concentrations of organic matter are best attacked initially with HNO_3 alone, before resorting to more powerful (and potentially explosive) oxidising agents. The oxidising properties of nitric acid are retained only in concentrated form, and when the acid is diluted below approximately 2 M these characteristics are lost.

Nitric acid matrices are generally regarded as being the best acid medium for ICP–MS analysis, since the constituent elements are already present in air entrained by the plasma. Hydrogen, N_2 and O_2 form a range of polyatomic ions (Figure 7.1) which (unlike most other mineral acids) is not increased significantly by the addition of an HNO_3 matrix (Gray, 1986b; Tan and Horlick, 1986; Horlick et al., 1987). Additionally, HNO_3 is one of the few acids which can be easily obtained commercially in very high purity form, making it ideal for ultratrace element work.

7.2.2.2 *Hydrochloric acid*   Concentrated hydrochloric acid (12 M, 36%) is an excellent solvent for many metal oxides and metals which are oxidised more easily than hydrogen, but is rarely used alone for the digestion of more complex matrices. Under elevated temperatures and pressures, many silicates and other refractory oxides, sulphates and fluorides are attacked by HCl to produce soluble salts. Unlike HNO_3, HCl is a weakly reducing agent and is not generally used to digest organic materials, although it is an effective solvent for basic organic compounds such as amines and alkaloids in aqueous solution, as well as some organometallic compounds. Hydrolysis of materials with hydrochloric acid is a routine preliminary procedure for the analysis of amino acids and carbohydrates.

Unlike atomic absorption techniques, where chloride matrices are preferred because of reduced interference effects, HCl is best avoided as a sample matrix in ICP–MS because Cl polyatomic ions (e.g. $ArCl^+, ClO^+, ClOH^+$) cause major interferences on the available isotopes of As and $V(^{75}As, ^{51}V)$ and to a lesser extent on many other trace elements (Cr, Fe, Ga, Ge, Se, Ti, Zn). Such problems, however, are generally only significant at masses below the Ar dimer ($^{40}Ar_2^+$) at m/z 80 (although Cl_3^+ species may potentially cause small interferences at higher masses), and do not compromise the determination of the many important groups such as Au (m/z 197) and the platinum group elements (m/z 96–110, 184–198), which are stable only in chloride matrices.

Since the boiling point of the HCl azeotrope with water is below that of the HNO_3 azeotrope, hydrochloric acid can be effectively removed from sample solutions by repeated evaporation to incipient dryness with HNO_3, the final solution being taken up in nitric acid. Remnant chloride ions do not produce significant polyatomic ion interferences after this procedure (Date et al., 1988). However, potential losses of volatile metal chlorides (As, Sb, Sn, Se, Ge, Hg) must be assessed if such a procedure is adopted.

7.2.2.3 *Hydrofluoric acid*   Hydrofluoric acid is the only acid which will readily dissolve silica-based materials, forming SiF_6^{2-} ions in acid solution. High purity concentrated HF (29 M, 48%) is commercially available and the acid forms an azeotrope with water of 38.3% HF (22 M) which has a boiling point of 112°C. This low boiling point and the acid's high vapour pressure, result in the acid being easily volatilised. Silicates are converted to volatile SiF_4, which will be lost in open vessel digestion procedures. This process may be expressed by the general simplified equations.

$$SiO_2 + 6HF \rightarrow H_2SiF_6 + 2H_2O \qquad (7.1)$$

$$H_2SiF_6 \rightarrow SiF_4 + 2HF \qquad (7.2)$$

Other volatile fluorides (and chlorides) of B, As, Sb and Ge may also be lost along with the HF, but their volatilities are highly dependent on oxidation states which are controlled by a combination of the acid mixture used and sample composition. Broad generalisations are not easily made (see Sulcek and Povondra, 1989 for discussion). Furthermore, HF is rarely used as the sole reagent because some salts (e.g. K, Ca) are poorly soluble in this acid. Oxidising acids such as HNO_3 and/or $HClO_4$ are generally used in combination with HF to ensure complete dissolution and to produce uniformly high oxidation states in final solutions.

Hydrofluoric acid, even at low concentrations, will etch glass, making plastic (preferably Teflon) containers essential. Additionally, any HF remaining in the final solutions will rapidly attack the glassware used in ICP–MS nebulisers, spray chambers and injectors. Inert alternatives such as PTFE nebulisers and spray chambers and sapphire injectors may be used to overcome these problems, but they do not remove the potential safety hazards involved in running solutions containing HF. Alternatively, HF may be complexed with saturated boric acid (H_3BO_3) prior to analysis. However, this significantly increases the total dissolved solids (TDS) in the solution, causing a serious deterioration in the limits of quantitation (section 7.2.3) obtained for the method. It is generally best, therefore, to remove all HF from the solutions by evaporation prior to analysis, although this precludes the determination of Si (and some other elements) in the final solutions.

Finally, it should be noted that HF is both highly corrosive and toxic, and is one of the most hazardous mineral acids used in the laboratory. Hydrofluoric acid will quickly cause irreparable damage to skin and eyes, and should on no account be used without adequate safety precautions. Full protective clothing and use within a fume cupboard are essential. Small contact areas should be drenched with water, and then calcium gluconate gel should be liberally applied to the wound. Immediate hospital treatment is recommended in all but the most trivial cases and always where there is any contact with eyes. An HF-specific first aid kit should be kept in the laboratory.

7.2.2.4 *Perchloric acid*   Perchloric acid is one of the strongest mineral acids known. The acid forms an azeotrope with water at 72.4% $HClO_4$ which boils at 203°C. The hot concentrated acid (i.e. the azeotrope) is a powerful oxidising agent which will react explosively with organic compounds, although cold or dilute solutions do not exhibit these properties. For this reason, organic samples (such as biological tissues) are best pre-treated with HNO_3 or an $HNO_3/HClO_4$ mixture (with HNO_3 in excess by at least four times) to destroy reactive phases prior to $HClO_4$ oxidation. Samples containing high proportions of fats and oils should not be attacked with $HClO_4$ until these compounds have been decomposed by other methods. Fats and oils will remain undigested in $HNO_3/HClO_4$ mixtures, and will react explosively following the evaporation of HNO_3. Many solid perchlorates are spontaneously explosive, so baking of samples should always be avoided. Heavy metal perchlorates are particularly unstable, and the use of perchloric acid for the digestion of heavy metals or their mineral concentrates is not recommended.

Most salts of $HClO_4$ are highly soluble and stable in aqueous solutions, and although some alkali (K, Rb, Cs) perchlorates are less soluble, this generally does not cause any problems in practical analysis. In addition to its use in dissolving organic compounds, $HClO_4$ is commonly used for the digestion of steels and other Fe alloys (Bock, 1979), because it not only dissolves samples rapidly but also simultaneously oxidises the common elements to their highest oxidation states. We use perchloric acid, in combination with HF, in several of our acid digestion procedures. The addition of $HClO_4$ has a number of advantages, including the rapid oxidation of organic compounds and the more efficient attack of refractory minerals by improving the efficiency of HF, due to the increased boiling temperature of the reaction mixture. Furthermore, the high boiling point of the acid ensures the effective removal of HF during evaporation stages. Silicon is lost by the evolution of volatile silica tetrafluoride, followed by evaporation of the low boiling point (112°C) HF-water azeotrope. Unlike HCl, however, chloride ions introduced during the digestion procedure (remnant $HClO_4$ and perchlorate species in sample solutions) are difficult to remove by evaporation (cf. Beauchemin et al., 1988a) and will compromise the determination of low levels of As and V (Figure 7.1), due to polyatomic ions interferences by $^{40}Ar^{35}Cl^+$ on ^{75}As and $^{35}Cl^{16}O^+$ on ^{51}V.

It must be noted that extreme safety precautions are required when using perchloric acid (Everett and Graf, 1971), which should only be used in a dedicated fume cupboard fabricated from inert plastics and equipped with fume-scrubbing and wash-down facilities. The interior surfaces of the fume cupboard should be washed regularly to prevent the build-up of potentially explosive anhydrous perchlorates. On no account should organic compounds be used in the same fume cupboard. Small spillages may be diluted and mopped up with paper towels, but these must be immediately rinsed with copious amounts of water, prior to disposal.

7.2.2.5 *Sulphuric acid* Concentrated sulphuric acid (18 M, 98%) is an effective solvent for a wide range of ores, metals, alloys, oxides and hydroxides, and is traditionally used in combination with HNO_3 (and other acids) for the digestion of organic materials. Concentrated H_2SO_4 is a near-anhydrous reagent with an extremely high boiling point (338°C), which prevents its use in PTFE vessels (which deform above 260°C and melt at 327°C). Sulphuric acid is not generally an oxidising agent, although the hot concentrated acid does have some oxidising ability. In addition to its use as a method of wet oxidising organic matter, H_2SO_4 has been used as a hydrolysing agent for esters, carbohydrates and proteins (Bock, 1979). Unfortunately, some inorganic sulphates have low solubilities (particularly Ba, Ca, Pb, Sr), and volatilisation of trace elements (Ag, As, Ge, Hg, Re, Se) may occur during the digestion of some samples (see Sulcek and Povondra, 1989 for a review).

Sulphuric acid matrices are viscous, producing transport effects during sample introduction which necessitate the use of matrix-matched standards to obtain accurate data. In addition, H_2SO_4 causes severe polyatomic ion interferences in ICP–MS (Gray, 1986b), notably on isotopes of Ti, V, Cr, Zn, Ga and Ge (e.g. $^{32}S\,^{16}O_2^+$, $^{34}S\,^{16}O_2^+$ on ^{64}Zn, ^{66}Zn; $^{34}S\,^{16}O_2\,^1H^+$ on ^{67}Zn), and prolonged aspiration of dilute H_2SO_4 leads to serious degradation of the nickel sampler cones commonly used in the interface regions of ICP–MS instruments. Finally, unlike HCl, the high boiling point of H_2SO_4 precludes its preferential removal by evaporation with other mineral acids. The use of H_2SO_4, therefore, is not recommended in sample preparation procedures for ICP–MS.

7.2.2.6 *Phosphoric acid*   Phosphoric acid is somewhat limited in its use in analytical chemistry because the presence of phosphate ions interferes with many of the subsequent steps of traditional analytical procedures. However, the acid has found some applications in the digestion of aluminium and iron-based alloys, ceramics, ores and slags (Bock, 1979). Possible difficulties in ICP–MS include the high viscosity of the acid which causes transport effects during sample introduction (see above), the formation of polyatomic species of P which produce interferences on isotopes of Cu, Ga, Ni, Se, Ti and Zn (e.g. $^{31}P\,^{16}O^+$ on ^{47}Ti, $^{31}P\,^{16}O\,^1H^+$ on ^{48}Ti), and very rapid erosion of the Ni sampler cone. The use of H_3PO_4 should, therefore, be avoided.

7.2.2.7 *Aqua regia*   A wide range of acid mixtures are used in sample digestion, and some of these are discussed in relation to specific sample preparation procedures. A mixture of one part 16 M HNO_3 with three parts 12 M HCl (by volume) produces a particularly useful reagent known as *aqua regia*. The effectiveness of this reagent is probably due to the complexing power of the Cl^- ion combined with the catalytic effects of molecular Cl_2 and nitrosyl chloride (NOCl). The reagent becomes more effective if allowed to stand for 10–20 min before use. On heating, NOCl dissociates to corrosive chlorine gas which also aids decomposition.

Aqua regia is traditionally used to attack metals, alloys, including steels, sulphides and other ores and is well known for its ability to digest Au, Pd and Pt. Aqua regia is not an ideal final matrix for ICP–MS analysis because of its high Cl ion content and its volatility. Nevertheless, the reagent can be extremely useful in 'opening-up' materials prior to evaporation to incipient dryness and final digestion in dilute HNO_3. As with HCl alone, no problems should be encountered in removing excess chloride ions by this procedure.

7.2.3 Limits of quantitative analysis

ICP–MS is characterised by greater sensitivity and lower instrumental detection limits than any other rapid multi-element technique (Gray, 1989a). The limitations of quoted instrumental detection limits, however, are not fully appreciated by many users of chemical data. In ICP techniques, detection limits are generally based on eleven determinations of the signal produced by a blank solution (commonly 1 M HNO_3). Two or three sigma standard deviations calculated from these data are expressed as an equivalent concentration for each analyte, and reported as the instrumental detection limits. In some cases, analytical conditions are optimised for measuring the detection limit of each element, so as to produce the most favourable figures. Clearly, these conditions are very different to those used for routine analysis, where compromise instrumental settings are used and more complex (and therefore less well behaved) solutions are analysed. Furthermore, even in the ideal case where the blank measurements follow a normal (gaussian) distribution, the three sigma detection limit represents only the lower limit of qualitative analysis; errors in determinations at this level are infinite (see Potts, 1987 for further discussion). The instrumental detection limits are useful, therefore, for comparing the performances of different instruments, but they provide the analyst with little information on the lowest levels which can realistically be measured in samples.

The statistical limitations of detection limit measurements can be overcome in part by using six or ten sigma standard deviation data. The latter has been proposed (American Chemical Society Committee on Environmental Improvement, 1980) as providing a good approximation to the lower limit of quantitative analysis, and has been termed the 'limit of quantitation' (LOQ). The ideal error in measurements at this signal level is 10% relative (one sigma). Two additional factors influence the way in which LOQ data are related to samples. Firstly, the magnitude of the standard deviation on the background is matrix dependent. ICP–MS mass spectra are characterised by very low background count rates which are not greatly affected by sample compositions. Nevertheless, the composition of the final acid medium (e.g. the presence of chloride, sulphate and other ions) and the levels of contaminants in reagent blank solutions may significantly increase LOQs for some elements. Secondly, the instrumental LOQ takes no account of dilution factors introduced during sample dissolution. This limitation is particularly relevant to ICP–MS analyses because the technique is relatively intolerant of high levels of total dissolved solids (TDS), which should generally be below 0.1% in the final solution. This necessitates dilution factors of several orders of magnitude for most dissolution methods, causing a corresponding deterioration in the LOQ for samples.

In practice, it is useful to report LOQs for individual dissolution procedures, calculating ten sigma standard deviation on reagent blank solutions and expressing the LOQ as a concentration in samples, thereby accounting for the dilution factors used. In practice, LOQs measured in this way provide a good indication of the lower limits of quantitative analysis for samples, although the error in measurement made close to the LOQ is generally in the order of 15–35%, rather than the 10% predicted by the ideal statistics.

7.2.4. *Precision and accuracy: assessing a digestion procedure*

Precision is an estimation of the analytical reproducibility of a measurement, and is normally calculated from the standard deviation (one sigma) of a replicate set of analyses. The precision is easily obtained for any analytical method. Accuracy, on the other hand, is a measure of how closely the analytical data lie to the 'true' composition of the sample, and is far more difficult to assess. In practice, accuracy is best estimated by the analysis of standard reference materials (SRMs). These materials comprise samples of proven homogeneity that have been analysed by a wide variety of techniques (commonly involving a large number of laboratories wordwide), and a measure of their 'true' composition obtained. SRMs are available from a number of distributors and for a wide range of matrices (e.g. see Gladney *et al.*, 1987; Govindaraju, 1989). Where SRMs of comparable compositions to samples are unavailable, accuracy can only be estimated by the use of techniques such as isotope dilution or standard addition.

Comparisons of measured and 'true' values for SRMs provide the means by which the accuracy of a method can be assessed. Where possible, therefore, ICP–MS data for selected SRMs are presented for each of the sample digestion procedures recommended here. In some cases our suggested methods incorporate modifications to previously published procedures, but the data presented should enable the reader to evaluate the accuracy and potential limitations of the different methods. Unfortunately, many SRMs remain very poorly characterised for most ultratrace elements, making an assessment of ICP–MS data more difficult than for many other less sensitive techniques. This limitation should be borne in mind when using the SRM data provided.

7.3 Digestion procedures

Digestion techniques may be subdivided into three basic types: (1) open-vessel and (2) closed-vessel digestion using mixed-acid attacks, and (3) fusion procedures, which generally employ alkali fluxes. Variations on these methods have been developed recently which use microwaves rather than conductive heating to provide energy for digestions. Microwave digestions employ many novel approaches, and therefore will be treated separately from more traditional procedures. A variety of methods and applications for each digestion method are presented, and the advantages and disadvantages are discussed with particular reference to ICP–MS analysis.

The presentation of digestion procedures has been arbitrarily subdivided into a number of categories, including plant and animal tissue, geological samples, environmental materials and metals. Many other possible categories exist, but the procedures described are generally applicable to the preparation of many comparable sample types. Archaeological materials such as flints, clays, pottery, bricks, tiles, glazes, slags and ores, for instance, may be successfully digested using procedures described in 'geological' applications. Similarly, metal analysis spans a large number of disciplines. Whatever the applications, it is hoped that the reader will be able to identify a matrix comparable to that being studied, and may use or modify the described procedure accordingly.

7.3.1   Open vessel digestions

Open vessel acid digestions are undoubtedly the most common method of sample decomposition used in chemical laboratories. Thousands of different methods and minor variations on these methods have been described in the literature, some of which are discussed here.

7.3.1.1   *Plant and animal tissue*   The low detection limits of ICP–MS make the method well suited to the analysis of animal and plant tissues, overcoming the sensitivity limitations of ICP–atomic emission spectrometry (ICP–AES) for biotically important trace elements such as Co and Cd. Some liquid samples such as blood serum and urine can be nebulised directly after suitable dilution (a 5- to 10-fold dilution with $0.14\,M\ HNO_3$ is suitable for serum; Vanhoe *et al.*, 1989; Allain *et al.*, 1990). However, the viscosity and nebulisation efficiency of undigested samples will differ from aqueous standards, and these differences, together with the high Na content of most biological fluids (which may cause signal suppression), necessitate the preparation of matrix-matched standards for quantitative analysis. Additionally, the high chloride content of undigested biological samples may preclude the determination of some trace elements, although chloride may be removed by a simple gel filtration procedure (Lyon *et al.*, 1988b), or precipitation by the addition of silver nitrate (Lyon *et al.*, 1988a).

The complete decomposition and dissolution of biological and botanical samples generally requires complete oxidation of the organic matrix. This may be accomplished by two main methods: (1) ignition in air (dry ashing), and (2) treatment with strongly oxidising acids (wet ashing). Details of such procedures may be found in Gorsuch (1970) and Bock (1979).

Dry ashing, generally using platinum or silica crucibles in a muffle furnace, is a cheap, simple and rapid method. Unfortunately, it has a number of disadvantages which make it poorly suited to ICP–MS analysis. One problem is that the high temperatures (450–600°C) necessary to complete the ashing process cause the partial (e.g. Cd, Pb, Zn) or complete (e.g. As, Hg, Se) loss by volatilisation of many trace elements. Even where the element itself is not volatile, a compound (e.g. chloride) may form during ashing which will be lost (most biological materials contain chloride species). Non-volatile elements, on the other hand, commonly exhibit the opposite behaviour, being absorbed onto the crucible walls and/or transformed into refractory phases which are difficult to dissolve subsequently. Most or all of these difficulties may be alleviated or removed by the addition of so-called ashing aids, such as calcium, magnesium and aluminium nitrates. These compounds assist the oxidation process, help retain volatiles and transform the analytes into readily soluble compounds (nitrates). Their addition, however, will enhance blank levels and significantly increase the TDS of the final solution, both of which are highly detrimental to the LOQs of the method.

Wet ashing is currently the most common method used to destroy organic matter. An enormous range of procedures are in general use, although most (Jolly, 1963; Gorsuch, 1970; Bock, 1979) use combinations of (1) sulphuric and nitric acids or (2) perchloric and nitric acids. The former are not recommended for ICP–MS studies because of problems with polyatomic ion interferences (see above), possible volatilisation of trace elements, the precipitation of insoluble sulphates, and because the high viscosity of H_2SO_4 may lead to differences in transport efficiency between sample and standard solutions.

Ridout et al. (1988) successfully digested 0.1 g of lobster hepatopancreas by prolonged heating with 1 ml 16 M HNO_3. Samples were diluted to 10 ml with deionised water and analysed by ICP–MS for a range of trace metals. However, significant signal suppression was observed for many elements, necessitating a further 10-fold dilution to achieve acceptable accuracy. Such increased dilutions clearly compromise the determination limits of the method. Suppression effects caused by the high levels of TDS in sample solutions were almost certainly due to the inadequate decomposition of organic compounds by nitric acid. In contrast, digestion procedures developed for the analysis of liquid milk (Emmett, 1988) and plant materials (McCurdy, 1990) by ICP–MS use HNO_3 and $HClO_4$ in combination with H_2O_2 to completely decompose organic samples, potentially enabling the analysis of more concentrated solutions. The proposed method is based on this work.

Method

Equipment

(1) 250°C hotplate, sited in a suitable fume cupboard.
(2) 50 ml Pyrex beakers with borosilicate watch-glass covers (PTFE vessels are unsuitable for this procedure because oils and fats liberated from samples adhere to the vessel walls and remain undigested).

(3) 50 ml Class A Pyrex volumetric flasks.
(4) Liquid dispensers for reagents.
(5) New 60 ml polypropylene bottles.

Reagents
(1) Deionised water (18–24 Mohm cm).
(2) Concentrated (68–71% m/m) Aristar nitric acid (16 M HNO_3).
(3) 100 volumes (30% m/v H_2O_2) Aristar hydrogen peroxide solution.
(4) Concentrated (70% m/m) Aristar perchloric acid (12 M $HClO_4$).

Procedure
(1) Air dry samples at 60°C for 48 h (note that a proportion of Hg, Se may be
    volatilised at this temperature; air drying at 25°C is preferred if volatile
    elements are sought). Samples which are difficult to dry by heating can be
    freeze-dried, or homogenised and weighed wet. Moisture contents can be
    determined separately.
(2) Weigh 0.500 g sample into a clean beaker; add 5.0 ml HNO_3, cover with a
    watch glass, and leave to cold soak for 7 days.
(3) Add a further 5.0 ml HNO_3, replace watch glass and heat at 80°C for 4 h,
    and then at 180°C for a further 5–6 h, or until solution volume is reduced
    to about 5 ml.
(4) Remove beaker from hotplate and allow to cool for 1 min. Add 1.0 ml H_2O_2
    dropwise to the solution. Care is essential at this stage as the reaction of
    the H_2O_2 in the hot acid is very vigorous.
(5) After the reaction has died down, replace watch glass and return beaker to
    hotplate. Allow the evolution of brown NO_2 fumes to die down, and then
    repeat the H_2O_2 addition a further four times, allowing solution to cool
    for 1 min before each addition.
(6) After the final addition of H_2O_2, allow the solution (which should be
    colourless or pale green to yellow) to cool. Ensure that no oil or fat globules
    are floating in the liquid. If any globules remain these must first be
    decomposed by additional HNO_3/H_2O_2 attacks before proceeding to the
    next stage (hot $HClO_4$ may react explosively with this material).
(7) Add 5.0 ml $HClO_4$ and heat at 180°C for 1 h. Remove watch glass and fume
    off $HClO_4$ until approximately 1 ml remains (approximately 1–2 h).
(8) Allow solution to cool for 1 min, then add 2.0 ml H_2O_2, 1 ml at a time. After
    each 1 ml addition, return to the hotplate until effervescence has ceased.
(9) Cool and add 15.0 ml deionised water, then heat gently for up to 5 min,
    until the solution clears.
(10) Allow solution to cool, pour into a volumetric flask and dilute to 50 ml.
    Transfer to a polypropylene bottle for storage. The sample may be analysed
    by ICP–MS without further dilution.

Siliceous material (e.g. silica phytoliths in grasses) remains undigested
during the above procedure and will remain as a fine residue. The cloudy
solution may be successfully aspirated using 'high dissolved solids' nebulisers
such as a De Galan V-groove, but less tolerant nebulisers (particularly
Meinhard designs) require the filtration of cloudy solutions prior to analysis
by ICP–MS.

Plant materials digested by the above procedure generally produce solutions containing total dissolved solid concentrations of 0.1% or less, making them suitable for analysis by ICP–MS without further dilution. McCurdy (1990) documented ICP–MS LOQs in the order of a few tens of $ng\,g^{-1}$ for 32 trace elements (including all of the naturally occurring rare earth elements) in solutions prepared by this technique. Data for a range of plant SRMs agreed well with reference values for most of the elements studied (Table 7.2), differences generally being attributable to inadequate characterisation of the SRMs rather than analytical error.

It should be noted that the use of $HClO_4$ in the digestion scheme compromises the determination of As and V at low levels, due to polyatomic ion interferences by Cl species on the main isotopes of these elements which require correction. Recent work in our laboratory indicates that the use of $HClO_4$ is not always necessary for the complete dissolution of samples, although H_2O_2 alone will not as effectively reduce TDS by oxidation of organic compounds. Siliceous material is also incompletely digested by the technique, preventing the determination of total Si. Within these limitations, however, the method should yield excellent results for most plant and animal materials.

7.3.1.2 *Geological samples* Open vessel acid digestions with HF, in combination with a range of other acids, are routinely used to digest geological materials for trace element determination (Bock, 1979; Jefferey and Hutchison, 1981; Johnson and Maxwell, 1981; Hickman *et al.*, 1986; Potts, 1987; Sulcek and Povondra, 1989; Thompson and Walsh, 1989; Jarvis and Jarvis, 1991, 1992). However, if silicon is to be determined, digestion must be undertaken in closed vessels and is normally followed by complexing the excess fluoride with boric acid prior to analysis. Unfortunately, the addition of boric acid increases the TDS of the solutions, causing a significant deterioration in the limits of quantitation for the method.

We prefer to use alkali fusion solutions for Si determinations, and the open vessel digestion given below for most trace elements. In the latter procedure both HF and silicon are removed as the volatile silicon tetrafluoride SiF_4, reducing the TDS and removing the necessity for an addition of boric acid to complex the HF.

Method

Equipment

(1) 250°C hotplate sited in a dedicated $HF/HClO_4$ fume cupboard equipped with fume-scrubbing and wash-down facilities.

(2) 60 ml or 100 ml PTFE breakers.

(3) 50 ml Class A Pyrex volumetric flasks.

(4) Liquid dispensers for acids.

(5) New 60 ml polypropylene bottles.

Reagents

(1) Deionised water (18–24 Mohm cm).

Table 7.2 Comparison of ICP–MS and reference values ($\mu g\,g^{-1}$) for some plant SRMs prepared by $HNO_3/H_2O_2/HClO_4$ open vessel digestion[a]

Element	Mass (m/z)	LOQ	NIST 1572 citrus leaves ICP–MS	NIST 1572 citrus leaves Ref.	NIST 1573 tomato leaves ICP–MS	NIST 1573 tomato leaves Ref.	NIST 1575 pine needles ICP–MS	NIST 1575 pine needles Ref.
Ba	137	0.038	19.5 ± 0.7	21.0 ± 3.0	64.1 ± 1.7	57.0 ± 9.0	8.0 ± 0.5	7.2 ± 0.8
Ce	140	0.021	0.377 ± 0.012	0.28	1.26 ± 0.02	1.60	0.166 ± 0.022	0.4
Cr	52	0.34	1.2 ± 0.1	0.8 ± 0.2	4.9 ± 0.3	4.5 ± 0.3	2.7 ± 0.2	2.6 ± 0.2
Cs	133	0.005	0.092 ± 0.006	0.098	0.047 ± 0.009	0.057 ± 0.008	0.130 ± 0.004	0.11 ± 0.01
Cu	63	0.47	16.4 ± 0.8	16.5 ± 1.0	12.5 ± 0.2	11.0 ± 1.0	2.7 ± 0.6	3.0 ± 0.3
Fe	56	1.5	98.1 ± 3.8	90.0 ± 10.0	761 ± 17	690 ± 25	197 ± 6	200 ± 10
La	139	0.008	0.161 ± 0.017	0.19	0.628 ± 0.007	0.9	0.116 ± 0.014	0.2
Li	7	0.068	0.191 ± 0.030	0.230 ± 0.105	0.658 ± 0.015	nd	0.136 ± 0.038	0.34
Mn	55	0.055	19.8 ± 1.3	23 ± 2	303 ± 13	238 ± 7	717 ± 13	675 ± 15
Mo	98	0.025	0.190 ± 0.029	0.17 ± 0.09	0.546 ± 0.039	0.53 ± 0.09	0.112 ± 0.008	0.15 ± 0.05
Pb	208	0.099	11.7 ± 0.5	13.3 ± 2.4	6.3 ± 0.1	6.3 ± 0.3	10.8 ± 0.1	10.8 ± 0.5
Rb	85	0.007	5.0 ± 0.2	4.8 ± 0.1	18.5 ± 0.2	16.5 ± 0.1	12.0 ± 0.5	11.7 ± 0.1
Sb	121	0.015	0.028 ± 0.004	0.04	0.039 ± 0.006	0.036 ± 0.007	0.132 ± 0.016	0.2
Sr	88	0.016	108 ± 4	100 ± 2	51.8 ± 0.3	44.9 ± 0.3	4.8 ± 0.1	4.8 ± 0.2
Th	232	0.006	0.010 ± 0.001	nd	0.152 ± 0.004	0.17 ± 0.03	0.030 ± 0.003	0.037 ± 0.003
Ti	47	0.16	3.4 ± 0.1	22.0 ± 0.4	37.3	56.0	6.5 ± 1.7	13.7 ± 39.0
U	238	0.009	0.037 ± 0.005	0.040 ± 0.002	0.038 ± 0.005	0.061 ± 0.003	0.016 ± 0.003	0.020 ± 0.004
Y	89	0.018	0.334 ± 0.004	nd	0.295 ± 0.014	nd	0.084 ± 0.016	nd
Zn	68	0.039	32.9 ± 1.7	29 ± 2	75.9 ± 1.2	62 ± 6	66.7 ± 3.6	67 ± 9
Zr	90	0.041	0.071 ± 0.008	nd	0.654 ± 0.037	nd	0.183 ± 0.009	nd

[a] ICP–MS data (mean and standard deviation of three replicates) from McCurdy (1990). LOQ = limit of quantitation; NIST = National Institute of Standards and Technology, USA. Reference values after Gladney et al. (1987), compilation; nd = not determined.

(2) Concentrated (48% m/m) Aristar hydrofluoric acid (23 M HF).

(3) Concentrated (70% m/m) Aristar perchloric acid (12 M $HClO_4$).

(4) Aristar nitric acid, 5 M (dilute 320 ml concentrated (68–71% m/m) HNO_3 to 1 l with deionised water).

(5) Aristar nitric acid, 1 M (dilute 65 ml concentrated (68%–71% m/m) HNO_3 to 1 l with deionised water).

Procedure

(1) Dry samples overnight at 105°C (note that labile Hg, As, Se will be volatilised at this temperature).

(2) Weigh 0.500 g into PTFE beaker and moisten with a few ml deionised water.

(3) Slowly add 10.0 ml HF and 4.0 ml $HClO_4$ (the effectiveness of the acid attack will be improved if the sample is allowed to stand for several hours in the acid mixture before the evaporation stage is initiated), and evaporate on a hotplate at 200°C until a crystalline paste is formed (2–3 h).

(4) Add a further two aliquots of HF and $HClO_4$, following each addition by evaporation to incipient dryness.

(5) Add 4.0 ml $HClO_4$ and evaporate to incipient dryness, to remove any remnant HF.

(6) Add 10.0 ml 5 M HNO_3 and warm gently, until a clear solution results. Inspect the solution for undigested material. If present, evaporate to incipient dryness and repeat the HF/$HClO_4$ additions until digested, followed by an $HClO_4$ evaporation and addition of 5 M HNO_3.

(7) Allow clear solution to cool, dilute to 50 ml in a volumetric flask (producing a 1 M HNO_3 solution) and transfer immediately to a new polypropylene bottle for storage.

(8) Dilute a 1.0 ml aliquot to 5.0 ml with 1 M HNO_3, immediately prior to analysis by ICP–MS.

Refractory minerals, such as chromite, garnet, magnetite and zircon, will be only partially attacked by this procedure. Samples containing large quantities of sulphide minerals are also not dissolved very efficiently by the HF/$HClO_4$ mixture, but can be 'opened up' by a preliminary attack with 10.0 ml Aristar-grade aqua regia. Samples should be decomposed slowly at room temperature until any effervescence ceases, followed by 1 h at 60°C, before evaporation to dryness at 150°C. Digestion can then be completed with the use of the HF/$HClO_4$ procedure.

Jarvis (1988, 1990) and Totland *et al.* (1991) recently assessed the applicability of the HF/$HClO_4$ open digestion procedure to ICP–MS. The latter authors determined selected major and trace elements by ICP–AES plus 38 trace elements (including all 14 of the naturally occurring rare earth elements, REEs) by ICP–MS. Good agreement with reference values was achieved for most major and trace elements in a wide range of SRMs (Table 7.3). Some elements (Cr, Hf, Mo, Sc, Zr), however, yielded inconsistent data in HF/$HClO_4$ digestions, with Cr, Hf and Zr in particular, displaying incomplete recovery in certain samples. The heavy REEs (Gd to Lu) may also be partly insoluble, particularly in zircon-bearing rocks (Gromet *et al.*, 1984; Sholkovitz, 1990). Other trace elements (Se, Hg, As, Ge, Te, Re, Os,

Table 7.3 Comparison of ICP–MS and reference values ($\mu g\,g^{-1}$) for some geological SRMs prepared by open vessel HF/HClO₄ digestion[a]

Element	Mass (m/z)	LOQ	USGS AGV-1 andesite		USGS BHVO-1 Hawaiian basalt		USGS SCo-1 Cody shale	
			ICP-MS	Ref.	ICP-MS	Ref.	ICP-MS	Ref.
Ba	135, 138	0.63	1100 ± 50	1226	i32 ± 2	139 ± 14	507 ± 2	570 ± 30
Ce	140	0.11	63.9 ± 2.8	67	38.7 ± 3.0	39 ± 4	65.4 ± 2.8	62 ± 6
Co	59	0.27	14.3 ± 0.7	15.3	48.6 ± 1.3	45 ± 2	9.58 ± 0.17	10.5 ± 0.8
Cs	133	0.25	1.12 ± 0.05	1.28	< 0.25	0.13 ± 0.06	7.07 ± 0.07	7.8 ± 0.7
Cu	63	0.47	55.4 ± 2.0	60	144 ± 1	136 ± 6	24.6 ± 0.5	28.7 ± 1.9
La	139	0.13	35.1 ± 1.7	38	15.8 ± 1.3	15.8 ± 1.3	34.4 ± 1.4	29.5 ± 1.1
Nb	93	0.33	12.5 ± 0.5	15	18.4 ± 0.7	19 ± 2	10.3 ± 0.2	11 ± 3
Ni	60	1.35	15.3 ± 0	16	124 ± 2	121 ± 2	26.0 ± 0.1	27 ± 4
Pb	207, 208	0.92	38.6 ± 6.8	36	1.89 ± 0.44	2.6 ± 0.9	29.1 ± 0.9	31 ± 3
Rb	85	0.48	70.7 ± 3.1	67.3	10.0 ± 0.3	11 ± 2	109 ± 2	112 ± 4
Sb	123	0.55	4.00 ± 0.21	4.3	< 0.55	0.159 ± 0.036	2.38 ± 0.14	2.50 ± 0.13
Sn	118, 120	0.55	4.40 ± 0.34	4.2	1.90 ± 0.05	2.1 ± 0.5	3.10 ± 0.25	2.7 ± 0.8
Sr	88	0.32	614 ± 29	662	398 ± 14	403 ± 25	162 ± 2	174 ± 16
Ta	181	0.15	0.87 ± 0.02	0.90	1.23 ± 0.05	1.23 ± 0.13	0.84 ± 0.05	0.92 ± 0.09
Th	232	0.12	5.80 ± 0.35	6.5	1.21 ± 0.02	1.08 ± 0.15	8.55 ± 0.10	9.7 ± 0.5
Tl	205	0.28	0.32 ± 0.04	0.34	< 0.28	0.058 ± 0.012	0.61 ± 0.01	0.72 ± 0.13
U	238	0.17	1.81 ± 0.12	1.92	0.43 ± 0.01	0.42 ± 0.06	2.92 ± 0.03	3.0 ± 0.2
W	186	0.52	0.54 ± 0.04	0.55	< 0.52	0.27 ± 0.06	1.49 ± 0.05	1.4 ± 0.2
Y	89	0.082	15.4 ± 1.1	20	23.9 ± 3.2	27.6 ± 1.7	24.2 ± 1.0	26 ± 4
Zn	66	5.3	86.5 ± 4.5	88	114 ± 4	105 ± 5	91.0 ± 4.1	103 ± 8

[a] ICP–MS data (mean and standard deviation of three replicates) from Totland *et al.* (1991). LOQ = limit of quantitation; USGS = US Geological Survey. Reference values after Govindaraju (1989), compilation; standard deviations for reference data after Gladney and Roelandts (1988).

Ru) may be potentially lost by this method (cf. Bock, 1979; Sulcek and Povondra, 1989), and as with other procedures which use $HClO_4$, quantification of low levels of As and V may be compromised by polyatomic ion interferences (Figure 7.1). Clearly, therefore, the procedure should be assessed carefully with respect to the sample types being studied.

7.3.1.3 *Environmental samples* The simplest type of environmental sample for ICP–MS analysis is freshwater, which may be analysed directly (Beauchemin *et al.*, 1987b; Garbarino and Taylor, 1987) or following evaporation preconcentration. During collection, samples should be filtered through a $0.45\,\mu m$ Micropore filter and acidified to $pH < 2$ with Aristar HNO_3 (Taylor, 1989; Long and Martin, 1991). Filtration separates the particulate and dissolved phases, while acidification is necessary to prevent the absorption of trace elements onto the container walls and the precipitation of trace metals or their co-precipitation with major constituents, once removed from ambient conditions. Particulate material can be discarded or dried and digested separately (using standard 'geological' dissolution techniques), as required. The high sensitivity of ICP–MS and the low levels of TDS in freshwaters make an almost ideal combination, eliminating the need for the complex pre-concentration procedures required when using most other analytical techniques.

Analysis of seawater is far less straightforward (chapter 9) and requires the chemical separation of trace elements from the sodium chloride-rich (3.5% TDS) matrix (e.g. McLaren *et al.*, 1985; McLaren, 1987; Beauchemin *et al.*, 1988c; Plantz *et al.*, 1989), or the use of alternative sample introduction strategies such as electrothermal vaporisation (ETV) and flow injection (Hutton and Eaton, 1988). Undiluted seawater samples cause rapid blocking of the sample cone and excessive signal suppression, while dilution leads to unacceptably high LOQs which are further degraded by polyatomic ion interferences from Na species (e.g. $^{23}Na_2{}^{16}O^+$ on ^{62}Ni, $^{23}Na^{40}Ar^+$ on ^{63}Cu). Consequently, McLaren *et al.* (1985) and Beauchemin *et al.* (1988c) employed a separation and pre-concentration procedure using silica-immobilized 8-hydoxyquinoline and ICP–MS to accurately determine trace metals (Cd, Co, Cr, Cu, Mn, Mo, Ni, Pb, U, Zn) at levels below $1\,ng\,ml^{-1}$ in National Research Council of Canada (NRCC) seawater SRMs CASS-1 and NASS-1.

The complete digestion of environmental samples such as soils, sediments, sewage sludges, domestic, road and industrial dusts, and domestic and industrial refuse may be accomplished using an $HF/HClO_4$ and HNO_3 attack similar to that described for geological materials. The high organic contents of most environmental materials, however, require that labile organic compounds are oxidised prior to the introduction of concentrated $HClO_4$ (this is particularly important when the presence of oils and fats is expected). This may be accomplished by prolonged digestion with concentrated HNO_3 (24 h at 100°C, followed by 10 h at 150°C), prior to the addition of an $HF/HClO_4$ or $HF/HClO_4/HNO_3$ mixture.

In many cases, the total metal contents of environmental samples are not sought, and the destruction of silicate phases (and therefore the use of HF) is unnecessary. For these applications HNO_3 alone, or in conjunction with $HClO_4$, is commonly used. The use of $HClO_4$ has the advantage of ensuring that oxidising conditions are maintained throughout the digestion, minimising (but not preventing totally) the losses of mercury compounds and the hydride-forming elements As, Bi, Se, Sn, Te. Such partial attacks have the advantage that many 'major' elements are not dissolved, enabling the use of lower dilution factors to produce enhanced LOQs by ICP–MS. In these cases, it is important to ensure the complete oxidation of organic constituents (finely dispersed colloids and dissolved compounds), which will increase the viscosity of the solution, and may lead to transport and aspiration problems during sample introduction. The following procedure (modified from Thompson and Walsh, 1989) is suggested:

Method

Equipment
(1) 250°C hotplate with aluminium heating block, or block bath, sited in a dedicated $HF/HClO_4$ fume cupboard equipped with fume-scrubbing and wash-down facilities.
(2) Medium-walled Pyrex test tubes and test-tube racks.
(3) Liquid dispensers for acids.

Reagents
(1) Deionised water (18–24 Mohm cm).
(2) Concentrated (68–71% m/m) Aristar nitric acid (16 M HNO_3).
(3) Concentrated (70% m/m) Aristar perchloric acid (12 M $HClO_4$).
(4) Aristar nitric acid, 5 M (dilute 320 ml concentrated (68–71% m/m) HNO_3 to 1 l with deionised water).
(5) Aristar nitric acid, 1 M (dilute 65 ml concentrated (68–71% m/m) HNO_3 to 1 l with deionised water).

Procedure
(1) Dry samples for 48 h at 60°C (note that a proportion of Hg, Se may be volatilised at this temperature; air drying at 25°C is preferred if volatile elements are sought).
(2) Weigh 0.100 g into a test tube.
(3) Slowly add 4.0 ml concentrated HNO_3.
(4) Place test tubes in the heating block and heat at 100°C for 24 h. Increase to 150°C and evaporate to dryness (approximately 10 h).
(5) Add 4.0 ml concentrated HNO_3 and 1.0 ml $HClO_4$, and heat at 150°C for 3 h and 190°C for 6 h, or until dry.
(6) Allow tubes to cool, add 2.0 ml 5 M HNO_3 and warm gently at 50°C for 1 h. Transfer to racks and allow to cool. Add 8.0 ml deionised water (to produce a 1 M HNO_3 solution) and mix contents thoroughly.
(7) Allow residues to settle overnight or centrifuge.
(8) Dilute a 1.0 ml aliquot to 5.0 ml with 1 M HNO_3, immediately prior to analysis by ICP–MS.

Environmental samples are characterised by varied compositions, and are often highly enriched in heavy metals (e.g. $\mu g\, g^{-1}$: 100–2000 Cu; 100–2000 Pb; 300–2000 Zn; 1–50 Cd). High concentrations of these and other metals may necessitate further dilution of solutions during analysis. Finally, an $HNO_3/HClO_4$ attack may not be suitable for studies where accurate data for highly volatile elements (e.g. Hg) are sought, and specialised digestion procedures should be considered for these applications.

7.3.1.4 *Metals* The compositional variations of metals is enormous, including such materials as aluminium and copper, and their alloys, ferrous metals, steels and ferro-alloys, boron-containing alloys, hardmetals, chromium metal, magnet alloys, refractory products and precious metals. A wide range of sample preparation techniques is required to cover such a wide range of compositions.

Samples are best reduced to a fine grain size prior to digestion. Fine turnings may be used for highly soluble materials, but grinding to a fine powder in an agate Tema mill is recommended for more refractory samples. A simple attack with 8 M HNO_3 will attack many metals (e.g. Longerich *et al.*, 1987a), although various HNO_3/HCl mixtures including aqua regia are also commonly used, sometimes in conjuction with HF. High carbon alloys require fuming with $HClO_4$ to break down carbides, although this may cause precipitation of Si, Nb, Ta, W (when present at high concentrations) which must be filtered off, fused with $LiBO_2$, and returned to the stock solution

Table 7.4 Comparison of ICP–MS and reference values (%) for some metallurgical SRMs prepared by open vessel HNO_3 digestion[a]

Element	Ma (m/z)	NIST 671 nickel oxide		NIST 365 electrolytic iron	
		ICP–MS	Ref.	ICP–MS	Ref.
Al	27	0.009	0.009	nd	
As	75	nd		0.0002	0.0002
Co	59	0.299	0.31	0.0069	0.007
Cr	52	0.013	0.025	0.0087	0.007
Cu	65	0.192	0.20	0.0063	0.006
Fe	56	0.370	0.39	nd	
Mg	24	0.032	0.030	nd	
Mn	55	0.129	0.13	0.0064	0.006
Mo	95	nd		0.0064	0.005
Ni	60	nd		0.0522	0.041
Si	28	0.031	0.047	nd	
Ti	47	0.023	0.024	nd	
V	51	nd		0.0005	0.0006

[a] ICP–MS data from Meddings and Ng (1989). NIST = National Institute of Standards and Technology, USA. Reference value from NIST certificates. nd = not determined.

(Walton, 1989). A range of metal dissolution procedures are outlined in Bock (1979), Ohls and Sommer (1987) and Thompson and Walsh (1989).

Traditional approaches to the digestion of metallurgical samples, many of which retain HCl or $HClO_4$ in the final matrix and/or utilise fusion with sodium compounds (e.g. Bozic et al., 1989) are generally not well suited to ICP–MS analysis (Vaughan and Horlick, 1989). Further work is required to modify these procedures to enable the greater use of nitric acid media with or without lithium-based fluxes (see below). Many methods which utilise hydrochloric or sulphuric acids, however, may be extended to include a final evaporation stage, with samples being finally taken up in 0.5 M HNO_3 (Meddings and Ng, 1989). Remnant chloride and sulphate ions will remain in these solutions, so the rigorous preparation of a procedural blank, in some cases combined with matrix-matched standards, may be necessary to compensate for polyatomic ion interferences. Matrix matching has the additional advantage of minimising errors caused by signal suppression from the heavy, essentially single-element, matrices characteristic of many metal digestions. Despite these difficulties, ICP–MS analysis of metallurgical samples in 0.5 M HNO_3 can produce excellent results (Table 7.4).

7.3.2 Closed vessel digestions

The use of sealed-vessel dissolution began in 1860 when Carius described the digestion of samples with concentrated HNO_3 in sealed, strong-walled glass vials (Carius tubes). Despite advances in vessel design (Bernas, 1968; Langmyhr and Paus, 1968; Uhrberg, 1982), such methods have generally remained time-consuming and potentially dangerous. Recently, however, closed-vessel systems designed to be rapidly heated by microwave energy have been developed (see below).

Closed-vessel digestion has a number advantages over other techniques:

(1) Digestions reach higher temperatures because the boiling point of the reagents is raised by the pressure generated within the vessel. Such increased temperatures and pressures may significantly decrease sample decomposition times and enable the dissolution of refractory phases.
(2) Volatile elements such as As, B, Cr, Hg, Sb, Se, Sn will remain within the vessel, allowing their retention in solution.
(3) Smaller reagent volumes are required because evaporation is avoided. This reduces consumable costs and minimises the amount of toxic fumes produced during the digestion procedure.
(4) Contamination is reduced by lowering reagent volumes and because the sealed system excludes the possible introduction of airborne particles during decomposition.

Modern conventional acid digestion bombs consist of a PTFE beaker and lid which fit tightly into an outer stainless steel jacket (Figure 7.2). The outer

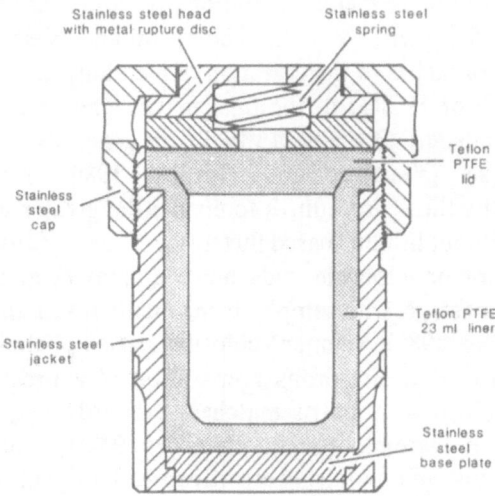

Figure 7.2 Diagrammatic cross-section through a Parr general purpose high-pressure digestion bomb. This model (Parr 4749) uses a 23 ml Teflon PTFE liner with a flanged closure. The stainless steel pressure casing is tightened onto the liner using a spanner, enabling the vessel to operate at temperatures of up to 250°C and internal pressures of 12.4 MPa (1800 psi). A safety disc will rupture if the internal pressure exceeds 24 MPa.

jacket has a screw-top or screw-studded lid, which when tightened forms a gas-tight high-pressure seal between the PTFE beaker and its lid. Following the addition of sample and acids, the bomb is sealed and placed in an oven at 110–250°C for periods of one to several hours.

Care is required when using digestion bombs, since vaporisation of the reaction mixture produces confined pressures in the order of 7–12 MPa. Samples and reagents should never constitute more than 10–20% of the volume of the liner; overfilling may generate pressures which (explosively) exceed the safety tolerances of the vessels. Similarly, organic materials should never be mixed with strong oxidising agents in sealed bombs, because the evolution of gases may also cause the explosive rupturing of the vessel. Bombs should always be allowed to return completely to room temperature prior to opening, and should be opened with great caution in a suitable fume cupboard. Despite these hazards, bomb techniques enable the rapid digestion of refractory minerals which are difficult or impossible to dissolve by other techniques.

Many biological samples may be successfully digested in Teflon-lined steel bombs using concentrated HNO_3. De Boer and Maessen (1983) accurately determined a number of transition metals by ICP–AES following the digestion of a 0.25 g sample of NIST SRM 1577 bovine liver with 3 ml 16 M HNO_3 in a 23 ml capacity bomb. The sample was heated for 2 h at 140°C, cooled, and diluted to 5 ml with deionised water prior to analysis.

Digestion of geological samples (Ito, 1961; Bernas, 1968; Hendel *et al.*, 1973; Price and Whiteside 1977; Potts, 1987; Sulcek and Povondra, 1989; Thompson and Walsh 1989; Jenner *et al.*, 1990) in steel digestion bombs typically involves weighing a 0.200 g aliquot into the PTFE liner and adding 0.5 ml 16 M HNO_3 (aqua regia is preferred by some authors), followed by 5 ml 29 M HF. The bomb is sealed, placed in an oven at 150°C for 1 h, and then removed and allow to cool to room temperature. The vessel may then be opened, and inspected for undigested material. Undecomposed samples should be returned to the oven for a further 1 h (or longer, as necessary). Boric acid (2.00 g H_3BO_3 dissolved in a small amount of water) is added to digested samples, which are resealed and reheated at 130°C for 15 min, and again returned to room temperature. Solutions should be diluted to 200 ml with 1 M HNO_3 in volumetric flasks, and transferred to new polypropylene bottles for storage. Samples require an additional 10-fold dilution with deionised water prior to analysis by ICP–MS.

Samples decomposed by the above method apparently display no significant elemental losses (Van Eenbergen and Bruninx, 1978). The addition of boric acid allows the retention of Si and other volatile fluorides, and prevents the precipitation of insoluble fluorides. However, the 'neutralised' HF will still attack glassware, and solutions should not be aspirated into delicate glass nebulisers such as Meinhard designs. Furthermore, the large quantities of H_3BO_3 introduced (producing levels of TDS which exceed those of many fusion techniques) are highly detrimental to LOQs, precluding the determination of lower abundance trace elements in many samples. The technique, therefore, may have some application for the determination of 'volatile' elements in refractory samples (e.g. Pb in zircon), but is not recommended for general ICP–MS work.

The use of boric acid in the bomb digestion procedure may be avoided if HF is retained or removed by evaporation with $HClO_4$ (or less effectively, HNO_3). Steels may be successfully digested using an $HNO_3/HCl/HF$ mixture and direct analysis of the diluted solution by ICP–MS (Vaughan and Horlick, 1989). However, the retention of chlorides in this method causes significant interference problems, while the measurement of low levels of Si, P and S are difficult if not impossible due to interferences inherent in the Ar plasma (e.g. $^{14}Na_2^+$, $^{12}C^{16}O^+$ on ^{28}Si). Removal of HF by evaporation may totally preclude the determination of Si and some other elements, but it avoids the addition of high levels of TDS (this approach is described in more detail in the discussion of microwave digestion techniques; see below).

Procedures using a combination of high-pressure bomb digestion followed by open evaporation have been successfully adopted for the complete dissolution of plant materials (Van Loon and Barefoot, 1989) and marine sediments (McLaren *et al.*, 1987a, b). Date *et al.* (1987a, 1988) used a multi-stage digestion procedure for the dissolution of limestones, phosphorites, manganese nodules and iron ores, to avoid the introduction of Cl ions prior to analysis by ICP–MS. These authors first used an open digestion with $HNO_3/HF/H_2O_2$ (HNO_3 alone for limestones) followed by a bomb-dissolution (overnight at 150°C) of the insoluble residue with an HNO_3/HF mixture. The HF was removed by multiple evaporations to incipient dryness with HNO_3, and the residue taken up in HNO_3. Finally, the two solutions were combined to produce an acid matrix of 0.2 M HNO_3. Good agreement with reference values was obtained for up to 42 trace elements in SRMs analysed by ICP–MS. Unfortunately, the time-consuming, labour intensive nature of this procedure makes it unsuitable for routine analysis.

Interestingly, Date et al. (1988) compared their HNO_3-based procedure with one using predominantly HCl, but including a final evaporation to incipient dryness prior to digestion in HNO_3. This study demonstrated that residual chlorides produced no serious polyatomic interferences from $^{35}ClO^+$ or $^{40}Ar^{35}Cl^+$, enabling the determination of ^{51}V and ^{75}As. However, some loss of As, Ba, Ge, Sb, W apparently occurred during the HCl-based digestion.

An allied procedure to the steel digestion bomb is the High Pressure Asher (Boorn et al., 1985; Knapp and Grillo, 1986), which has been successfully used for the digestion of botanical and biological materials. This system (Figure 7.3) consists of 2–70 ml quartz glass closed digestion vessels and a heating block sited in a stainless steel programmable autoclave. Venting of the vessels is prevented by a 10 MPa confining pressure of nitrogen, while the impermeable nature of quartz glass (unlike Teflon PTFE and PFA) enables the retention of Hg, Os and the halides.

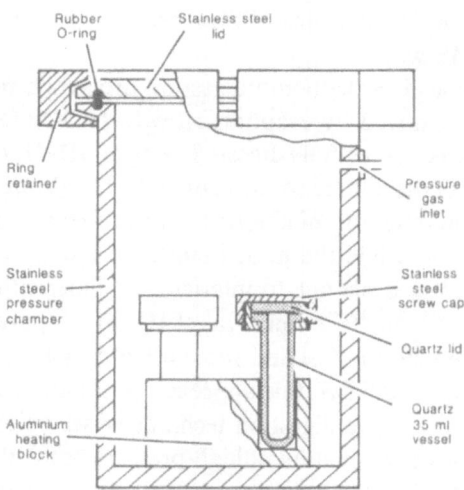

Figure 7.3 Schematic diagram of a high-pressure asher (after Knapp and Grillo, 1986). The sample is digested in closed quartz vessels at temperatures of up to 320°C. Venting of the vessels and loss of volatiles is prevented by a 10 MPa confining pressure of nitrogen. The impermeable nature of the quartz glass vessels ensures the retention of Hg, Os and halides.

Boorn *et al.* (1985) have described a method for the digestion of organic samples using a High Pressure Asher. In their procedure, 2.0 ml aliquots of 16 M Aristar-grade HNO_3 are added to 0.500 g samples in 25 ml digestion vessels, which are placed in the heating block in the autoclave and pressurised to 10 MPa with nitrogen. Samples are ramped from 70 to 120°C over 15 min and then rapidly up to 220°C, which is maintained for 2 h. After cooling the nitrogen pressure is released, the autoclave opened, and samples made to 10.0 ml with deionised water. Analysis by ICP–MS of solutions produced from NIST SRMs 1567 wheat flour, 1568 rice flour and 1572 citrus leaves, and yielded fair agreement with reference values for As, Cd, Ce, Co, La, Mo, Ni, Pb, Sn, Tl (Boorn *et al.*, 1985).

7.3.3 *Alkali fusions*

Alkali fusions are rarely used for the decomposition of botanical or biological materials because the high levels of flux required produce unacceptably high limits of quantitation. Similarly, partial acid digestions in open vessels prove adequate for most environmental applications. In contrast, a wide range of alkali fluxes have been used for the decomposition of geological and metallurgical samples, particularly lithium metaborate ($LiBO_2$), lithium tetraborate ($Li_2B_4O_7$), sodium carbonate (Na_2CO_3), sodium hydroxide (NaOH), sodium peroxide (Na_2O_2), the equivalent potassium salts and alkali fluorides (particularly KHF_2). Fusion procedures may be used to effectively digest even the most refractory phases, although particular fluxes may be more effective for certain sample types.

The major disadvantage of all fusion procedures is the high level of total dissolved solids introduced during sample preparation. These necessitate the use of large dilutions prior to analysis, with a consequent deterioration in the quantitation limits for the method. Alkali fluxes have two further disadvantages. Firstly, it is well documented (Gregoire, 1987c; Gray, 1989a) that the presence of excess amounts of easily ionisable elements, particularly Na, K, Ca, can cause severe matrix effects (e.g. signal suppression) in ICP–MS, beyond those caused by the TDS content. Such effects need to be investigated for the particular compounds and instrument conditions employed, and may require dilution of samples well below 0.1% TDS. Secondly, the addition of large quantities of an element may cause severe polyatomic ion interferences (e.g. $^{23}Na\,^1H^+$ on ^{24}Mg, $^{23}Na\,^{16}O^+$ on ^{39}K, $^{23}Na_2\,^{16}O^+$ on ^{62}Ni, $^{23}Na\,^{40}Ar^+$ on ^{63}Cu), which will cause deteriorated LOQs, and may restrict (Table 7.5) the range of trace elements determinable.

Lithium metaborate fusions have been widely used in sample preparation for atomic absorption spectrometry and ICP–AES (Ingamells, 1964, 1970; Walsh, 1979; Brenner *et al.*, 1980; Walsh and Howie, 1980; Thompson and Walsh, 1989; Van Loon and Barefoot, 1989; Jarvis and Jarvis, 1992), predominantly because the method enables the determination of all major

Table 7.5 Potential polyatomic ion interferences produced by alkali fluxes[a]

Element	Isotope	Abundance (%)	MH$^+$	MO$^+$	MOH$^+$	MO$_2^+$ (mass)	MO$_2$H$^+$	M$_2^+$	M$_2$O$^+$	MAr$^+$
Lithium	6	7.5	7	22	^{23}**Na**	38	^{39}**K**	12	^{29}**Si**	46
	7	92.5	8	^{23}**Na**	^{24}**Mg**	^{39}**K**	40	14	30	^{47}**Ti**
Boron	10	20	11	^{26}**Mg**	^{27}**Al**	42	43	20	37	50
	11	80	12	^{27}**Al**	^{28}**Si**	43	^{44}**Ca**	22	38	^{51}**V**
Sodium	23	100	^{24}**Mg**	^{39}**K**	40	^{55}**Mn**	^{56}**Fe**	46	^{62}**Ni**	^{63}**Cu**
Potassium	39	93.3	40	^{55}**Mn**	^{56}**Fe**	^{71}**Ga**	^{72}**Ge**	78	94	^{79}**Br**
	40	0.01	41	56	57	^{72}**Ge**	^{73}**Ge**	80	^{95}**Mo**	80
	41	6.7	42	^{57}**Fe**	58	^{73}**Ge**	^{74}**Ge**	^{82}**Se**	96	^{81}**Bi**

[a] Serious interferences are in bold; after Date and Stuart (1988) and Date and Jarvis (1989). Abundance = relative abundances of naturally occurring isotopes.

elements (including sodium and potassium) in a single solution. The LiBO$_2$ flux is mixed with samples in a ratio of between 3 and 7:1, depending on their composition. We have found that a ratio of 5:1 is applicable to the majority of sample types, and a 3:1 mixture can be used in many cases (cf. Crock *et al.*, 1984). Relatively low flux to sample ratios are particularly useful when it is considered that matrix suppression effects are minimal for low mass elements such as lithium (Gray, 1989a), thereby enabling more concentrated solutions to be analysed. These characteristics, plus the small number of polyatomic interferences on trace elements which LiBO$_2$ generates (Figure 7.4, Table 7.5), and the ready availability of the high purity reagent at reasonable cost (e.g. Johnson Matthey Spectroflux 100A), make lithium metaborate the best alkali flux for ICP–MS work.

Samples are fused at 900–1050°C in non-wetting platinum-gold or graphite crucibles in a muffle furnace, or over a Meker burner, and the melts dissolved in dilute nitric acid. If poured directly into the acid, the molten bead is shattered by instantaneous cooling and is rapidly dissolved. Lithium metaborate effectively digests all major rock-forming minerals and most accessory phases. Samples containing large proportions of oxides or sulphides, however, are less effectively attacked (Cremer and Schlocker, 1976; Feldman, 1983), and may require longer fusion times, larger flux to sample ratios, or the use of other (less suitable) fluxes.

Platinum or platinum-gold crucibles have traditionally been used for lithium metaborate fusions, but the enormous capital investment involved now precludes their routine use in many laboratories. Vitreous carbon crucibles are also expensive and may retain some of the melt. Carbon crucibles are the most acceptable alternative but have a rather limited life (generally 5–6 fusions per crucible). They should be lightly brushed-out to remove excessive loose carbon particles between runs, and immediately after each fusion, crucibles should be inspected carefully for any remnant glass. Small beads can sometimes be transferred subsequently to the sample solution and

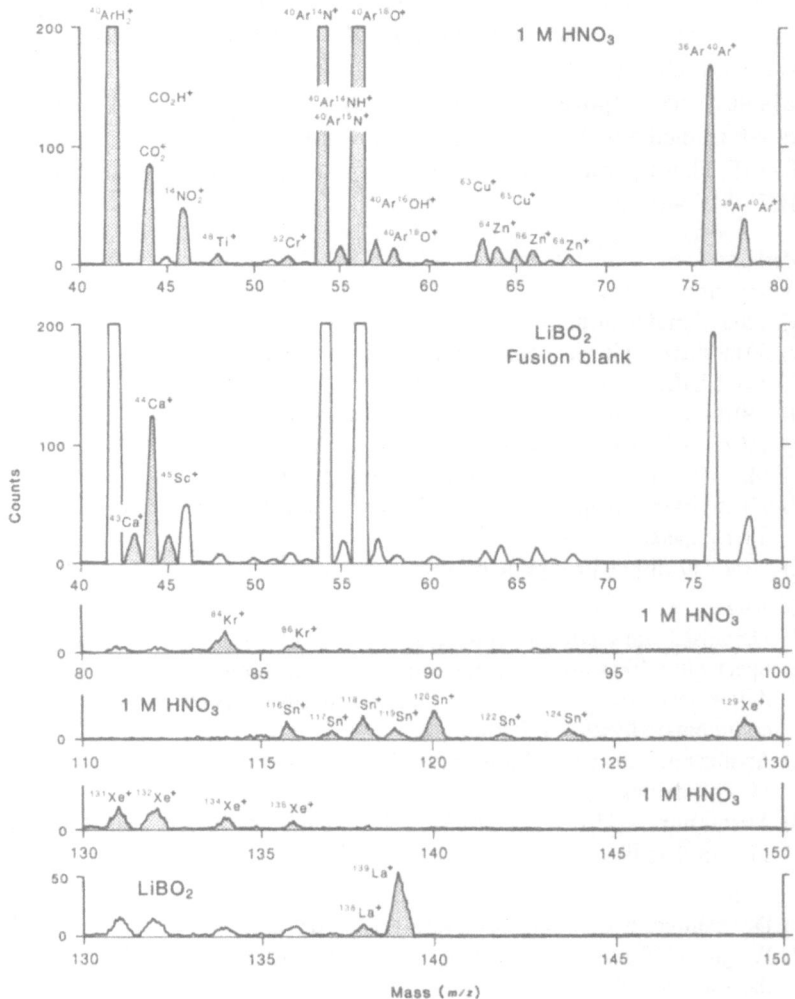

Figure 7.4 Spectra of the mass ranges 41–79, 81–100 and 110–150 m/z for BDH Aristar 1 M HNO$_3$ compared with those for a Johnson Matthey Spectroflux 100 A LiBO$_2$ fusion blank (section 7.3.3; 5000-fold dilution) prepared in the same matrix. Only parts of the LiBO$_2$ spectra which differ from those of 1 M HNO$_3$ are shown; the same vertical scale is used throughout. In addition to gas and gas polyatomic species, the HNO$_3$ spectra display contamination from small amounts of Cr, Cu, Sn, Ti and Zn. The LiBO$_2$ spectra are essentially identical to those for HNO$_3$ with the exception of the additional presence of Ca, La and Sc. The relatively high levels of La contamination make Spectroflux 100 A unsuitable for ultratrace level REE determinations; purer (more expensive) LiBO$_2$ is available from other manufacturers.

dissolved, but strongly adhered material will necessitate the disposal of both the sample and crucible, and the preparation of a duplicate using a higher flux to sample ratio. 'Difficult' (particularly Fe-rich) samples are best fused in old degraded crucibles which display a lower tendancy to retain material, and require high flux to sample ratios (typically 7:1). Metal-rich samples (e.g. Cu, Pb) which form alloys with Pt must always be fused in carbon.

The characteristics of lithium metaborate fusions of selected geological SRMs have been reported recently by Jarvis (1990) and Totland et al. (1991), who determined a wide range of major and trace elements by ICP–AES and ICP–MS. These studies clearly demonstrate the applicability of $LiBO_2$ fusions to ICP–MS work.

Method

Equipment
(1) 1200°C muffle furnace.
(2) 30 ml disposable plastic weighing boats, and camel hair brush.
(3) 30 ml UltraCarbon graphite crucibles.
(4) 250 ml polythene dispenser bottles with slopping shoulders and lids.
(5) 150 ml polythene measuring cylinder.
(6) Magnetic stirrers and PTFE-coated stirring bars.
(7) 250 ml Pyrex volumetric flasks, polythene filter funnels and Whatman No. 41 filter papers.
(8) New 250 ml polypropylene bottles.

Reagents
(1) Deionised water (18–24 Mohm cm).
(2) Spectroflux 100A lithium metaborate (Aldrich high purity $LiBO_2$ is preferred if low levels of La are to be determined since Spectroflux is generally contaminated with La; Figure 7.4).
(3) Aristar nitric acid, 0.8 M (dilute 50 ml concentrated (68–71% m/m) HNO_3 to 1 l with deionised water).
(4) Aristar nitric acid, 0.5 M (dilute 32.5 ml concentrated (68–71% m/m) HNO_3 to 1 l with deionised water).

Procedure
(1) Dry samples and $LiBO_2$ flux overnight at 105°C.
(2) Weigh 0.250 g into a weighing boat, followed by 1.250 g $LiBO_2$. Mix thoroughly with a plastic spatula.
(3) Transfer mixture to a carbon crucible, using a clean camel hair brush to remove any remnant powder from the weighing boat.
(4) Fuse at 1050°C for 20 min in a muffle furnace (allow the furnace to return to operating temperature before timing the fusion). Note: refractory samples e.g. those rich in zircon or chromite) will benefit from a gentle swirling once or twice during the fusion, and will require up to 45 min to be fully digested.
(5) Measure 150 ml 0.8 M HNO_3 into a polythene bottle, add a stirring bar, and place on a magnetic stirrer.
(6) Rapidly pour the melt into the stirring HNO_3, ensuring that no material remains in the crucible (swirling the melt in the crucible during removal from the furnace will facilitate the pouring process). Cap the bottle and stir solution until dissolved (30 min to 1 h).
(7) Filter (to remove carbon particles) into a volumetric flask. Make to volume with deionised water (to produce a final 0.5 M HNO_3 solution), and transfer immediately to a new polypropylene bottle for storage.

(8) Dilute a 1.0 ml aliquot to 5.0 ml with 0.5 M HNO_3, immediately prior to analysis by ICP-MS. Analyse solutions as soon as possible after preparation (silicic acid is metastable and may hydrolyse and precipitate if solutions are stored too long).

In addition to the determination of Si, Totland et al. (1991) demonstrated that lithium metaborate fusions rather than open vessel digestions are required to obtain quantitative data (Table 7.6) for Cr, Hf and Zr in many geological samples. Some other elements, such as the higher mass rare earth elements (REEs), yielded marginally more accurate fusion results for some materials, but other samples produced equally acceptable data by open digestions techniques. The high temperature of the fusion procedure, however, precludes the determination of Pb, Sb, Sn, Zn and other volatile elements, while the high LOQs for the method prevent the quantification of low-abundance trace elements in some samples.

A further study of fusion with an alkali flux combined with ICP-MS was provided by Date and Stuart (1988), who demonstrated the feasibility of the simultaneous determination of chlorine, bromine and iodine in environmental samples by ICP-MS, following a simple fusion with sodium carbonate (Na_2CO_3) containing 10% zinc oxide (ZnO). The method uses 0.100 g of sample to 0.300 g of flux, which are mixed in an agate mortar and pestle and transferred to a platinum crucible. A further 0.250 g of flux is used to clean the agate, and this flux is added to the crucible to form a thin covering layer over the sample mixture. The sample is slowly ignited over a Meker burner; the temperature being increased slowly until a maximum is reached. The crucible is cooled, rinsed externally with deionised water, and the crucible and its contents are transferred to a 125 ml beaker containing 50 ml deionised water. The beaker is warmed until the fusion cake has completely disintegrated, and the solution is then filtered through a Whatman No. 42 filter paper into a 100 ml polypropylene volumetric flask. The residue is washed carefully with further deionised water, and the solution is then made to volume.

Successful application of the method requires the use of high purity (i.e. low halogen) reagents, which may be difficult to obtain, and the use of matrix-matched standards to compensate for suppression and polyatomic ion interferences from the sodium-rich (0.5% Na_2CO_3, 2170 $\mu g\,ml^{-1}$ Na) matrix. Limits of quantitation calculated from Date and Stuart's (1988) data are approximately 970 $\mu g\,g^{-1}$ Cl, 14 $\mu g\,g^{-1}$ Br and 2.3 $\mu g\,g^{-1}$ I (the high ionisation energy of fluorine realistically precludes its determination in an Ar plasma). These relatively high values are limited by the dilution factor and the relative impurity of the reagent blanks. Furthermore, the high levels of sodium remaining in the diluted samples would be expected to cause significant instrumental drift, and for the analysis of batches of samples an additional four-fold dilution would be preferred. Unfortunately, such a dilution would cause a corresponding deterioration of the LOQ, and would

Table 7.6 Comparison of ICP–MS and reference values ($\mu g\,g^{-1}$) for some geological SRMs prepared by $LiBO_2$ fusion[a]

Element	Mass (m/z)	LOQ	USGS AGV-1 andesite		USGS BHVO-1 Hawaiian basalt		USGS SCo-1 Cody shale	
			ICP–MS	Ref.	ICP–MS	Ref.	ICP–MS	Ref.
Ba	135,138	4.2	1140 ± 20	1226	127 ± 7	139 ± 14	542 ± 11	570 ± 30
Ce	140	3.2	72.5 ± 1.9	67	40.6 ± 1.4	39 ± 4	62.2 ± 2.0	62 ± 6
Co	59	5.7	13.7 ± 0.4	15.3	48.0 ± 1.4	45 ± 2	10.5 ± 0.7	10.5 ± 0.8
Cr	52	8.7	< 8.7	10.1	321 ± 25	289 ± 22	66.5 ± 2.2	68 ± 5
Cu	63	32	56.2 ± 1.4	60	141 ± 4	136 ± 6	< 32	28.7 ± 1.9
Hf	177,178	3.5	5.36 ± 0.33	5.1	4.68 ± 0.46	4.38 ± 0.22	4.44 ± 0.46	4.6 ± 0.3
La	139	2.2	40.9 ± 1.1	38	16.6 ± 1.3	15.8 ± 1.3	32.0 ± 1.2	29.5 ± 1.1
Nb	93	6.7	12.9 ± 0.2	15	17.2 ± 0.9	19 ± 2	10.6 ± 0.2	11 ± 3
Ni	60	17	< 17	16	124 ± 10	121 ± 2	32.9 ± 7.8	27 ± 4
Rb	85	6.7	65.1 ± 1.4	67.3	9.35 ± 1.3	11 ± 2	106 ± 2	112 ± 4
Sr	88	14	612 ± 10	662	395 ± 19	403 ± 25	149 ± 6	174 ± 16
Th	232	2.2	5.91 ± 0.55	6.5	< 2.2	1.08 ± 0.15	8.68 ± 0.11	9.7 ± 0.5
U	238	1.1	1.90 ± 0.23	1.92	< 1.1	0.42 ± 0.06	2.93 ± 0.23	3.0 ± 0.2
Y	89	2.2	17.8 ± 1.2	20	25.2 ± 1.1	27.6 ± 1.7	24.2 ± 1.2	26 ± 4
Zr	90	4.3	208 ± 4	227	164 ± 8	179 ± 21	152 ± 3	160 ± 30

[a] ICP–MS data (mean and standard deviation of three replicates) from Totland et al. (1991). LOQ = limit of quantitation; USGS = US Geological Survey. Reference values after Govindaraju (1989), standard deviations for reference data after Gladney and Roelandts (1988).

only be feasible for samples containing high concentrations of halogens.

Fair agreement with reference values (in parentheses) were obtained for NIST Urban Particulate SRM 1648: 4140 (4500) $\mu g\,g^{-1}$ Cl, 485 (505) $\mu g\,g^{-1}$ Br and 14.6 (?20) $\mu g\,g^{-1}$ I, particularly given that the reference values themselves are constrained by few independent determinations. The method clearly has considerable potential for the analysis of halogens in environmental samples.

7.3.4 Microwave digestion

Most routine decomposition procedures involve heating samples and reagents over flames in open vessels, or electrically on hotplates or in furnaces. Such procedures originate from pre-classical times. Recently, however, methods have been developed which use microwave energy as a means of heating (see Matusiewicz and Sturgeon, 1989 for a review). During irradiation by microwave energy (generally at 2450 MHz; wavelengths of about 12 cm), polar molecules and ions are energised through the mechanisms of dipole rotation and ionic conductance (Neas and Collins, 1988; Gilman and Engelhart, 1989) respectively, causing rapid heating of the aqueous phase. However, microwaves not only affect the digestion medium (generally a mixture of mineral acids), they are also absorbed by sample molecules. This increases the kinetic energy of the matrix and causes internal heating and differential polarisation, which expand, agitate and rupture surficial layers of the solid material, exposing fresh surfaces to acid attack. These effects may be highly significant, producing rates of dissolution which are much greater than those predicted by the temperature of the acids alone.

Microwave heating is particularly applicable to closed-vessel techniques (Kingston and Jassie, 1988a) since solutions may be heated directly in low conductivity microwave-transparent (e.g. Teflon PFA) vessels (Figures 7.5, 7.6), allowing very rapid heating and cooling times compared with conventional-oven steel-jacketed digestion bombs. Teflon PFA (perfluoroalkoxy) is generally favoured for microwave digestion systems because it is transparent to microwave radiation, inert to mineral acids at temperatures of > 200°C, is an extremely poor conductor of heat, and has a high tensile strength. Teflon, however, is a semi-permeable membrane which allows the passage of some gases, enabling small quantities of water and mineral acids (including HNO_3, HCl, HF) to diffuse into the walls and escape during decomposition. Metal cations and most anions generally remain in solution, but elements which can pass through Teflon PFA include metallic Hg, OsO_4, and the hydrogen halides (Kingston and Jassie, 1988b). The use of Teflon may also restrict the range of reagents which can be used, H_2SO_4 must be avoided because its high boiling point (338°C) is above the melting point of the plastic (this acid is, in any case, not recommended for ICP–MS applications, section 7.2.2.5).

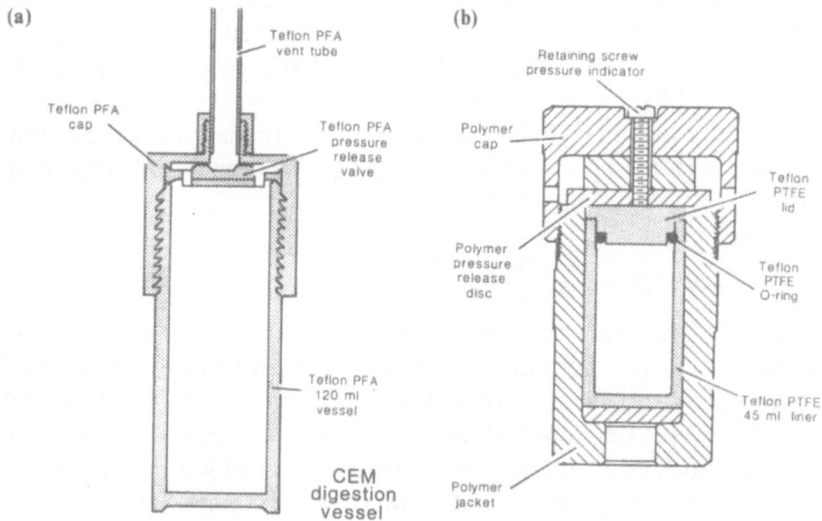

Figure 7.5 Diagrammatic cross-sections through two common types of microwave closed digestion vessel. (a) The CEM system is designed to operate with an internal temperature of up to 200°C and will remained sealed if the internal pressure does not exceed 830 kPa (120 psi). If this pressure is exceeded, gases are vented automatically by the pressure release system which then reseals the vessel. (b) The Parr bomb will tolerate an internal temperature of 250°C and a pressure of 8.3 MPa. The PTFE O-ring will blow out if the pressure exceeds 10 MPa. The high tolerances of the Parr system are similar to those of conventional oven digestion bombs.

The performance of nitric acid (the best acid for ICP–MS work) is enhanced considerably in closed microwave vessels. The boiling point reaches 176°C at about 520 kPa (75 psi), substantially increasing its oxidation potential and decreasing reaction times. For the same temperature, pressures in excess of 930 kPa will be reached with HCl, and 830 kPa with HF (Kingston and Jassie, 1988b). The absorption of microwave energy by a solution is controlled by its concentration, ionic strength and the molecular species present. At any temperature the pressure developed in the vessel depends on the partial pressure of the solvent(s) under non-equilibrium conditions, and the gaseous decomposition products formed during decomposition. These conditions are complex and it is difficult to exactly predict reaction mechanisms. Consequently, method development requires the acquisition of empirical data for representative sample/reagent combinations, although the recent introduction of interactive pressure controllers on some commercial microwave digestion systems (Gilman and Engelhart, 1989; Grillo, 1990) greatly facilities this work.

7.3.4.1 *Plant and animal tissue* The first application of microwave digestion (Abu-Samra *et al.*, 1975) was the decomposition of biological samples in open glass vessels using HNO_3 and $HClO_4$, or HNO_3 with H_2O_2. The decomposition by nitric acid of carbohydrates, proteins and lipids (the basic

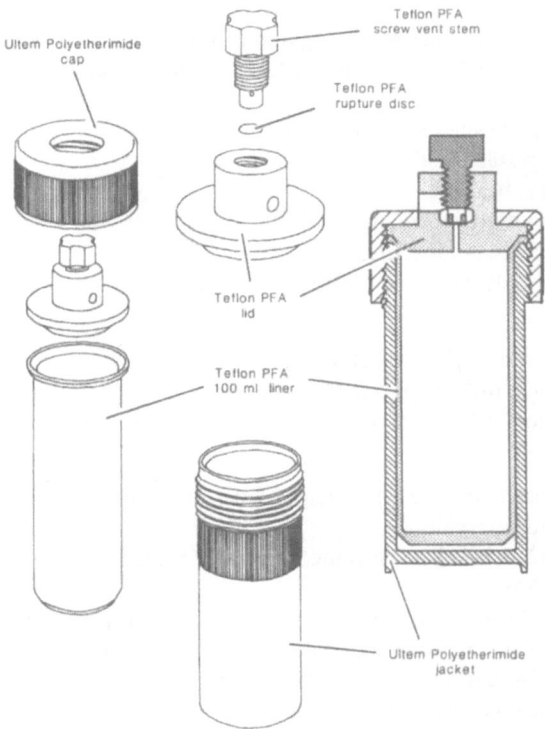

Figure 7.6 Schematic diagram and cross-section through a CEM lined microwave digestion vessel. This newly released (1990) vessel will operate with an internal temperature of 250°C and a pressure of 1.4 MPa. The Teflon PFA rupture disc will fail if this pressure tolerance is exceeded. The vessel is designed to operate in conjunction with a programmable pressure controller which maintains the digestion conditions within the limitations of the system.

constituents of biological and botanical samples), however, requires temperatures in excess of 140–160°C (Kingston and Jassie, 1986), which can only be achieved in closed vessels. Most of these molecular species break down in a few minutes under these conditions, leaving only nitroaromatics undigested, the presence of minute quantities of which will not adversely effect ICP–MS determinations.

Many elements have been successfully determined by GF–AAS and/or ICP–AES in a range of organic samples following closed-vessel microwave digestion with HNO_3 alone (Patterson et al., 1988), HNO_3/H_2O_2 (Matusiewicz et al., 1989) or mixed acid combinations (De Boer and Maessen, 1983; Schelkoph and Milne, 1988; Bettinelli et al., 1989). Recently, microwave digestion methods using HNO_3 and analysis by ICP–MS were used successfully by Beauchemin et al. (1988a) and Friel et al. (1990) who accurately determined a wide range of trace elements in biological materials. The suggested procedure is based on these studies.

Method

Equipment
(1) CEM MDS-81D microwave digestion systems with pressure monitoring system, including 12 Teflon PFA 120 ml digestion vessels, turntable and vessel capping station.
(2) 250°C hotplate sited in a fume cupboard.
(3) 30 ml disposable plastic weighing boats.
(4) 10 ml PTFE beakers.
(5) 25 ml Class A Pyrex volumetric flasks.
(6) Liquid dispensers for acids.
(7) New 30 ml polypropylene bottles.

Reagents
(1) Deionised water (18–24 Mohm cm).
(2) Concentrated (68–71% m/m) Aristar nitric acid (16 M HNO_3).
(3) 100 volumes (30% m/v H_2O_2) Aristar hydrogen peroxide solution.

Procedure
(1) Prepare partially defatted protein powders by acetone extraction. Spray dry or freeze dry residue.
(2) Weigh 0.250 g into a weighing boat, and transfer both to a microwave digestion vessel.
(3) Slowly add 6.0 ml concentrated HNO_3, washing any remnant powder from the surface of the weighing boat and the sides of the digestion vessel with the acid. Discard the weighing boat.
(4) Locate the relief valve on top of the vessel, cap and screw finger tight. Using the capping station, apply a torque of 16 Nm to the cap. Place a total of four vessels (including samples), equally spaced, in the sample turntable.
(5) Set the exhaust fan to maximum, heat at 100% power until a pressure of 550 kPa (80 psi) is attained, and maintain at this level for 25 min.
(6) Transfer the vessels and turntable to a cold water bath and allow to return to room temperature. Vent by hand any remaining vapours into a fume cupboard, and loosen the caps with the capping station.
(7) Carefully remove caps and transfer the vessel contents to a PTFE beaker, washing residual liquid from cap and vessel interiors with deionised water.
(8) Evaporate on a hotplate at 90°C to incipient dryness.
(9) Add 2.0 ml 16 M HNO_3 and warm gently. Carefully add 1.0 ml H_2O_2 dropwise to the solution, allowing effervescence to die down between additions. Evaporate to near dryness and repeat the HNO_3/H_2O_2 addition and evaporation twice more.
(10) Add 0.5 ml 16 M HNO_3 to the residue and warm gently until digested. Allow solution to cool, dilute to 25 ml with deionised water in a volumetric flask (producing a 0.3 M HNO_3 solution) and transfer immediately to a new polypropylene bottle for storage. The sample may be analysed by ICP–MS without further dilution.

The evaporation steps are incorporated to more completely digest organic compounds, to allow a reduction in the acid concentration of the final solution, and to decrease Cl levels by the formation and evaporation of HCl.

Removal of Cl sufficiently reduces Cl-polyatomic ion interferences to enable the accurate determination of As, Cr and V, following corrections based on interference standards (Friel *et al.*, 1990). Unfortunately, Hg, I and Br are lost along with Cl.

Beauchemin *et al.* (1988a) used a procedure similar to the above (but without hydrogen peroxide) for the digestion of SRM shellfish tissue NOAA-I and cod liver tissue NOAA-K, followed by isotope dilution ICP–MS analysis. These authors obtained good agreement with trace-element determinations obtained by graphite furnace atomic absorption spectrometry (GF–AAS), and with results for a range of digestion and analytical techniques provided by other laboratories. Friel *et al.* (1990) digested their samples in a Parr microwave bomb (Figure 7.5), rather than a CEM system, and used ICP–MS with multiple internal standards (to correct for matrix and drift effects) and external calibration to analyse three biological SRMs. After correcting for chloride ($^{35}Cl\,^{16}O^+$, $^{37}Cl\,^{16}O^+$, $^{40}Ar\,^{35}Cl^+$) polyatomic ion interferences on ^{51}V, ^{53}Cr and ^{75}As, and matrix interferences ($^{27}Al\,^{16}O^+$, $^{40}Ca\,^{16}OH^+$, $^{44}Ca\,^{16}O^+$, $^{48}Ca\,^{16}OH^+$) on ^{43}Ca, ^{57}Fe, ^{60}Ni and ^{65}Cu, the latter authors obtained excellent agreement with reference values for these materials (Table 7.7).

The proposed method should prove suitable for a wide range of biological and botanical samples, the use of HNO_3 under pressure enabling the breakdown of organic compounds without resorting to the addition of $HClO_4$. However, siliceous material such as silica phytoliths in grasses, will remain undigested. This insoluble residue should not compromise the determination of most trace elements but may necessitate the filtration of samples prior to analysis, in order to avoid the blockage of concentric glass nebulisers.

7.3.4.2 *Geological samples*   The wide range of sample compositions represented by geological materials preclude the use of any one digestion procedure. Consequently, a number of methods have been developed which utilise microwave heating for the dissolution of rocks, minerals and ores (Lamothe *et al.*, 1986; Matthes, 1988; Nakashima *et al.*, 1988c; Rantala and Loring, 1989; Kemp and Brown, 1990; Nölter *et al.*, 1990). Few microwave digestion procedures have yet been used in combination with ICP–MS. However, closed-vessel acid attacks, followed by an open evaporation were used successfully by Nölter *et al.* (1990) and Totland *et al.* (1991) to rapidly digest a range of geological materials, prior to analysis by ICP–MS. The suggested procedure is based on that of the latter workers.

Method

Equipment

(1) CEM MDS-81D microwave digestion system, including 4 Teflon PFA 120 ml digestion vessels, turntable and vessel capping station. The exhaust should be vented into a dedicated fume extraction system.

Table 7.7 Comparison of ICP–MS and reference values ($\mu g\,g^{-1}$) for some biological SRMs prepared by closed-vessel microwave digestion[a]

Element	Mass (m/z)	LOQ	NIST 1566 oyster tissue		NIST 1577a bovine liver		IAEA CRM H4 animal muscle	
			ICP-MS	Ref.	ICP-MS	Ref.	ICP-MS	Certified
Al	27	0.67	136 ± 9	255 ± 23	1.1 ± 1.7	3.4	12 ± 21	(10)
As	75	0.067	13.2 ± 0.5	13.0 ± 1.2	0.079 ± 0.010	0.048 ± 0.008	< 0.067	0.006
Ba	137	0.033	3.7 ± 0.6	5.18	0.09 ± 0.06	nd	0.13 ± 0.10	nd
Ca	43	36	1600 ± 150	1400 ± 120	143 ± 9	121 ± 5	187 ± 10	188
Cd	111	0.17	3.4 ± 0.1	3.43 ± 0.16	0.39 ± 0.01	0.455	< 0.17	0.0049
Ce	140	0.002	0.39 ± 0.04	0.42	0.027 ± 0.009	nd	0.009 ± 0.008	0.0023
Cr	53	0.33	0.9 ± 0.4	0.65 ± 0.08	< 0.33	1.0	< 0.33	0.009
Cs	133	0.002	0.027 ± 0.001	0.0405	0.014 ± 0.001	nd	0.14 ± 0.01	0.12
Cu	65	0.17	64 ± 2	63 ± 2	154 ± 5	149 ± 14	3.80 ± 0.60	3.65
Fe	57	4	188 ± 10	195 ± 11	188 ± 7	155 ± 17	48 ± 6	49
La	139	0.002	0.297 ± 0.022	0.37	0.013 ± 0.001	nd	0.012 ± 0.001	nd
Li	7	0.13	0.293 ± 0.018	0.323	0.21 ± 0.04	nd	< 0.13	nd
Mg	25	0.067	1330 ± 30	1330 ± 100	618 ± 45	612 ± 36	1150 ± 121	1050
Mn	55	0.013	17.6 ± 0.8	17.0 ± 1.2	10.2 ± 0.4	9.9 ± 0.4	0.494 ± 0.052	0.466
Mo	95	0.003	0.32 ± 0.17	0.14 ± 0.04	3.8 ± 0.2	3.43	0.050 ± 0.007	0.41
Ni	60	0.067	1.04 ± 0.38	1.01 ± 0.09	0.8 ± 0.4	nd	0.251 ± 0.410	(0.006)
Rb	85	0.003	4.52 ± 0.13	4.5 ± 0.5	13.2 ± 0.7	12.2	21 ± 1	19
Sb	121	0.003	0.08 ± 0.11	0.19 ± 0.20	0.003 ± 0.002	0.031	0.022 ± 0.032	nd
Se	82	0.2	2.6 ± 2.3	2.08 ± 0.20	0.82 ± 0.04	0.78 ± 0.20	0.46 ± 0.08	0.28
Sr	88	0.002	18.0 ± 0.3	10.1 ± 0.7	0.152 ± 0.006	nd	0.076 ± 0.007	nd
Tl	205	0.003	0.007 ± 0.001	< 0.005	0.004 ± 0.001	nd	< 0.003	(< 0.005)
V	51	0.067	2.4 ± 0.1	2.7 ± 0.2	0.146 ± 0.026	0.097	< 0.067	(0.003)
Y	89	0.002	0.432 ± 0.023	nd	0.011 ± 0.013	nd	0.006 ± 0.008	nd
Zn	66	0.13	863 ± 34	854 ± 24	119 ± 8	122 ± 4	79 ± 6	86

[a] All samples air-dried at 55°C overnight prior to dissolution with HNO_3/H_2O_2. ICP–MS data (mean and standard deviation of four replicates) from Friel et al. (1990). LOQ = limit of quantitation; NIST = National Institute of Standards and Technology, USA; IAEA CRM = International Atomic Energy Authority Certified Reference Material. Reference values from Gladney et al. (1987); compilation. Certified values from originator, values in parentheses are uncertified and provided for reference only; nd = not determined.

(2) 250°C hotplate sited in a dedicated $HF/HClO_4$ fume cupboard equipped with fume-scrubbing and wash-down facilities.
(3) 30 ml disposable plastic weighing boats.
(4) 100 ml PTFE beakers.
(5) 50 ml Class A Pyrex volumetric flasks.
(6) Liquid dispensers for acids.
(7) New 60 ml polypropylene bottles.

Reagents
(1) Deionised water (18–24 Mohm cm).
(2) Concentrated (68–71% m/m) Aristar nitric acid (16 M HNO_3).
(3) Concentrated (48% m/m) Aristar hydrofluoric acid (29 M HF).
(4) Concentrated (70% m/m) Aristar perchloric acid (12 M $HClO_4$).
(5) Aristar 5 M nitric acid (dilute 320 ml concentrated (68–71% m/m) HNO_3 to 11 with deionised water).
(6) Aristar nitric acid, 1 M (dilute 65 ml concentrated (68–71% m/m) HNO_3 to 11 with deionised water).

Procedure
(1) Dry samples overnight at 105°C (note that labile Hg, As, Se will be volatilised at this temperature).
(2) Weigh 0.500 g into a weighing boat, and transfer both to a microwave digestion vessel.
(3) Slowly add 8.0 ml concentrated HNO_3, 4.0 ml HF and 2.0 ml $HClO_4$, washing any remnant powder from the surface of the weighing boat and the sides of the digestion vessel with the acid. Discard the weighing boat and allow any effervescence to cease (the effectiveness of the acid attack will be improved if the sample is allowed to stand for several hours in the acid mixture before microwave heating is initiated).
(4) Locate the relief valve on top of the vessel, cap, and screw finger tight. Using the capping station, apply to torque of 16 Nm to the cap. Place a total of four vessels (including samples), equally spaced, in the sample turntable.
(5) Set the exhaust fan to maximum, heat for 1 min at 100% power, followed by 60 min at 50% power.
(6) Transfer the vessels and turntable to a cold water bath and allow to return to room temperature. Vent by hand any remaining vapours into a fume cupboard, and untighten the caps with the capping station.
(7) Carefully remove caps and transfer the vessel contents to PTFE beakers, washing residual liquid from cap and vessel interiors with deionised water.
(8) Evaporate on a hotplate at 200°C to incipient dryness. To ensure complete removal of HF, add two further 2.0 ml aliquots of $HClO_4$, following each addition by further evaporation to a crystalline paste.
(9) Add 10.0 ml 5 M HNO_3 and warm gently, until a clear solution results.
(10) Allow solution to cool, dilute to 50 ml in a volumetric flask (producing a 1 M HNO_3 solution) and transfer immediately to a new polypropylene bottle for storage.
(11) Dilute a 1.0 ml aliquot to 5.0 ml with 1 M HNO_3, immediately prior to analysis by ICP-MS.

The evaporation stages are included to ensure the complete removal of HF, and to ensure the dissolution of potentially insoluble fluorides. Larger batches of samples can be prepared using the procedure but require longer heating times in the microwave. Using the above acid volumes, batches of 12 samples have been successfully digested using 8 min at 100% power, followed by 180 min at 75% power.

Totland et al. (1991) reported results for geological SRMs prepared by the above microwave method, which are comparable to $HF/HClO_4$ open digestions. These authors were additionally able to obtain quantitative data for Cr in an ultrabasic rock using a modified acid mixture (8 ml HF, 4 ml HNO_3, 2 ml $HClO_4$). Microwave methods have the advantages of more rapid (unattended) dissolution and reduced reagent consumption, but do not always allow the determination of Cr, Hf or Zr. Difficulties with incomplete digestion of refractory phases may be circumvented, however, by using higher pressure digestion vessels (cf. Nölter et al. 1990), such as Parr microwave bombs (Figure 7.5). The effectiveness of microwave procedures for the retention of volatile elements has not yet been adequately tested.

7.3.4.3 *Environmental samples*   Closed-vessel microwave digestion of soils and sediments using nitric and hydrochloric acids in combination with H_2O_2 may be used to rapidly (15 min) partially extract environmentally sensitive elements (Kammin and Brandt, 1989a, b) in proportions which are compatible with established environmental protection guidelines (e.g. US EPA method 3050, normally requiring 1–2 h). Good recovery of mercury, for example, can be achieved by such procedures (van Delft and Vos, 1988; Millward and Kluckner, 1989). These methods, however, retain HCl in the final sample matrix, preventing the determination of As and V at low concentrations by ICP–MS, and compromising the LOQs for some other trace elements. Protocols involving the use of nitric acid alone are currently being developed (Anon, 1991).

7.3.4.4 *Metals*   Microwave methods developed for metal digestion (Matthes, 1988) generally retain HCl in their final matrices. Clearly, new microwave digestion schemes need to be developed in conjunction with ICP–MS to better combine the inherent strengths of the two techniques.

7.4 Separation and pre-concentration methods

In many cases, the exceptionally low detection limits of ICP–MS are not translated into significantly improved limits of quantitation for samples, compared with other analytical techniques. This is primarily due to the limited tolerance of ICP–MS to high levels of total dissolved solids, and to matrix effects caused by the presence of elevated concentrations of a single, or a small number of matrix elements. One approach to solving this problem is to remove a suite of analytes from their major-element matrix. Separation not only removes possible matrix effects but, more importantly, also enables the pre-concentration of solutions, commonly by several orders of magnitude, producing corresponding improvements in the LOQ. Such an approach, however, is only viable for a multi-element technique such as ICP–MS if the separation can be made as specific as possible for the major elements. Highly selective separation procedures for only one, two or three elements are far better suited to much cheaper single-element techniques such as GF–AAS.

Separation and pre-concentration procedures fall into three main categories: (1) solvent extraction, (2) ion exchange chromatography and (3) co-precipitation/adsorption. In addition, classical fire assay methods may be used to separate the platinum group elements (PGEs). Most solvent extraction and ion-exchange techniques were originally developed for AAS, and many are too 'element specific' to be of general application to ICP–MS. Co-precipitation methods have an even longer history, forming the basis of classical wet chemistry, but again they were generally designed to be highly focused in nature. Clearly, therefore, much work needs to be done to modify existing procedures or to develop new methods which better suit the multi-element capabilities of ICP–MS. The robustness of ICP–MS, however, may in fact extend the application of less efficient separations, since in many cases it is no longer necessary to totally remove matrix elements from sample solutions; a significant reduction in TDS or the partial removal of an interfering element may lower the LOQ sufficiently for the trace elements sought.

The separation and pre-concentration of trace metals from seawater using silica-immobilised 8-hydroxyquinoline (oxine) was an early application of such methods to ICP–MS (McLaren et al., 1985; Beauchemin et al., 1988c), and similar techniques were used by Hall et al. (1987) for the determination of low levels of W and Mo in geological materials. Work on separation procedures is continuing in many laboratories, and only a few general methods are described here. It is likely, however, that new techniques, particularly those employing commercial high performance ion chromatographs, will be developed in the near future.

7.4.1 Rare earth elements

The rare earth elements (REEs) were one of the first element groups to be successfully determined in a range of geological materials by ICP–MS (Date and Gray, 1985; Doherty and Vander Voet, 1985). The group is well suited to determination by ICP–MS, consisting of 15 (14 naturally occurring) elements with atomic numbers between 57 (La) and 71 (Lu) and ranging in mass from ^{136}Ce to ^{176}Lu and ^{176}Yb. Concentrations in geological materials typically range from $> 50\,\mu g\,g^{-1}$ for La and the lighter REEs to $< 1\,\mu g\,g^{-1}$ for Lu. The determination of the REEs by most other techniques, such as

AAS, ICP–AES and X-ray fluorescence (XRF) is hindered by large numbers of spectral interferences which necessitate the separation of the group from their matrix. In contrast, the REEs lie in an area of the mass spectrum characterised by minimal interferences and greatest sensitivity for ICP–MS. All of the REEs have at least one isotope free from isobaric overlap, and consequently the entire group is easily determined in most geological samples (Date and Hutchison, 1987; Lichte et al., 1987; Jarvis, 1988, 1989a, b).

In cases such as pure limestones, ultrabasic rocks, some mineral separates, and many botanical, biological and environmental materials, levels of the REEs are too low to determine the low abundance members of the group in normal sample solutions. The determination of such ultratrace levels may be readily accomplished by the separation and pre-concentration of the group using a simple cation-exchange separation (Jarvis, 1988, 1989a, b), based on that previously developed for ICP–AES analysis (Strelow and Jackson, 1974; Walsh et al., 1981; Jarvis and Jarvis, 1985) and described by Jarvis and Jarvis (1985). A similar approach was adopted by Hirata et al. (1988) in their ICP–MS study.

The digestion procedures used for the determination of the REE require some comments. Comparison of ICP–MS data for solutions prepared by standard $HF/HClO_4/HNO_3$ open digestions with those obtained by lithium metaborate fusions, indicate incomplete recovery of Zr in some cases (e.g. Totland et al., 1991), which may be associated with low recovery of the heavy REEs (HREEs, Gd–Lu; Sholkovitz, 1990). This observation suggests the incomplete digestion of refractory zircons (and in some cases sphene and garnet), which are known to be highly enriched in the REEs, particularly the HREEs. It is essential, therefore, that the digestion procedure should ensure the complete dissolution of these phases.

A number of possibilities exist. Walsh et al. (1981) and Jarvis and Jarvis (1985) used open vessel digestions followed by 'minifusions' of visible insoluble residues with an alkali flux (KHF_2 is most effective for large amounts of zircon, but $LiBO_2$ or NaOH are equally suitable in most cases). Date and Hutchison (1987) used a high-pressure closed vessel digestion to dissolve their insoluble residues. In the light of more recent evidence (Sholkovitz, 1990), which suggests that even minute (essentially invisible) residues may lead to biased HREE determinations, it is considered that routine filtration and fusion or high pressure digestion are an essential precaution (only some rock types will contain insoluble phases), if open vessel digestions are to be used for REE determinations. Alternatively, Watkins and Nolan (1990, 1991) successfully used a lithium metaborate fusion (0.500 g samples plus 1.500 g $LiBO_2$) in conjunction with a standard cation-exchange procedure to accurately determine the REEs in a wide range of rock types. In all cases a final matrix of 100 ml 0.5 M HCl is required for the cation-exchange procedure (contrary to the opinion of some authors, e.g. Thompson and Walsh (1989), no problem should be encountered in dissolving $LiBO_2$ in hydrochloric acid). Our method (Figure 7.7) is adapted from that of Jarvis and Jarvis (1985).

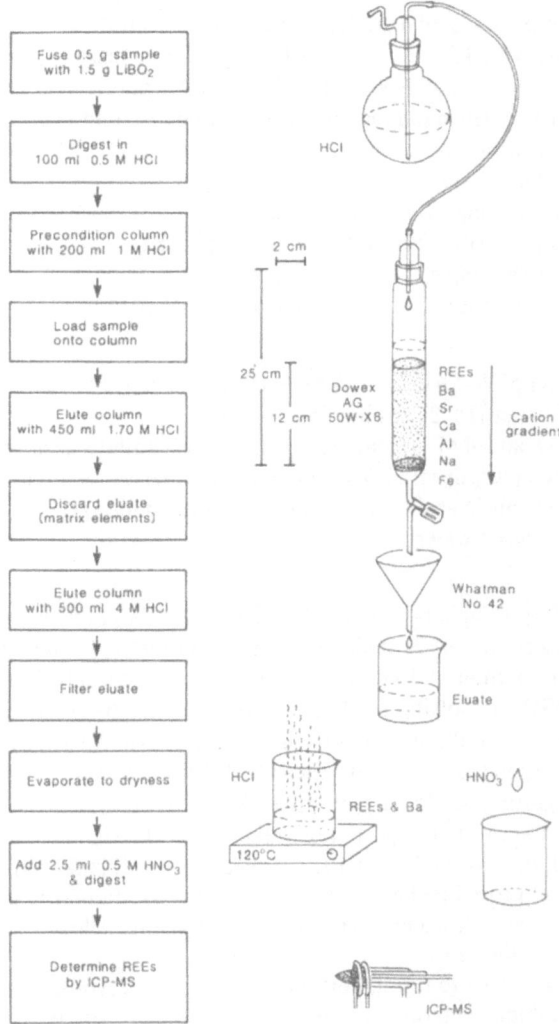

Figure 7.7 Schematic diagram of the apparatus and procedure used for the separation of the rare earth elements by cation-exchange chromatography. Fusion of samples with $LiBO_2$ is preferred for silicate rocks and sediments containing refractory minerals, but organic samples and many other materials are better suited to open vessel acid decomposition. The ion-exchange separation may be applied to any sample prepared in a 0.5 M HCl matrix.

Method
Equipment
(1) 500 ml round bottomed Pyrex flasks with Quickfit necks, corresponding Quickfit Drechsel heads, and PVC tubing to connect to column adapter heads.
(2) Chromatographic glass columns, 25 cm long, 2 cm internal diameter with

glass porosity O sintered discs, PTFE Rotaflo stopcocks, Quickfit necks and corresponding tubing adapter heads with central glass tube.
(3) Pyrex glass Quickfit powder funnels to fit column necks.
(4) Pyrex filter funnels with Whatman No. 42 filter papers.
(5) Plastic or wooden stand and support assembly to hold glassware.
(6) 600 ml low form Pyrex beakers.
(7) 50 ml tall form Pyrex beakers.
(8) 1 l glass measuring cylinder for acids.
(9) 250°C hotplate sited in a suitable fume cupboard.
(10) 5 ml tall glass sample vials with snap tops.
(11) 2 l Pyrex beaker for storage of homogenised resin.

Reagents
(1) Bio Rad AG 50W-X8 (200–400 mesh) cation-exchange resin.
(2) Deionised water (18–24 Mohm cm).
(3) 1 M, 1.70 M and 4 M Analar hydrochloric acid (dilute accurately titrated concentrated (36% m/m) HCl with deionised water as required).
(4) 0.5 M Analar nitric acid (dilute 32 ml concentrated (68–71% m/m) HNO_3 to 1 l with deionised water).

Procedure
(1) Digest 0.500 g sample by $HF/HClO_4/HNO_3$ (Analargrade acids) open acid digestion (section 7.3.1.2), making up final solution in 100 ml Analar 0.5 M HCl. Filter solution and perform minifusion on any insoluble residue with Aldrich $LiBO_2$, dissolving bead in an aliquot of the open digestion solution. Return the dissolved minifusion to the main solution. Alternatively, prepare sample by fusion (section 7.3.3) with Aldrich $LiBO_2$ (0.500 g sample + 1.500 g $LiBO_2$), digesting bead in 100 ml 0.5 M Analar HCl.
(2) Homogenise resin in a large Pyrex beaker with 1 M HCl (new resin should be washed in 1 M HCl and then subjected to a blank run of the full ion-exchange procedure before use). Load resin into columns to give a settled height of 12 cm, discarding excess acid after loading. Allow 1 cm of acid to remain above the resin.
(3) Add 20 ml 1 M HCl to round bottomed flask acid reservoirs, locate Drechsel heads and tubing adaptors, and rinse through columns with HCl (again retain 1 cm of acid above the resin to avoid disturbing the resin bed) into 600 ml beakers, discarding eluates.
(4) Remove tubing adapters and locate the powder funnels into the columns. Carefully pour samples into funnels, washing sample vessel sides with deionised water to ensure complete transfer of samples. Load samples onto resin, discarding eluates, which contains only anionic species. Rinse funnels with deionised water and remove.
(5) Relocate tubing adapters and rinse columns with 450 ml 1.70 M HCl, discarding eluates, which contain major and most trace elements.
(6) Place filter funnels below column outlets (to eliminate traces of resin from the final solutions), and rinse columns with 500 ml 4 M HCl into washed 600 ml beakers, these eluates contain the REEs, Y, Sc, Ba and some Sr, Zr, Hf.
(7) Evaporate eluates on hotplate at 120°C until approximately 10 ml remains, and transfer to 50 ml beakers. Evaporate to dryness and allow to cool.

(8) Add 2.5 ml 0.5 M HNO_3 and digest residues with gentle warming. Transfer solutions to glass sample vials for storage. Samples may be analysed by ICP–MS without further dilution.

(9) Following elution of the REEs, wash the resin from the columns into a large Pyrex beaker, homogenise by stirring with a PTFE rod, and store in 1 M HCl.

A fully automated version of a scaled-down cation-exchange procedure for the REEs is described by Govindaraju and Mevelle (1987), who used conveyor techniques and laboratory robots to performs sample digestions, REE separations and ICP–AES analyses. Separations are performed manually in our laboratory, in batches of nine samples plus one blank. The blank column is used to assess possible contamination from the previously homogenised batch of resin, and generally only yields measurable quantities of REEs (commonly La, Ce) after extended non-use of the resin or following a batch of samples containing very high levels of the REEs.

The cation-exchange procedure provides quantitative recovery of the REEs and eliminates the bulk of the major and trace elements. This enables a sample to be analysed by ICP–MS with only a 5-fold dilution factor (compared with the 500-fold typical of most acid digestion procedures), producing limits of quantitation of around $2 \, ng \, g^{-1}$ for most REEs (Jarvis, 1988). The small amount of other elements which are eluted with the REEs do not generally pose any analytical problems, and good agreement with reference values was obtained by Jarvis (1988) for separated samples of a range of geological standard reference materials (Table 7.8).

Unfortunately, barium causes some difficulties in Ba-rich samples (e.g. barite) because Ba is eluted with the REEs. During ICP–MS analysis BaO^+ and $BaOH^+$ formation (typically 1–2% of the parent ion) produce significant isobaric interferences ($^{134}Ba^{16}O^{1}H^+$, $^{135}Ba^{16}O^+$, $^{136}Ba^{16}O^{1}H^+$, $^{137}Ba^{16}O^+$) on ^{151}Eu and ^{153}Eu, one of the low abundance REEs which may display anomalous behaviour in natural systems. Where Ba levels are not too high, the Ba interference can be corrected by running a Ba interference standard, but in some cases unacceptably high correction factors (> 90% of the value) result from this approach. These problems were addressed by Jarvis et al. (1989), who avoided the spectral interferences on Eu by reconfiguring the ICP–MS to enable the sensitive measurement of Eu^{2+} ions. This doubly charged ion mode of measurement provided more accurate data for Eu than that produced by interference correction.

Several other methods of ion-exchange separation of the REEs have been published, including those of Bolton et al. (1982) and Crock and Lichte (1982) who used a two-stage procedure involving the use of both cationic and anionic resins. Such a procedure might be used to more effectively separate Ba from the REEs. Other possible approaches to REE separation include organic solvent extraction (Weiss et al., 1990), although this has not yet been coupled with ICP–MS analysis.

Table 7.8 Comparison of ICP–MS and reference values (ng g^{-1}) for the REEs and Y in some geological SRMs after separation using cation-exchange chromatography[a]

Element	Mass (m/z)	LOQ	NIST 70a potash feldspar		ANRT UB-N serpentine		USGS PCC-1 peridotite	
			ICP-MS	Ref. 1	ICP-MS	Ref. 2	IPC-MS	Ref. 2
La	139	1.25	161 ± 19	420	750 ± 240	500	175 ± 67	52
Ce	140	1.7	402 ± 30	590	1140 ± 440	1000	364 ± 9	100
Pr	141	1.5	59 ± 3	<130	135 ± 3	nd	48 ± 1	13
Nd	146	3.5	229 ± 5	250	644 ± 125	600	135 ± 4	42
Sm	147	2.5	74 ± <1	70	219 ± 33	200	32 ± 9	6.6
Eu	153	1	397 ± 10	520	80 ± 18	80	22 ± 7	1.8
Gd	157	1.5	81 ± 3	100	283 ± 40	300	20 ± 3	14
Tb	159	0.5	20 ± <1	nd	60 ± 8	60	3 ± <1	1.5
Dy	163	2	117 ± 1	140	377 ± 49	380	15 ± 4	10
Ho	165	0.5	27 ± <1	30	86 ± 9	nd	3 ± <1	2.5
Er	166	1	81 ± 3	90	255 ± 28	280	12 ± 3	12
Tm	169	0.2	14 ± 1	nd	42 ± 6	nd	3 ± 1	2.7
Yb	172	1	81 ± 3	90	249 ± 30	250	20 ± <1	24
Lu	175	1	12 ± <1	20	38 ± 39	40	4 ± 1	5.7
Y	89	2	630 ± 3	840	2190 ± 360	2500	105 ± 18	100

[a] ICP–MS data (mean and standard deviation of two replicates) from Jarvis (1988). LOQ = limit of quantitation; NIST = National Institute of Standards and Technology, USA; ANRT = Association Nationale de la Recherche Technique, France; USGS = US Geological Survey. Reference values after: (1) Jarvis and Jarvis (1988), ICP–AES with REE separation; (2) Govindaraju (1989), compilation. nd = not determined.

7.4.2 *Precious metals*

The precious metals, gold, silver and the platinum group elements (PGEs), fall
into two groups, each containing four elements: atomic numbers 44–47
(Ru, Rh, Pd, Ag) and 76–79 (Os, Ir, Pt, Au). Like the REEs, the high masses of
the precious metals (^{96}Ru–^{110}Pd and ^{184}Os–^{198}Pt) make them well suited
to determination by ICP–MS. Unlike the REEs, however, the precious metals
occur in most natural materials at very low concentrations (10–100 ng g^{-1}
Ag; 1–10 ng g^{-1} Au, Pd, Pt, Ru; < 1 ng g^{-1} Ir, Os, Rh) and are very hetero-
geneously distributed, commonly being concentrated in discrete particles or
phases. Such low abundances and, more importantly, sample inhomogeneity,
require well-designed sampling strategies and sample preparation protocols
which incorporate sufficient mixing and grinding to homogenise the material,
to ensure that a representative aliquot is analysed. Cross-contamination
provides a constant hazard during such work.

In classical fire assay remains the commonest sample preparation procedure
for precious metal analysis (Bacon *et al.*, 1989; van Loon and Barefoot, 1989,
1991). One basic reason for the continued use of this technique is the relatively
large sample size which can be treated, typically one assay ton (29.1667 g),
which to some extent compensates for nugget effects during sample preparation.
Smaller sample sizes are generally considered to be inadequate for assessing
low-grade ores. In classical fire assay (e.g. Beamish, 1966), the sample is fused
with a flux of litharge (PbO), silica (SiO$_2$), sodium carbonate (Na$_2$CO$_3$), lime
(CaO), fused borax (Na$_2$B$_4$O$_7$), potassium nitrate (KNO$_3$) and wheat flour,
and the precious metals are concentrated and collected in a lead button. This
button is cupelled (generally with silver) in a furnace to give a noble metal
prill which is dissolved for analysis (e.g. by flame atomic absorption
spectrometry, FAAS, ICP–AES or graphite furnace atomic absorption
spectrometry, GFAAS). The method enables the accurate determination of
Au, Pd, Pt and Rh (although low recovery of Rh is commonly encountered),
but more complex procedures are required to quantify Ir, Ru and Os which
are partially lost by volatilisation or cupel absorption during cupellation of
the lead button.

In the neo-classical nickel sulphide fire assay technique (Robert *et al.*, 1971;
Hoffman *et al.*, 1978; Haines and Robert, 1982), a flux of fused borax, sodium
carbonate, nickel powder and flowers of sulphur is used, the cooled button
being ground and dissolved to leave insoluble noble metals. These may be
determined directly after filtration (e.g. by instrumental neutron activation
analysis, INAA) or dissolved prior to analysis (e.g. by GFAAS). Nickel
sulphide fire assay is gaining in popularity because it provides efficient
collection for the complete group of noble metals, although the procedure
is less efficient for gold collection and special techniques are necessary to
avoid the loss of volatile Os compounds (particularly OsO$_4$) at the button
dissolution stage. Fire assay procedures vary considerably from laboratory
to laboratory and their reliability depends to a large extent on the skill of

the analyst, the exact composition of the sample and the fusion flux mixture used. Achieving efficient recovery for all of the noble metals generally requires extensive experimentation with flux compositions and assay conditions.

The use of a nickel sulphide fire assay as a sample preparation technique for ICP–MS has been described by Date *et al.* (1987b), Gregoire (1988) and Jackson *et al.* (1990), and yields determination limits which are similar to or better than those achieved using INAA or GFAAS (Denoyer *et al.*, 1989). The ICP–MS technique has the advantage of rapidity of analysis, and avoids the need for sophisticated hardware (i.e. the nuclear reactor of INAA) and complex solution chemistry (necessary for GFAAS). Consequently, ICP–MS is now used routinely by a number of commercial assay laboratories. As already emphasised, however, the success of fire assay depends on a large number of factors which are sample dependent. Reliable results will be obtained only if all possible precautions are taken, such as re-fusion of the slags, duplicate (and preferably replicate) analyses, and the use of blank buttons. It is suggested, therefore, that the reader should consult more detailed literature (e.g. Robert *et al.*, 1971; Haines and Robert, 1982; Moloughney, 1986; Robert, 1987; Asif and Parry, 1989; Van Loon and Barefoot, 1989, 1991) before attempting to use fire assay techniques with ICP–MS; no universal recipe is possible. The following procedure is based on Robert *et al.* (1971), incorporating modifications from Date *et al.* (1987b), Gregoire (1988) and Jackson *et al.* (1990). Cleanliness is essential to avoid cross-contamination and all glassware should be boiled in aqua regia and rinsed thoroughly with deionised water between samples.

Method
Equipment
 (1) 1200°C muffle furnace.
 (2) Large fireclay crucibles, Battersea round size J or K.
 (3) Conical cast iron or steel mould.
 (4) Steel hydraulic press.
 (5) Tungsten carbide Tema mill.
 (6) Tall-form 1 l Pyrex beakers with watch-glass covers.
 (7) 250°C hotplate sited in a fume cupboard.
 (8) Micropore filter system with 0.45 μm cellulose nitrate membrane filter, pre-washed with 1 M HCl.
 (9) 30 ml reflux test tubes with condensers.
 (10) 100 ml class A Pyrex volumetric flasks.
 (11) New 125 ml polypropylene bottles.

Reagents
 (1) Carbonyl nickel powder 5 μm, 99.8% Ni.
 (2) Precipitated sulphur powder.
 (3) Analar anhydrous sodium carbonate (Na_2CO_3).
 (4) Analar fused borax (sodium tetraborate, $Na_2B_4O_7$).
 (5) Silica, SiO_2 floated powder 62 μm.

(6) Deionised water (18–24 Mohm cm).
(7) Concentrated (36% m/m) Analar hydrochloric acid (12 M HCl).
(8) Analar HCl, 1 M (dilute 85 ml concentrated (36% m/m) HCl to 1 l with deionised water).
(9) Concentrated (68–71% m/m) Analar nitric acid (16 M HNO_3).
(10) Tellurium solution, 2000 $\mu g\,ml^{-1}$ Te in 1 M HCl (dissolve Specpure 99.999% Te metal in aqua regia, followed by two evaporations with 12 M HCl, make up to volume with 1 M HCl).
(11) Stannous chloride solution (dissolve 100 g $SnCl_2.2H_2O$ in 150 ml 12 M HCl and dilute to 500 ml with deionised water; solution must be prepared weekly).

Procedure
(1) Dry samples and flux components for 48 h at 60°C.
(2) Weigh 30.0 g sample, 10.0 g Ni, 6.0 g S, 20.0 g Na_2CO_3, 40.0 g $Na_2B_4O_7$, 6.0 g SiO_2 (10.0 g SiO_2 for ultrabasic rocks) into a fireclay crucible, and mix thoroughly with a spatula.
(3) Fuse the mixture in a preheated furnace at 1050°C for 1.25 h. Remove crucible and pour contents into a cast iron mould.
(4) After cooling for 30 min, remove the charge and separate the NiS button from the slag. Crush the slag in a hydraulic press, place the fragments in a Tema mill and grind for 30 s. Return the ground slag to the original crucible, add 2.5 g Ni, 1.5 g S. Repeat the fusion process and separate the second NiS button.
(5) Weigh both buttons, crush in a hydraulic press and grind to a powder. Weigh powder to allow for crushing loss and transfer to a 1 l Pyrex beaker.
(6) Add 600 ml 12 M HCl and cover beaker with a watch-glass. Heat for 2–3 h on a hotplate set at 200°C until dissolution is complete. A black noble metal-bearing precipitate may be visible.
(7) Allow solution to cool until warm, add 3.5 ml Te solution. Dilute with deionised water to approximately 1 l and add 12 ml stannous chloride. Bring the solution to the boil over 30 min to coagulate the Te precipitate. Allow to cool.
(8) Filter solution, discarding the filtrate. Wash filter and precipitate thoroughly with 1 M HCl to remove traces of Ni.
(9) Place filter in a reflux tube, add 5 ml nitric acid, and attach condenser. Allow filter to dissolve (no heat should be applied if Os is sought). Add 5 ml 12 M HCl through the top of the condenser and warm at 95°C for 20–30 min, until precipitate is dissolved.
(10) Allow to cool, wash condenser with deionised water and transfer solution to a volumetric flask, making to 100 ml with deionised water. Store solution in a polypropylene bottle. This solution is analysed directly by ICP–MS.

If stable isotope spike solutions are to be used for calibration (cf. Date *et al.*, 1987b), these may be added to the charge prior to the initial fusion. It should be emphasised, however, that the very different nature of the spike compared with noble metal-bearing minerals may lead to different behaviours during fusion (Gregoire, 1988), and may not always compensate for elemental losses.

The fire assay procedure described above is applicable to a wide range of sample types, but a number of problems may arise; chromite-rich samples may be difficult to fuse completely, and are more effectively attacked if lithium tetraborate is used in place of borax (see Robert, 1987 for details). Samples containing > 1% Ni or S must have the proportions of those elements adjusted in the fusion mixture to prevent the formation of insoluble NiS_2, while samples rich in Cu and/or Zn introduce a number of analytical problems which must also be addressed (e.g. Jackson et al., 1990).

Five PGEs and gold were determined in two standard reference materials by Date et al. (1987b) using a nickel sulphide fire assay ICP–MS procedure and external calibration. Reasonable agreement with reference value for Au, Ir, Pd, Pt, Rh, Ru were obtained for South African Merensky Reef standard reference platinum ore SARM-7. Fair agreement with INAA data for Ir, Pd, Pt, Ru has been reported (Gregoire, 1988) in a number of samples prepared by nickel sulphide fire assay and analysed by both solution nebulisation and electrothermal vaporisation (ETV) ICP–MS. The ETV–ICP–MS technique yielded lower determination limits than solution nebulisation but required the use of isotope dilution to obtain reproducible data. More recently, all of the noble metals except Ag were determined in three SRMs (Table 7.9) by Jackson et al. (1990), who incorporated a tellurium co-precipitation step during dissolution of the nickel sulphide button. Despite this additional precaution, recoveries of only around 90% are reported for most elements, which may be attributed in part to the lack of a second 'cleaning charge' in their procedure. Isotope dilution techniques might improve the accuracy of the technique but only if isotopic equilibration can be assured. It should be noted that although Ag was not determined because of limited solubility in the final $HCl–HNO_3$ matrix, it is recovered quantitatively by the fire assay and could be determined in a modified procedure.

The low detection limits and multi-element capability of nickel sulphide fire assay combined with ICP–MS represent a major advance which should allow geologists to routinely study precious metal distributions in unmineralised materials. For the future, laser ablation ICP–MS of the undissolved sulphide button provides an exciting alternative approach for the direct determination of the PGEs, which should obviate the recovery problems introduced by the sample dissolution procedure. Silver determinations should be readily obtained by this technique.

Other common approaches to the separation and pre-concentration of the noble metals are ion-exchange chromatography and tellurium co-precipitation. The feasibility of anion-exchange separation has been demonstrated recently by Chung and Barnes (1988) who determined Ag, Au, Pd, Pt with reasonable accuracy in SARM-7 by ICP–AES, following fusion with 4:1 lithium metaboratelithium tetraborate, chelation of the precious metals by a poly-dithiocarbamate resin, decomposition of the resin with hydrogen peroxide, and digestion of the residue in 1 M HNO_3. Sen Gupta (1989) adopted the

Table 7.9 Comparison of ICP–MS and reference values (ng g^{-1}) for the noble metals in some SRMs after separation using nickel sulphide fire assay[a]

Element	Mass (m/z)	LOQ	SACCRM SARM-7 platinum ore		CANMET PTC-1 sulphide concentrate		CANMET SU-1a nickel-copper-cobalt ore	
			IPC–MS	Ref. 1	ICP–MS	Ref. 2	ICP–MS	Ref. 2
Ru	101	1.9	397 ± 56	430 ± 57	442 ± 46	nd	44 ± 4	nd
Rh	103	0.47	212 ± 18	240 ± 13	606 ± 30	620 ± 70	56 ± 4	nd
Pd	105	5.6	1350 ± 140	1530 ± 30	11400 ± 400	12700 ± 700	327 ± 27	370 ± 30
Os	189	0.87	53 ± 10	63 ± 7	303 ± 84	nd	nd	nd
Ir	193	0.33	71 ± 7	74 ± 12	165 ± 9	nd	24 ± 2	nd
Pt	195	4.9	3400 ± 460	3740 ± 50	2580 ± 290	3000 ± 200	367 ± 40	410 ± 60
Au	197	29	253 ± 50	310 ± 15	466 ± 124	650 ± 100	126 ± 27	nd

[a]ICP–MS data (mean and standard deviation of 5–29 replicates) from Jackson et al. (1990). LOQ = limit of quantitation based on routine reagent blanks and a 15 g sample size (better values are attainable); SACCRM = South African Committee for Certified Reference Materials; CANMET = Canadian Centre for Mineral and Energy Technology. Reference values after (1) Steele et al. (1975) and (2) Bowman (1990). nd = not determined.

alternative approach of separating matrix elements by absorption onto a cationic resin (Dowex 50W-X8), and successfully determined all of the noble metals excluding Os in a range of SRMs by this method. Acceptable data have also been obtained by tellurium co-precipitation for the same materials (Sen Gupta, 1989; Sen Gupta and Gregoire, 1989). Work on ion-exchange separation procedures for the noble metals and their determination by ICP–MS is currently being undertaken in our laboratory.

7.4.3 *Petrogenetic discriminations: Hf, Nb, Ta, Zr*

The accurate determination of low abundances of Hf, Nb, Ta, Zr in geological materials has been found highly informative in petrogenetic studies (Pearce *et al.*, 1984) and in diamond exploration (Van Wambeke, 1960). Alkali fusions are necessary to ensure the complete dissolution of Hf- and Zr-bearing minerals (e.g. zircon $ZrSiO_4$, baddeleyite ZrO_2) and the limits of quantitation provided by these methods (e.g. $4\,\mu g\,g^{-1}$ for a $LiBO_2$ fusion) are insufficiently low to enable the determination of these elements in some sample types (see Totland *et al.*, 1991). Lower levels of Nb and Ta may be determined successfully in open acid digestions ($0.2\text{–}0.3\,\mu g\,g^{-1}$) but again the quantities present in some materials may be too low to quantify. These limitations have been addressed recently by Hall *et al.* (1990a) and Hall and Pelchat (1990) who described a procedure for the analysis of Hf, Nb, Ta, Zr in rocks, soils, sediments and plants, utilising an analyte separation procedure with cupferron prior to analysis by ICP–MS. This method, as described below, yields quantitation limits of $1.2\,\mu g\,g^{-1}$ for Zr and $0.06\text{–}0.08\,\mu g\,g^{-1}$ for Hf, Nb, Ta.

Method

Equipment
(1) 1200°C muffle furnace.
(2) 30 ml disposable plastic weighing boats, and camel hair brush.
(3) 30 ml Ultra Carbon graphite crucibles.
(4) 100 ml PTFE beakers.
(5) Magnetic stirrer and PTFE-coated stirring bars.
(6) 0.45 µm Millipore HAWP cellulose membrane filters and holders.
(7) 6000 rev min^{-1} centrifuge and 100 ml polypropylene centrifuge tubes.
(8) Water bath.
(9) 25 ml polypropylene volumetric flasks, new 25 ml polypropylene bottles.

Reagents
(1) Deionised water (18–24 Mohm cm).
(2) Spectroflux 100A lithium metaborate.
(3) Concentrated (36% m/m) Analar hydrochloric acid (12 M HCl).
(4) Concentrated (48% m/m) Analar hydrofluoric acid (29 M HF).
(5) 6% m/v cupferron (the ammonium salt of nitrosophenylhydroxylamine) solution at 10°C.
(6) Concentrated (68–71% m/m) Analar nitric acid (16 M HNO_3).

(7) 100 volumes (30% m/v H_2O_2) Analar hydrogen peroxide.

(8) Analar nitric acid, 1 M (dilute 65 ml concentrated (69–71% m/m) HNO_3 to 1 l with deionised water).

Procedure

(1) Dry samples and $LiBO_2$ flux overnight at 105°C.

(2) Weigh 0.400 g into a weighing boat, followed by 2.00 g $LiBO_2$. Mix thoroughly with plastic spatula.

(3) Transfer mixture to a carbon crucible, using a clean camel hair brush to remove any remnant powder from the weighing boat.

(4) Fuse at 1050°C for 20 min in a muffle furnace (allow the furnace to return to operating temperature before timing the fusion). Note: refractory samples (e.g. those rich in zircon or chromite) will benefit from a gentle swirling once or twice during the fusion, and will require up to 45 min to be fully digested.

(5) Add 70 ml deionised water, 10 ml HCl, 3.5 ml HF to each PTFE beaker, add a stirring bar, and place on the magnetic stirrer.

(6) Rapidly pour the melt into the stirring acid, ensuring that no material remains in the crucible (swirling the melt in the crucible during removal from the furnace will facilitate the pouring process). Stir solution until dissolved (30 min to 1 h).

(7) Filter (to remove carbon particles) into a centrifuge tube, and cool in a refrigerator to 10°C.

(8) Add 6 ml of cold cupferron solution and store in a refrigerator at 10°C for 10 min to complete precipitation.

(9) Centrifuge at 2800 rev min^{-1}. Decant off and discard liquid phase.

(10) Add 2 ml 16 M HNO_3 and 1 ml H_2O_2 to residue. Allow reaction to subside and immerse centrifuge tube in a water bath at 90°C. Progressively add 2 ml portions of H_2O_2, mixing well between additions, until a clear solution is obtained (10–12 ml H_2O_2 may be required).

(11) Make to volume with deionised water in a 25 ml volumetric flask, and transfer to a polypropylene bottle for storage.

(12) Dilute a 1.0 ml aliquot to 100 ml (for Zr) and a second 1.0 ml aliquot to 5.0 ml (for Hf, Nb, Ta) with 1 M HNO_3, immediately prior to analysis by ICP–MS.

The concentration of HCl used in the cupferron precipitation step is not critical for full recovery of Hf, Nb, Ta, Zr but the temperature must be kept at or below 10°C to prevent the decomposition of cupferron. The amount of cupferron used is also not critical and is in great excess. The addition of 4–6% HF was found to be necessary to prevent the precipitation of silica which leads to low recoveries of analytes by occlusion. Most matrix elements except Fe are effectively removed by the cupferron separation, although Fe, Mo, V are recovered quantitatively along with some Cu, W and small amounts of Al. A final chloride matrix was avoided to prevent the possible polyatomic interference of $^{56}Fe^{37}Cl^+$ on mono-isotopic $^{93}Nb^+$; the small amount of HCl remaining after centrifuging does not produce any measurable interference.

Table 7.10 Comparison of ICP–MS and reference values ($\mu g \, g^{-1}$) for Hf, Nb, Ta, Zr in some geological SRMs after separation using cupferron[a]

Element	Mass (m/z)	LOQ	USGS MAG-1 marine sediment		USGS RGM-1 rhyolite		CCRMP SY-2 syenite	
			ICP-MS	Ref.	ICP-MS	Ref.	ICP-MS	Ref.
Hf	178	0.06	4.08 ± 0.11	3.7 ± 0.5	6.47 ± 0.15	6.2 ± 0.3	8.18 ± 0.43	7.7
Nb	93	0.06	14.8 ± 1.3	12 ± 2	9.50 ± 0.21	8.9 ± 0.6	28.8 ± 1.5	29
Ta	181	0.08	1.08 ± 0.11	1.11 ± 0.22	0.89 ± 0.02	0.95 ± 0.10	1.78 ± 0.06	2.01
Zr	90	1.2	124 ± 2	126 ± 13	218 ± 3	219 ± 20	278 ± 6	280

[a] ICP–MS data (mean and standard deviation of five replicates) from Hall et al. (1990a). LOQ ± limit of quantitation; USGS = US Geological Survey; CCRMP = Canadian Certified Reference Materials Project. Reference values after Govindaraju (1989), compilation, standard deviations for reference data from Gladney and Roelandts (1988).

Data for rock standard reference materials prepared by the cupferron separation method yield excellent agreement with reference values (Table 7.10) and RSDs is the order of 2% for Zr and 3–5% for Hf, Nb, Ta. These results demonstrate that the method should be applicable to a wide range of petrogenetic and biogeochemical studies.

7.5 Conclusions and overview

There is little doubt that many current methods of sample preparation restrict the enormous power of ICP–MS for rapid multi-element analysis. Much work remains to be done to free the instrumentalist from the shackles of existing preparation chemistry. New dissolution techniques need to be designed which address the limitations of the instrumentation and maximise its potential. Recent advances in closed-vessel microwave digestion are already allowing new approaches to be made to the decomposition of many sample types. The development of separation and pre-concentration procedures will continue, enabling some of the instrumental limitations to be overcome. Methods of solid sample analysis have considerable potential, but its is unlikely that solid sample introduction will ever totally replace solution techniques because of sampling problems associated with the former and the latter's inherent suitability for chemical manipulation.

A major limitation of existing sample preparation schemes is that they are slow, labour intensive and require dedicated highly skilled staff. These limitations are generally considered unimportant in academic laboratories, but they are highly significant to commercial operations where labour costs limit profitability. The technology already exists to solve many of these problems, and there is little doubt that we will see increasing automation in sample preparation during the next decade. Such automation will include the use of laboratory information management systems combined with flexible, programmable robots, continuous flow procedures and intelligent instrumentation including computer-controlled sample digestors. Automation will revolutionise the analytical laboratory, freeing trained staff to concentrate on analytical problem solving and data interpretation rather than performing repetitive mundane tasks, thereby improving rates of sample throughput, and ultimately leading to better analytical reproducibility.

8 Elemental analysis of solutions and applications

8.1 Introduction

In the seven years since the introduction of the first commercial ICP–MS systems, the technique has been put to a wide variety of applications in the fields of geological, environmental, industrial, food, medical and nuclear sciences. Due to the often confidential nature of the work carried out, published data for nuclear and industrial applications are sparse although ICP–MS is widely used in both of these areas. The information given below attempts to summarise the range of applications in each of the main fields of interest and, where possible, to give recommendations for instrument operation and to highlight specific problem areas. The analysis of natural waters is dealt with separately in chapter 9.

8.2 Multi-element determinations

The range and combination of elements which can be analysed by ICP–MS is very broad. Modern instruments provide the possibility for the simultaneous determination of most elements in the periodic table at major, trace and ultratrace levels. However, there are a number of limitations which, in practice, restrict those elements which can be measured simultaneously in a single solution. For example, the precision required for many major elements in geological samples is between 0.5 and 1% RSD, a figure which is not routinely achievable on current ICP–MS instruments. Some volatile trace elements may be lost during sample preparation and many may therefore not be available for determination. Thus in practice, analytical methodology is often best developed for specific elemental groups, taking into account the matrix and method of sample preparation. The preferred isotopes for elemental determination in a wide range of sample types are shown in Table 8.1. After selection of appropriate isotopes, either a scanning or peak hopping mode of measurement should be chosen for analysis using the criteria discussed in chapter 6. This choice will generally depend on the number of isotopes to be measured and their distribution across the mass range. Prior to quantitative determination, samples of unknown composition should be qualitatively scanned to identify major and trace components and to identify possible interferences, particularly from dissolution acids. Final sample dilutions

Table 8.1   Preferred isotopes for elemental analysis

Element	Preferred (m/z)	Abundance (%)	Comments and interferences	Digest	Fusion
Li	7	92.5	Most abundant free isotope	Yes	Yes[a]
Be	9	100	Mono-isotopic	Yes	Yes
B	11	80	Most abundant free isotope	No	Yes[a,b]
Na	23	100	Mono-isotopic	Yes	Yes[c]
Mg	24	79	Most abundant free isotope	Yes	Yes
Al	27	100	Mono-isotopic	Yes	Yes
Si	29	4.7	NH	No	Yes
P	31	100	NO, NOH	Yes	Yes
S	33	0.75	O_2H	?	?
Cl	35	75.8	Poor sensitivity	No	Yes
K	39	93.3	Good resolution needed	Yes	Yes[d]
Ca	44	2.08	N_2O	Yes	Yes
Sc	45	100	CaH	Yes	Yes
Ti	47/49	7.3/5.5	SOH, CaH	Yes	Yes
V	51	99.7	ClO	Yes[e]	Yes
Cr	52/53	83.8/9.5	ClOH	?	Yes
Mn	55	100	Mono-isotopic	Yes	Yes
Fe	56/57	91.7/2.2	ArO, ArOH	Yes	Yes
Co	59	100	Mono-isotopic	Yes	Yes
Ni	60	26.1	Most abundant free isotope	Yes	Yes
Cu	63/65	69.2/30.8	Average isotopes	Yes	Yes
Zn	66/68	27.9/18.8	S_2	Yes[f]	Yes
Ga	71	39.9	^{69}Ga coincident with ^{138}Ba^{2+}	Yes	Yes
Ge	72/73	27.4/7.8	FeO, FeOH, FeO	Yes	Yes
As	75	100	ArCl	Yes[e]	?
Se	77	7.6	ArCl	?	?
Br	79/81	50.7/49.3	Average	?	Yes
Rb	85	72.2	Most abundant free isotope	Yes	Yes
Sr	88	82.6	Most abundant free isotope	Yes	Yes
Y	89	100	Mono-isotopic	Yes	Yes
Zr	90	51.4	Most abundant free isotope	No	Yes
Nb	93	100	Mono-isotopic	Yes	Yes
Mo	95	15.9	Most abundant free isotope	?	?
Ru	99/101	12.7/17.0	Average	Yes	Yes
Rh	103	100	Mono-isotopic	Yes	Yes
Pd	105	22.3	Most abundant free isotope	Yes	Yes
Ag	107/109	51.8/48.2	Average; caution chemistry!	Yes[e]	Yes
Cd	111	12.8	Most abundant free isotope	?	?
In	115	95.7	Isobaric overlap with Sn	Yes	Yes
Sn	118	24.2	Most abundant free isotope	Yes	Yes
Sb	121	57.3	Most abundant free isotope	Yes	Yes
Te	125	7.1	Most abundant free isotope	?	?
I	127	100	Mono-isotopic	?	Yes
Cs	133	100	Mono-isotopic	Yes	Yes
Ba	137	11.2	Most abundant free isotope	Yes	Yes
La	139	99.9	Most abundant free isotope	Yes	Yes
Ce	140	88.5	Most abundant free isotope	Yes	Yes
Pr	141	100	Mono-isotopic	Yes	Yes
Nd	143/145/146	44.4	Average	Yes	Yes
Sm	147/149	28.8	Average	Yes	Yes
Eu	151/153	47.8/52.2	Average	Yes	Yes
Gd	157	15.7	Most abundant free isotope	Yes	Yes
Tb	159	100	Mono-isotopic	Yes	Yes

Table 8.1   (*Contd.*)

Element	Preferred (m/z)	Abundance (%)	Comments and interferences	Digest	Fusion
Dy	163	24.9	Most abundant free isotope	Yes	Yes
Ho	165	100	Mono-isotopic	Yes	Yes
Er	166/167	56.3	Average	Yes	Yes
Tm	169	100	Mono-isotopic	Yes	Yes
Yb	172/173	38.1	Average	Yes	Yes
Lu	175	97.4	Most abundant free isotope	Yes	Yes
Hf	178	27.2	Most abundant free isotope	No	Yes
Ta	181	99.9	Most abundant free isotope	Yes	Yes
W	182	26.3	Most abundant free isotope	?	Yes
Re	185	37.4	Most abundant free isotope	Yes	Yes
Os	189	16.1	Most abundant free isotope	Yes	Yes
Ir	193	62.7	Most abundant free isotope	Yes	Yes
Pt	194/195	32.9/33.8	Average	Yes	Yes
Au	197	100	Mono-isotopic	Yes	?
Hg	202	29.8	Most abundant free isotope	Yes	?
Tl	205	70.5	Most abundant free isotope	Yes	Yes
Pb	206/207/208	98.6	Sum isotopes	Yes	No
Bi	209	100	Mono-isotopic	Yes	Yes
Th	232	100	Mono-isotopic	Yes	Yes
U	238	99.3	Most abundant free isotope	Yes	Yes

[a] Not $LiBO_2$.
[b] Low temperature fusion, e.g. sodium peroxide.
[c] Not Na flux.
[d] Not K flux.
[e] No Cl.
[f] No S.

should only be made once the decision has been taken to include or exclude one or more internal standards. With sample preparation concluded, the ICP–MS instrument can be optimised either for general multi-element analysis or for the determination of only a few specific elements.

Data aquisition times should be long enough to ensure that the counting statistics obtained are adequate for the elemental concentration range expected in the samples. Standard calibration solutions may need to be matrix matched for some analytical tasks (see below), although simple aqueous standards are often sufficient. An analytical procedure should then be set up which will determine the run order of the standard solutions, blanks, unknown samples, quality control solutions, signal drift monitors, etc. A suggested procedure for multi-element determination is given in Table 8.2. Unknown samples should be analysed with up to three replicates, and repeat analysis of the same sample can provide useful additional data. If a signal drift monitor is included, it must be first analysed prior to any of the standards or unknown samples. Repeated analyses should be made at a relatively high frequency, e.g. every 5 or 10 samples, throughout the run, to monitor and correct for

Table 8.2 A suggested procedure for multi-element determination

Approx. time[a] (s)	Solution type	Number of runs
180	Standard blank	3
180	Procedural blank	3
90	Wash	
60	Uptake	
60	(Drift monitor)[b]	1
90	Wash	
60	Uptake	
60–180	Standard 1	1–3
90	Wash	
60	Uptake	
60–180	Standard 2	1–3
90	Wash	
60	Uptake	
60–180	Sample 1	1–3
90	Wash	
60	Uptake	
60–180	Sample 2	1–3
90	Wash	
60	Uptake	
60–180	Sample 3	1–3
90	Wash	
60	Uptake	
60–180	Sample 4	1–3
90	Wash	
60	Uptake	
60–180	Sample 5	1–3
90	Wash	
60	Uptake	
60	(Drift monitor)[b]	1
90	Wash	
60	Uptake	
60–180	Sample 6	1–3	etc.

[a] Total time 41–57 min.
[b] May be omitted if internal standards are used.

changes in analyte signal. If internal standards are used, then the monitor solution may be omitted.

8.3 Geological applications

Of the many scientific fields in which ICP–MS has been utilised for elemental analysis, geological applications have received the most attention. Early in the development of the technique its application to the earth sciences was demonstrated (Date and Gray, 1985). Since this time the technique has been applied to a wide variety of geochemical topics, elemental groups and sample types.

8.3.1 *Rare earth elements*

In many areas of science, the term rare earth elements (REE) has become synonymous with 'lanthanides'. However, the lanthanide series are strictly defined as rare earth elements of atomic number 57 to 71, which have chemical properties similar to lanthanum (Weast, 1987). In addition, the REE also include Y and Sc although these are often not analysed along with the other REE. Before considering the determination of the REE, it is useful to understand something of the chemical and physical behaviour which gives the group a special place in geochemical studies. Furthermore, a basic knowledge of the group can provide a basis for critical assessment of data quality.

REE are predominantly trivalent in terrestrial rocks and show a progressive decrease in ionic radii (Shannon, 1976) from 1.03 Å for La^{3+} to 0.861 Å for Lu^{3+}. This can lead to the preferential uptake by some minerals of heavy REE (HREE) relative to light REE (LREE), or vice versa. In addition, Ce occurs in the tetravalent form under oxidising conditions, while Eu^{3+} may be reduced to Eu^{2+}. This can lead to extensive fractionation of Ce and Eu relative to the other REE. With this knowledge to hand, one can, to some extent, predict the relative behaviour of the group. Since neighbouring REE display striking differences in abundance (Oddo Harkins rule) a simple plot of concentration against atomic number produces a jagged pattern. For comparative purposes therefore, concentrations are divided by the elemental abundance in 'average' chondritic meteorite (e.g. Nakamura, 1974) or for sedimentary and related rocks, by values in 'average' shale (e.g. Piper, 1974). Plotting REE concentrations on a chondrite or shale normalised diagram should result in a smooth pattern (with the possible exception of Ce and Eu). A smooth profile can therefore be used as an additional assessment of relative analytical accuracy. This method of data assessment and comparison is widely used by many workers in the field.

The REE are a difficult elemental group to quantify by many instrumental techniques and consequently their easy determination by ICP–MS has received much attention. The ICP–MS spectra are simple to interpret, each REE has at least one isotope free from isobaric overlap and sensitivity is relatively uniform from ^{139}La to ^{175}Lu (Figure 8.1). The major potential analytical problem encountered on some ICP–MS instruments, is the level of refractory oxide formation (e.g. Longerich *et al.*, 1987c). Since the REE form a continuous group from 139 to 175 m/z, the formation of LREE oxide species could produce significant interferences on the middle to HREE. In many sample types, the concentration of the LREE is considerably higher than the HREE and the potential for serious interferences is increased. The

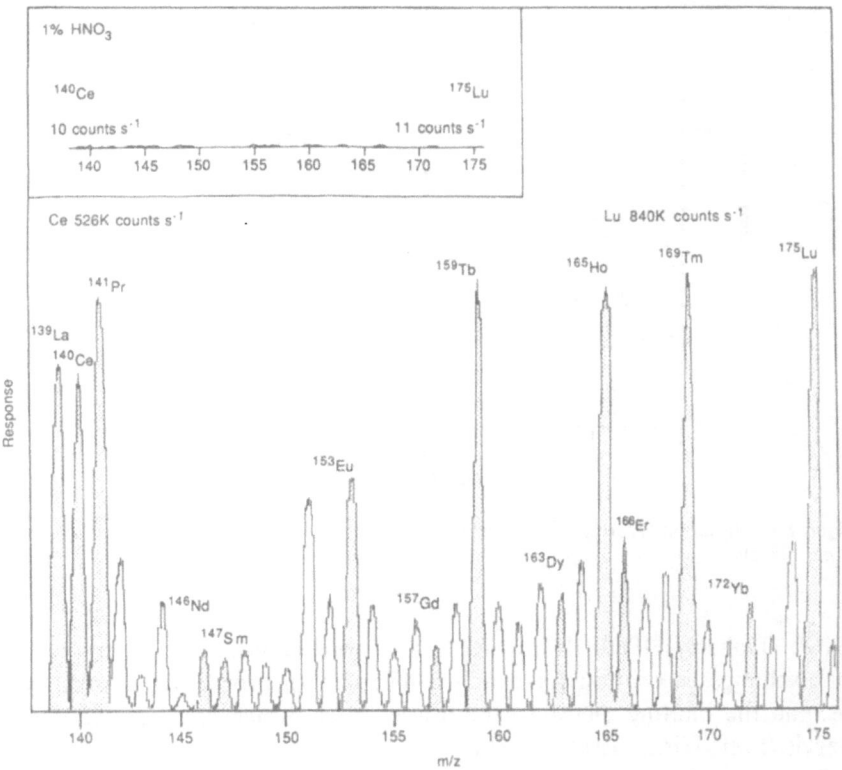

Figure 8.1 ICP–MS spectrum showing the general characteristics of the REE for a $1\,\mu g\,ml^{-1}$ mixed REE solution (after Jarvis, 1989b).

relative level of oxide formation is, in general, a direct function of the oxide bond strength of the parent element (Figure 8.2). However, instrumental operating conditions and basic instrument design can significantly affect the oxide level. Reports of oxide levels $(MO^+/M^+ \times 100)$ on instruments of similar vintage, but of different designs, vary considerably from as low as 0.05–0.5% (Date and Hutchison, 1987) to 0.1–70% (Longerich *et al.*, 1987c). On most modern instruments, oxide levels are now specified by the manufacturers to be in the region of 1% for LaO^+.

It is clear therefore, that the extent of correction which will be required, if any, is dependent on a number of factors including the instrument used, instrument optimisation and the relative concentration of LREE to HREE in the sample analysed. Despite these potential problems, accurate REE data have been reported for bulk rock samples which have been taken through a mixed acid digestion (chapter 7) with no separation of the REE from their matrix (e.g. Doherty and Vander Voet, 1985; Longerich *et al.*, 1987c; Date and Hutchison, 1987; Lichte *et al.*, 1987; Hirata *et al.*, 1988b; Jarvis, 1988,

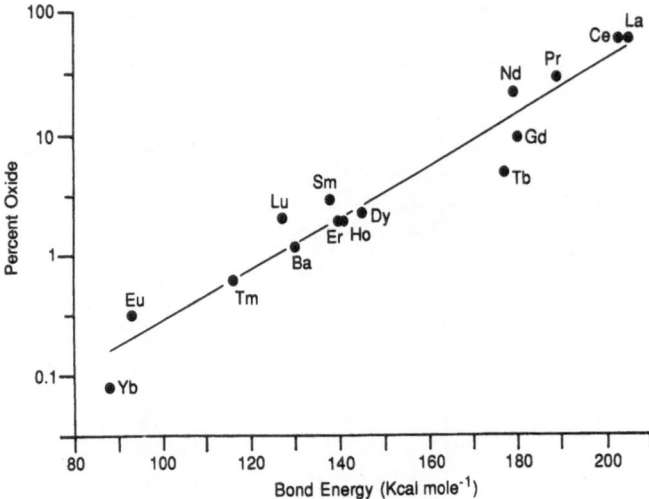

Figure 8.2 Variation of percent oxide as a function of metal-oxide bond strength (after Longerich *et al.*, 1987c).

1989b; Doherty, 1989). If the REE are isolated from the matrix using ion-exchange separation for example (chapter 7), procedural blanks then become the limiting factor to the lowest levels which can be determined (Jarvis, 1988; Hirata *et al.*, 1988b).

Recommended instrumental operating conditions for REE determination using a scanning mode of operation, are given in Table 8.3. Whether the instrument is operated in the scanning or peak hopping mode, data are collected only across the mass range of interest. If Y data are also required a measurement

Table 8.3 Instrumental operating conditions for REE determination using the scanning mode

Forward power	1300 W
Reflected power	< 5 W
Nebuliser gas flow rate	$0.75\,l\,min^{-1}$
Coolant gas flow rate	$13\,l\,min^{-1}$
Auxiliary gas flow rate	$0\,l\,min^{-1}$
Sample uptake rate	$0.5\,ml\,min^{-1}$
Nebuliser	De Galan/Meinhard concentric
Spray chamber	Water cooled
Ion lenses	Optimised on ^{140}Ce
Scan range	$88-176\,m/z$
Skip scan	$91-130\,m/z$
Dwell time	$80\,\mu s$
Channels	2048
Sweeps	400
Analysis time	Approximately 60 s

can be made at 89 m/z and the region from 90 to 130 m/z omitted. The isotopes of Ba (130–138 m/z) are included in order to monitor the level of Ba (and the potential interference from BaO on Eu). Under these operating conditions, fully quantitative data can be obtained down to about 10 × chondritic levels (3 $\mu g\,ml^{-1}$ La to 0.3 $\mu g\,ml^{-1}$ Lu).

The isotopes used for the determination of the REE are shown in Table 8.4. In general, the most abundant isotope is used where it is free from isobaric overlap. Primary calibration should be carried out using mixed multi-element standards. The number of standard solutions used is a matter of preference. For example Doherty (1989) used two solutions containing 80 $\mu g\,ml^{-1}$ of either Y, La, Ce, Pr, Nd, Sm, Eu, Yb, Ru and Re or Gd, Tb, Dy, Ho, Er, Ru and Re (Ru and Re are used as internal standards) while other authors have used a single solution standard with an acid blank (e.g. Hirata et al., 1988b; Jarvis, 1988, 1989b). The use of a mixed concentration standard is recommended (Jarvis, 1988, 1989b) rather than one in which all elements are present at the same concentration (Table 8.5). In this case the distribution of concentrations

Table 8.4 Choice of isotopes for REE determination[a]

Element	Isotope	Abundance	Comment
La	139	99.9	Most abundant free isotope
Ce	140	88.5	Most abundant free isotope
Pr	141	100	Mono-isotopic
Nd	143, 145, 146	Σ44.4	Average value used
Sm	147, 149	Σ28.8	Average value used
Eu	151	47.8	Less prone to interference
Gd	157	15.7	Most abundant free isotope
Tb	159	100	Mono-isotopic
Dy	163	24.9	Most abundant free isotope
Ho	165	100	Mono-isotopic
Er	166, 167	Σ56.3	Average
Tm	169	100	Mono-isotopic
Yb	172, 173	Σ38.1	Average
Lu	175	97.4	Most abundant free isotope

[a] From Jarvis (1990).

Table 8.5 Rare earth element concentrations in an ideal synthetic standard solution[a]

Element	Concentration	Element	Concentration
La	0.8	Dy	0.08
Ce	1.2	Ho	0.08
Pr	0.16	Er	0.08
Nd	0.8	Tm	0.08
Sm	0.16	Yb	0.08
Eu	0.08	Lu	0.08
Gd	0.16	Y	0.8
Tb	0.08		

[a] From Jarvis (1989a).

Table 8.6 A comparison of REE detection limits for ICP–MS and some other analytical techniques[a]

Element	ICP–MS Detection limit (ng ml^{-1})	Quantitation limit[b] (μg g^{-1})	Quantitation limit[c] (μg g^{-1})	Element	ICP–AES[d] Quantitation limit[g] (μg g^{-1})	INAA[e] Quantitation limit (μg g^{-1})	XRF[f] Quantitation limit (μg g^{-1})
^{139}La	0.075	0.125	0.00125	La	0.044	0.83	1.17
^{140}Ce	0.1	0.17	0.0017	Ce	0.145	2.50	1.10
^{141}Pr	0.09	0.15	0.0015	Pr	0.065	–	1.03
^{146}Nd	0.2	0.35	0.0035	Nd	0.085	7.7	0.87
^{147}Sm	0.2	0.25	0.0025	Sm	0.031	0.17	0.77
^{153}Eu	0.06	0.10	0.001	Eu	0.005	0.08	0.73
^{157}Gd	0.1	0.15	0.0015	Gd	0.035	6.5	0.67
^{159}Tb	0.03	0.05	0.0005	Tb	–	0.15	0.67
^{163}Dy	0.1	0.20	0.002	Dy	0.011	–	0.63
^{165}Ho	0.04	0.05	0.0005	Ho	0.010	2.83	0.60
^{167}Er	0.06	0.10	0.001	Er	0.019	–	0.57
^{169}Tm	0.01	0.02	0.0002	Tm	–	0.57	0.57
^{172}Yb	0.06	0.10	0.001	Yb	0.004	0.23	0.57
^{175}Lu	0.05	0.10	0.001	Lu	0.004	0.17	0.50
^{89}Y	0.10	0.20	0.002	Y	0.025	–	1.63

[a] From Jarvis (1989a).
[b] 500 × dilution.
[c] 5 × dilution.
[d] Jarvis and Jarvis (1988).
[e] Potts et al. (1981).
[f] Robinson et al. (1986).
[g] 10 × dilution.

within the group in both samples and standard are similar and this can, to some extent, compensate for low levels of LREE oxides which may form.

The detection limits for all fourteen naturally occurring REE are typically between 0.01 and 0.1 ng ml^{-1}. A more useful comparative measurement for the lower level of measurement in real samples is the limit of quantitation or LOQ (chapter 7) which takes into account the dilution factor used. The LOQs assuming either a × 5 or × 500 dilution factor are shown in Table 8.6 with comparative figures for some other analytical techniques.

In practical analysis on many instruments, the extent of oxide formation is not a serious problem. With CeO$^+$ levels at about 1% the LREE: HREE ratio (Ce:Yb) in geological samples is rarely large enough to pose a serious problem, except perhaps in the analysis of REE minerals such as apatite or monazite where the Ce:Yb ratio can approach 10 000. A more serious practical problem is the formation of BaO$^+$ which can produce a significant interference on ^{151}Eu and ^{153}Eu when the Ba:Eu concentration ratio is > 200:1. This is frequently the situation in geological samples, particularly in those samples prepared using a cation-exchange separation procedure when Ba may be eluted along with the REE. The signal which results due to the oxide species is, in general, relatively stable and a correction for the interference can readily be made if necessary. However, if the oxide signal contribution is very large, an alternative approach may be used. A method

Table 8.7 REE data determination by ICP–MS[a] for SRM's granite NIM-G and basalt NIST-688

	Concentration (μg g^{-1})			
	NIM-G		NIST-688	
Element	Measured	References[b]	Measured	Reference
La	108 ± 2.9	109	2.60 ± 0.3	5.3
Ce	205 ± 5	195	13.1 ± 0.6	13.3
Pr	22.3 ± 0.3	–[c]	2.00 ± 0.09	–
Nd	71.6 ± 1.6	72	9.20 ± 0.3	9.6
Sm	14.8 ± 0.2	15.8	2.70 ± 0.05	2.79
Eu	0.39 ± 0.005	0.35	1.10 ± 0.05	1.07
Gd	13.7 ± 0.3	14	3.30 ± 0.3	3.2
Tb	2.3 ± < 0.001	3	0.50 ± 0.05	0.45
Dy	17.2 ± 0.2	17	3.60 ± 0.12	3.4
Ho	3.70 ± 0.1	–	0.80 ± 0.05	0.81
Er	11.9 ± 0.3	–	2.10 ± 0.09	2.11
Tm	1.70 ± < 0.001	2	0.30 ± < 0.001	0.29
Yb	12.2 ± 0.6	14.2	1.90 ± 0.07	2.09
Lu	1.60 ± 0.05	2	0.30 ± < 0.001	0.34

[a] ICP–MS data from Jarvis (1988).
[b] Reference values from Govindaraju (1989).
[c] –, no data available.

Table 8.8 Concentrations of the REE and Y in SRM feldspar FK-N determined after ion exchange separation

Element	Concentration ($\mu g\,g^{-1}$)		
	ICP–MS[a]	Reference[b]	Reference[c]
La	1.000 ± 0.305	0.900	1
Ce	0.691 ± 0.029	1.04	1
Pr	0.072 ± 0.001	< 0.13	nd[d]
Nd	0.231 ± 0.009	0.18	0.3
Sm	$0.058 \pm\, < 0.001$	< 0.06	0.06
Eu	0.302 ± 0.002	0.40	0.42
Gd	$0.047 \pm\, < 0.001$	0.07	0.05
Tb	0.009 ± 0.006	nd	0.01
Dy	0.046 ± 0.001	0.06	0.06
Ho	0.011 ± 0.001	0.02	nd
Er	0.032 ± 0.001	0.05	0.04
Tm	0.007 ± 0.005	nd	nd
Yb	0.034 ± 0.001	0.05	0.04
Lu	$0.006 \pm\, < 0.001$	0.01	0.01
Y	0.310 ± 0.010	< 0.05	0.3

[a] ICP–MS data from Jarvis (1988).
[b] From Jarvis and Jarvis (1988).
[c] From Govindaraju (1989).
[d] nd = not determined.

developed by Jarvis *et al.* (1989) uses the doubly charged Eu^{2+} ion for quantitative determination of Eu in the presence of high concentrations of Ba.

The data shown in Table 8.7 were determined on a VG Elemental PlasmaQuad of an early vintage. The samples were analysed in triplicate and the precision is better than 3% RSD with an accuracy which is generally better than 5% when compared to the compiled values of Govindaraju (1989). The elemental levels in NIST-688 are approximately 10 × chondrite. If the REE are separated from their matrix the LOQ can be reduced (Table 8.6). An example of the data which can then be obtained is shown for feldspar FK-N (Table 8.8). The concentrations of the HREE are only a few $ng\,g^{-1}$.

8.3.2 *Platinum group metals*

Although strictly confined to Ru, Rh, Pd, Os, Ir and Pt, the determination of this group by ICP–MS can also include Re and Au. Gold, Pd, Pt and Rh have traditionally been determined using classical fire assay. However, the neo-classical nickel sulphide fire assay technique (Robert *et al.*, 1971: Haines and Robert, 1982) is gaining popularity because it provides an effective collection for the whole group with the exception of Au and Os (see chapter 7). This collection technique has been adopted for ICP–MS determination with some success by a number of workers (Date and Hutchison, 1987; Gregoire, 1988; Jackson *et al.*, 1990). A review of methods (including ICP–MS) for the determination of Au, Pt and Pd in production orientated geochemistry laboratories is given by Hall and Bonham-Carter (1988) with an assessment of statistical bias between techniques.

The platinum group metals (PGM) lie in two distinct parts of the mass range from ^{96}Ru–^{110}Pd and ^{184}Os–^{198}Pt, and each has at least one isotope free from isobaric overlap. Due to their position in the mass spectrum, the PGM are relatively free from interference effects and all have detection limits of between 0.4 and 0.05 ng ml^{-1} (Denoyer et al., 1989; Totland, pers. commun.).

Even with these impressively low detection limits, most geological samples require the separation and pre-concentration of the PGM in order to increase the concentration in solution to above the LOQ. Date et al. (1987a) successfully determined Ru, Rh, Pd, Ir, Pt and Au in two reference materials (SARM7 and PTC-1) after separation by Ni-sulphide fire assay.

All of the PGMs can be determined by external calibration using a mixed element standard at 100 ng ml^{-1} in 1% HCl. If isotope dilution is employed, then the spike solution should be added to the samples in the middle of the fire assay charge, prior to fusion. The determination of all the PGM can be carried out in one analysis. The instrument should be optimised in the middle of the mass range of interest (e.g. ^{140}Ce) to ensure a relatively uniform response. Providing that the sample has been pre-concentrated and the PGMs separated from the matrix, all elements can be quantitatively determined in about 1 min per sample. If the scanning mode of operation is used, a skip scan region should be set which excludes 112–183 m/z. If peak hopping is used then the specific isotopes of interest should be selected. The data obtained using external calibration for two SRMs are shown in Table 8.9. The two separate fire assay preparations for SARM7 are in good agreement with reference values and the precision is better than 5% RSD. The measured data

Table 8.9 ICP–MS data[a] for the platinum group elements and gold in SARM-7 and PTC-1

| | Concentration (μg g^{-1}) | | | | |
| | SARM-T | | | PTC-1 | |
Element	ICP–MS(1)	ICP–MS(2)	Reference[b]	ICP–MS(1)	Reference[c]
Ru	0.41 ± 0.01	0.47 ± 0.04	0.43	0.37 ± 0.01	–
Rh	0.19 ± 0.01	0.22 ± 0.01	0.24	0.43 ± 0.01	0.62
Pd	1.34 ± 0.05	1.29 ± 0.07	1.53	6.87 ± 0.18	12.7
Ir	0.06 ± 0.01	0.08 ± 0.01	0.074	0.12 ± 0.01	–
Pt	3.04 ± 0.13	3.27 ± 0.11	3.74	1.60 ± 0.06	3.0
Au	0.57 ± 0.02	0.38 ± 0.02	0.31	0.25 ± 0.01	0.65

[a] ICP–MS data from Date et al. (1987b).
[b] From Steele et al. (1975).
[c] From Steger (1983).

for PTC-1 are low compared with reference values and the authors offered no explanation for this.

Gregoire (1988) also used a Ni-S fire assay for pre-concentration followed by determination of Ru, Pd, Ir and Pt using isotope dilution and electrothermal vaporisation (ETV) for sample introduction. If isotope dilution is used, it is necessary to carry out a 'rough' analysis prior to quantitative determination of the unknown samples to assess the approximate levels of each element of interest. The final solutions can then be spiked with appropriate amounts of ^{101}Ru, ^{105}Pd, ^{193}Ir and ^{194}Pt and should be allowed to equilibrate for 3 h prior to analysis. For sample introduction by ETV, the ICP–MS should be optimised using a 'dry' plasma. This may be achieved by placing a small vessel containing Hg onto the graphite filament and gently heating it to produce a steady stream of Hg vapour (Gregoire, 1988). The operating conditions used are shown in Table 8.10. A high temperature of about 3000°C is required to volatilise the PGMs from the graphite rod. During each vaporisation cycle, eight isotopes are monitored. Analyte signal pulses are measured by monitoring the centre of each peak (1–3 measurements per peak) for each isotope of interest. The transient analyte pulse last for 3–5 s and requires frequent count rate measurements to ensure that the signal is accurately measured. The instrument must therefore be operated using a

Table 8.10 Instrumental operating conditions for the determination of Ru, Pd, Ir and Pt using solution nebulisation and ETV ICP–MS[a]

Plasma conditions for solution nebulisation	
Forward power	1.0 kW
Reflected power	< 5 W
Coolant gas flow rate	12.5 l min^{-1}
Nebuliser gas flow rate	1.1 l min^{-1}
Auxiliary gas flow rate	2.2 l min^{-1}
Plasma conditions for ETV introduction	
Forward power	0.9 kW
Reflected power	5 W
Coolant gas flow rate	12.5 l min^{-1}
Nebuliser gas flow rate	0.7 l min^{-1}
Auxiliary gas flow rate	2.0 l min^{-1}
Electrothermal vaporiser conditions	
Carrier gas flow rate	1.0 l min^{-1}
Sample volume	2 μl
Matrix modifier	500 μg ml^{-1} Ni
Heating cycle	
Dry	10 s ramp to 0.5 V
	20 s hold at 0.5 V
Char	10 s ramp to 1.2 V
	10 s hold at 1.2 V
High temperature	5 s at 1.8 V (no ramp)

[a] After Gregoire (1988).

rapid scan or peak hopping routine. On the Sciex Elan used by Gregoire (1988), count rates were measured sequentially at each mass for 5 ms until a total measurement time of 50 ms had elapsed. This was repeated continuously at intervals of 500 ms throughout the vaporisation cycle. On systems with a multi-channel analyser, the instrument should be scanned as rapidly as possible, typically 80 μs dwell time per channel. Gregoire (1988) found that the presence of Ni in the fire assay solutions acted as a matrix modifier and caused an enhancement of the analyte signal. Therefore, to ensure that standard and sample solution behave alike, Ni (as $NiNO_3$; 500 μg ml^{-1}) can be added to all solutions. The degree of signal enhancement is typically between 4 and 9% depending on the element and the amount of Ni added (Table 8.11). Measured data from ID–ETV–ICP–MS compares well with values obtained by solution nebulisation ICP–MS and NAA.

Alternative pre-concentration procedures such as tellurium co-precipitation have also been used with some success for Ru, Pd and Ir in iron formation rocks, ores and related materials (Sen Gupta and Gregoire, 1989). The technique is suitable for determination at the μg ml^{-1} and ng ml^{-1} level. Twenty-two international reference rocks were determined by Sen Gupta and Gregoire (1989) using ETV for sample introduction with good accuracy at concentrations from 0.5 to 24 ng ml^{-1}.

In some analytical applications, data may be required for specific members of the PGM group, particularly Os and Re (Hirata *et al.*, 1989; Hirata and Masuda, 1990) and Au (Bakowska *et al.*, 1989). The determination of Os in geological samples is usually carried out in conjunction with the determination of Re and Os isotope ratios (see chapter 11). Osmium readily forms a volatile species, OsO_4, and is lost during pre-concentration procedures such as fire assay. However, the volatile properties of Os tetroxide can be used to advantage by the direct measurement of the Os vapour. A technique

Table 8.11 Effect of nickel (as the nitrate) on the ion count rate of some platinum group elements determined by electrothermal vaporisation ICP–MS[a]

Ni(μg g^{-1})	Recovery		
	^{106}Pd	^{103}Pd	^{195}Pt
	1.0	1.0	1.0
100	5.8	7.6	9.0
500	4.6	8.1	9.4
1000	4.6	7.4	8.3
3000	3.9	5.5	4.7
5000	4.0	3.7	2.6
7000	4.0	3.3	2.3
10000	4.4	2.3	1.7

[a] From Gregoire (1988).

described by Hirata *et al.* (1989) uses a small merging chamber in which the sample solution and a micro-heater are placed. OsO_4 is evaporated from the sample solution and is carried into the ICP using a mist produced by conventional nebulisation. The signal generated by this technique is a transient one but with careful heating control, a 20 µl sample produces a signal of approximately 60–120 s duration. The significant advantage of this mode of sample introduction over conventional nebulisation, is that sensitivity is increased by more than a factor of 10, with little change in background intensity. The absolute 3 sigma detection limit can therefore be reduced from 1.0 pg using solution nebulisation to 0.085 pg using the micro-heater merging chamber (Hirata *et al.*, 1989).

Ultratrace level gold determination has been carried out by Bakowska *et al.* (1989) in solutions pre-concentrated from seawater. The concentration range of gold in seawater is typically between 0.02 and 0.2 pg g^{-1} and contamination is therefore a potentially serious problem in such sample types. In the work by Bakowska *et al.* (1989) samples were spiked with [195]Au prior to separation and pre-concentration, to monitor losses during the separation procedure. Using this technique, the final volume available for analysis is 1 ml of 5% aqua regia. Using the instrumental operating conditions given in Table 8.12, the calibration is linear from 0.10 to 2.94 ng ml^{-1}. At these concentrations and low uptake rates, gold retention on the plastic walls of the peristaltic pump tubing was initially a problem resulting in anomalous calibration curves. However, providing a rigorous washout procedure was adopted using 5% aqua regia, the problem was eliminated. Samples were subsequently analysed using the procedure outlined in Figure 8.3 (Bakowska *et al.*, 1989).

Table 8.12 Instrumental operating conditions for the determination of gold in seawater[a]

ICP–MS instrument	Sciex Elan 250
Forward power	1.2 kW
Reflected power	< 5 W
Coolant gas flow rate	11.4 l min^{-1}
Nebuliser gas flow rate	1.10 l min^{-1}
Auxiliary gas flow rate	1.39 l min^{-1}
Sampling distance	27 mm
Torch	Sciex long
Spray chamber	Double pass
Nebuliser	Meinhard type C
Signal optimisation	[197]Au for maximum signal
Resolution	Low
Measurement mode	Sequential
Measurements per peak	3
Measurement time	0.8 s
Repeats per integration	5

[a] After Bakowska *et al.* (1989).

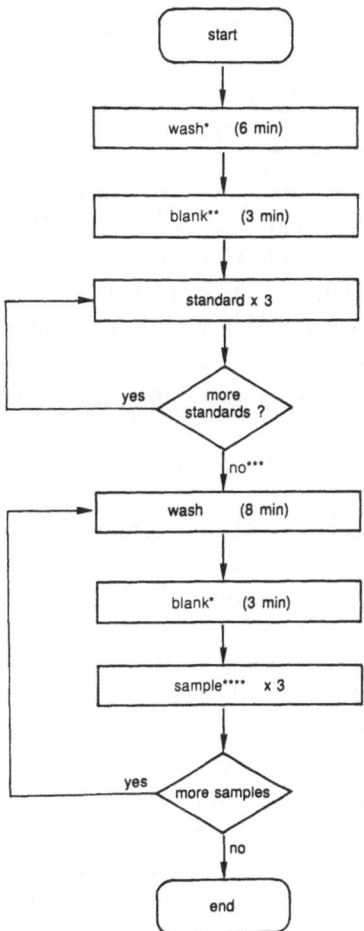

Figure 8.3 Block diagram of the procedure for gold determination. (*) Flexible tubing must be pre-conditioned; (**) all consecutive measurements are blank subtracted; (***) calibration curve is calculated, (****) or standard (after Bakowska *et al.*, 1989).

Unfortunately all but one (with a concentration of $0.542\,\mathrm{ng\,ml^{-1}}$) of the test seawater samples measured were below the limit of detection for the method.

8.3.3   *Zirconium, niobium, hafnium, tantalum, thorium and uranium*

The accurate measurement of low levels of highly incompatible elements is an essential requirement for many geochemical studies. The elements of interest vary, depending on the rock suite under investigation, but frequently include the heavier elements, e.g. Zr, Nb, Hf, Ta, Th and U. Some of these elements are partitioned into resistant mineral phases, which can most easily be brought into solution using a fusion technique (e.g. Kantipuly *et al.*, 1988;

Hall et al., 1990a; Jarvis, 1990: see chapter 7). Hall et al. (1990a) compared a number a fluxes ($LiBO_2$, KHF_2 and $K_2B_4O_7$) and found $LiBO_2$ to be the most effective. The fusion stage in this work was followed by separation of Zr, Nb, Hf and Ta from the matrix by precipitation with cupferron. Kantipuly et al. (1988) found $K_2B_4O_7 \cdot H_2O$ to be a suitable flux for the preparation of tourmaline for the determination of Th and U. Jarvis (1990) compared a $LiBO_2$ fusion with open acid digestion using $HF/HClO_4$. Using either preparation technique Th and U could be determined down to about $20 \, ng \, ml^{-1}$ with good accuracy indicating that, in the sample analysed, either preparation method was acceptable. However, the results obtained for some other elements was less conclusive. In general, quantitative data were obtained for Zr, Hf, Nb and Ta in the fused samples providing that the levels were significantly above the detection limit. Some of the elements showed quantitative recovery after acid digestion, but this was critically dependent on sample type.

To overcome the problems of low concentrations, Hall et al. (1990a) used a cupferron precipitation technique to separate Zr, Hf, Ta and Nb from their fused matrix (see chapter 7 for method). This allowed the use of lower dilution factors (approx. 300 for Nb, Ta and Hf) and consequent reduction in the LOQ (Table 8.13). A comparison of instrumental operating conditions for the determination of Zr, Hf, Nb, Ta, Th and U are shown in Table 8.14. If the scanning mode of operation is used, a skip scan region should be set from 94 to $176 \, m/z$. If peak hopping is used, only the isotopes of interest are counted in order to optimise counting times. Signal drift may occur and should be monitored and corrected for. This may be done by running the calibration standard solution after every five samples and applying a linear drift correction to the raw count integrals. Alternatively internal standards

Table 8.13 Detection limit data for some elements used as petrogenetic indicators

	Detection limit[a] (ng ml^{-1})	Quantitation limit ($\mu g \, g^{-1}$)		
		Acid digest[b]	Fusion[c]	Cupferron[d]
Zr	0.11	0.183	1.83	1.33
Nb	0.03	0.050	0.50	0.067
Hf	0.02	0.033	0.33	0.067
Ta	0.01	0.017	0.17	0.067
Th	0.01	0.017	0.17	nd[e]
U	0.01	0.017	0.17	nd

[a] 3 sigma in solution.
[b] 500 × dilution, 10 sigma $HF/HClO_4$ (Jarvis, 1990).
[c] 5000 × dilution, 10 sigma $LiBO_2$ fusion (Jarvis, 1990).
[d] Zr = 6250 × dilution; Nb, Ta, Hf = 313 × dilution cupferron separation (Hall et al., 1990a).
[e] nd, not determined.

Table 8.14 Operating conditions for the determination of Zr, Nb, Hf, Ta on two ICP–MS instruments

	PlasmaQuad	Elan[a]
Forward power (kW)	1.3	1.1
Reflected power	< 5	< 5
Coolant gas flow rate (l min^{-1})	12	14
Nebuliser gas flow rate (l min^{-1})	0.75	1.1
Auxiliary gas flow rate (l min^{-1})	0.5	2.2
Nebuliser	De Galan	Meinhard
Torch	Fassel	Sciex Long
Spray chamber	Surrey design single pass, water cooled 5°C	Scott, no cooling
Sample uptake rate (l min^{-1})	1.5	1.5
Cone (mm)	Ni 1.0	Ni 1.14
Skimmer (mm)	Ni 0.7	Ni 0.89
Ion lens optimisation	^{59}Co and ^{238}U	Hf

Data aquisition parameters

PlasmaQuad: Scanning mode of data aquisition, scan range 89–182 m/z, skipped masses 94–176 m/z, dwell time 80 µs, sweeps 200, channels 4096, total analysis time 60 s

Elan: Peak jumping (multi-channel) mode of data aquisition, measurement time 1 s, dwell time 50 µs, isotopes measured ^{90}Zr, ^{93}Nb, ^{99}Ru, ^{178}Hf, ^{181}Ta and ^{185}Re

[a] Elan data from Hall et al. (1990a).

can be used. ^{99}Ru can be used to correct for changes in signal for Zr and Nb, while ^{185}Re is used to correct for Hf and Ta. An example of the data given by Hall et al. (1990a) is shown in Table 7.10 and in general measured and reference values are in excellent agreement.

8.3.4 Molybdenum, tungsten and thallium

8.3.4.1 *Molybdenum and tungsten* These two elements have received some attention in geological samples. The average content of both Mo and W in most rock types lies in the range 1–2 µg g^{-1}. Although alternative analytical techniques are available for the determination of both W and Mo at these levels (NAA and spectrophotometry) these are often time consuming and require complex separation procedures to achieve the desired detection limits. Hall et al. (1987) have developed a method for trace level determination of both elements using either conventional solution nebulisation or ETV (Park and Hall, 1987) for sample introduction. For the former, the sample is first decomposed by an alkaline fusion with NaCO$_3$, the elements are then leached from the melt as their soluble oxy-anions, and then taken through a separation procedure based on selective formation of their oxinate complexes with subsequent absorption on activated charcoal. The separation of W and Mo from the matrix has considerable benefits. Significant amounts of Na, Pb and Cu ($100\,\mu g\,ml^{-1}$) have been shown to cause serious matrix effects under a range of operating conditions on the instrument used for this work (Hall *et al.*, 1987). Although the authors demonstrated that internal standards could be used to compensate effectively for these matrix effects across a range of operating conditions, separation of the elements was thought to be a more satisfactory alternative. Internal standards (Re for W, Ru for Mo) were still used, however, to improve the precision. The ICP–MS instrument may be operated in either the scanning or peak hopping mode. If scanning is used the scan range should be set from 93 to $186\,m/z$ with a skipped region from 101 to $180\,m/z$. External drift correction should be carried out if significant changes in analyte signal occur. Alternatively, internal standards may be used (^{99}Ru for Mo and ^{185}Re for W). When only a few elements are to be determined, such as in this particular case, analysis times per sample may be reduced to about 30 s. Determination limits of 0.07 and $0.08\,\mu g\,ml^{-1}$ for W and Mo, respectively, can be achieved using a common set of operating conditions (Hall *et al.*, 1987). A comparison of measured data using isotope dilution ICP–MS and internal standardisation ICP–MS for the determination of Mo are shown in Table 8.15. The precision is comparable using either method at the sub $\mu g\,g^{-1}$ level.

As an alternative to separation of the elements from the matrix, samples may be introduced into the ICP–MS using ETV (Park and Hall, 1987). Sample introduction by ETV rather than conventional nebulisation in the determination of Mo results in improved productivity and a superior determination limit of $0.03\,\mu g\,ml^{-1}$. However, problems are reported for the determination of W in a high salt matrix (see chapter 4). Although good data can be obtained with repeated measurements, the necessity of incorporating an ashing step in the ETV program causes a loss in signal during vaporisation and memory

Table 8.15 A comparison of precision measurements for molybdenum by ICP–MS and ICP–AES for stream sediment GSD-4 ($n = 5$)[a]

	ICP–MS isotope dilution		ICP–MS internal correction[b]		ICP–AES direct calibration	
	Mo ($\mu g\,g^{-1}$)	%RSD	Mo ($\mu g\,g^{-1}$)	%RSD	Mo ($\mu g\,g^{-1}$)	%RSD
	0.86	1.02	0.89	1.32	0.79	5.6
	0.82	0.91	0.79	1.91	0.92	31
	0.84	0.94	0.80	0.94	1.07	23
	0.81	1.22	0.78	1.52	1.00	1.7
	0.85	1.43	0.81	1.43	0.89	6.9
Mean	0.83	2.38	0.81	5.42	0.93	12

[a] From Hall *et al.* (1987).
[b] Using ^{99}Ru.

effects between samples (Park and Hall, 1987). The use of ETV sample introduction for W is therefore not recommended.

8.3.4.2 *Thallium* The average concentration of Tl in the earth's crust is 0.3–$0.5\,\mu g\,g^{-1}$ which is close to the limit of quantitation for Tl by solution nebulisation ICP–MS assuming a sample dilution factor of 500. The trace level determination of Tl may be carried out using ETV for sample introduction following the method of Park and Hall (1988). Samples were prepared using an $HF/HClO_4$ digestion and calibration was achieved by isotope dilution. Rather poor abundance sensitivity on the instrument used by Park and Hall (1988) necessitated the separation of Tl from matrices which contained $>500\,\mu g\,g^{-1}$ Pb. Overlap from the skirt of the poorly resolved ^{205}Pb peak caused considerable error in the analytical data since both of the Tl isotopes were required to carry out the isotope dilution calculations. Tl was extracted from the matrix where required using isobutyl methyl ketone (IBMK).

8.3.5 *Analysis of specific sample types*

The accurate and precise simultaneous determination of a wide range of elements has become a reality in many sample types. The analysis of marine sediments, particularly for the characterisation of new reference materials is a case in point. The applicability of ICP–MS to the determination of 11 trace elements (Cr, Ni, Zn, Sr, Mo, Cd, Sn, Sb, Tl, Pb and U) was demonstrated by McLaren *et al.* (1987a, b) using two Canadian SRMs, BCSS-1 and MESS-1. The use of standard additions for calibration was found to overcome some of the matrix effect problems which were experienced using external calibration procedures. Calibration by isotope dilution is shown to provide the most accurate and precise data compared with other calibration strategies, providing that isotopic equilibrium is achieved and the isotopes used for the ratio measurements are free from interference by molecular species. The measurement of the isotope ratios on unspiked samples, provides a sensitive diagnosis of such interferences. Regardless of the calibration technique used, samples are prepared by a similar method using $HF/HClO_4/HNO_3$ closed low-pressure digestion. If isotope dilution is being employed, the isotopic spike is added prior to the first digestion. The reference/spike isotope pairs used by McLaren *et al.* (1987b) were $^{52}Cr/^{53}Cr$, $^{60}Ni/^{61}Ni$, $^{68}Zn/^{67}Zn$, $^{88}Sr/^{86}Sr$, $^{98}Mo/^{100}Mo$, $^{114}Cd/^{111}Cd$, $^{118}Sn/^{117}Sn$, $^{121}Sb/^{123}Sb$, $^{205}Tl/^{203}Tl$, $^{208}Pb/^{207}Pb$, $^{238}U/^{235}U$. Corrections for a number of isobaric overlaps may be necessary for the determination of these isotopic pairs including ^{86}Kr on ^{86}Sr (corrected by blank subtraction), ^{98}Ru and ^{100}Ru on ^{98}Mo and ^{100}Mo, ^{123}Te on ^{123}Sb and ^{114}Sn on ^{114}Cd, depending on the specific sample matrix.

A comparison of results obtained by external calibration and isotope dilution for marine sediment PACS-1 is shown in Table 8.16. Where no

Table 8.16 Analysis of PACS-1 by ICP–MS[a]

Element	No internal standardisation	$^{40}Ar_2$ internal standardisation	Isotope dilution	Certified value
	External calibration			
V	136 ± 3	139 ± 2	–	127 ± 5
Cr	105 ± 2	107 ± 2	–	113 ± 8
Co	20.6 ± 0.4	21.1 ± 0.4	–	17.5 ± 1.1
Ni	43.5 ± 1.9	44.5 ± 2.6	44.7 ± 1.6	44.1 ± 3.0
Cu	451 ± 15	446 ± 24	447 ± 15	452 ± 16
Zn	911 ± 23	932 ± 16	836 ± 16	824 ± 22
As	195 ± 12	204 ± 11	–	211 ± 11
Sr	250 ± 5	264 ± 5	275 ± 2	277 ± 11
Mo	11.8 ± 0.5	15.7 ± 0.7	12.7 ± 0.4	12.3 ± 0.9
Cd	2.42 ± 0.16	–	2.23 ± 0.14	2.38 ± 0.20
Sn	42 ± 2	–	41.0 ± 2.5	41.1 ± 3.1
Sb	195 ± 8	–	206 ± 9	171 ± 14
Hg	–	–	4.56 ± 0.09	4.57 ± 0.16
Tl	–	0.74 ± 0.05	0.77 ± 0.05	–
Pb	401 ± 18	–	402 ± 10	404 ± 20
U	–	2.9 ± 0.5	3.02 ± 0.04	–

[a] From McLaren *et al.* (1988). Results in $\mu g/g$, with precision of ICP–MS analyses expressed as the standard deviation of six replicate analysis. Precision data for certified values are 95% tolerance limits.

internal standardisation is used, no attempt has been made to correct for signal drift during the run. Results are compared with certified values. The accuracy and precision for all elements is reasonably good and a slight improvement in precision is seen for the isotope dilution results.

8.3.5.1 *Tourmaline* The analysis of tourmaline samples for Th and U was carried out by Kantipuly *et al.* (1988) after fusion with $K_2B_4O_7$. Potassium was removed from the solutions as a precautionary measure by precipitation as $KClO_4$. This did not apparently co-precipitate the analytes of interest. The concentration of Th and U determined in a number of samples ranged from 1 to $40\,\mu g\,ml^{-1}$ and 0.5–$5\,\mu g\,ml^{-1}$, respectively.

8.3.5.2 *Calcium-rich samples* Rock types containing Ca as the major component are relatively common, e.g. limestones and dolomites, and their analysis by some analytical techniques such as ICP–AES is hampered by severe interference problems. The potential interference effects from a Ca matrix in ICP–MS was investigated by Date *et al.* (1987a). The response for a number of trace elements was recorded in the presence of $500\,\mu g\,ml^{-1}$ Ca and compared with a Ca-free acid matrix. The data indicate that significant matrix effects (both suppression and enhancements) could occur. However, data from our laboratory suggest that on more modern instruments, non-spectroscoic matrix effects are not severe for most elements. The effect on analyte signal of increasing calcium concentration is illustrated in Figure 8.4. For all elements except B, there is no evidence of systematic signal suppression even at 2000 μg ml^{-1} Ca. At high matrix element concentrations, precision is poorer, and the data are more scattered. It is only at about 2000 μg ml^{-1} Ca, that there appears to be a significant suppression of the B signal of about 10%. For a relatively pure limestone containing 37% Ca, a sample dilution factor of 1000 would give a solution containing less than 500 μg ml^{-1} Ca, a level at which signal suppression is not a serious problem.

For the determinatioin of trace metals in hydrothermal calcites where Ca concentrations are between 1300 and 1600 μg ml^{-1} Ca, Luck and Luders (1989) used matrix matched standards with respect to Ca content and acid concentration, to overcome possible matrix effects.

Isobaric overlaps should also be considered for Ca-rich matrices. Trace level Ti determinations may be subject to considerable error in Ca matrices since the most abundant Ti isotope has a direct overlap with ^{48}Ca. However,

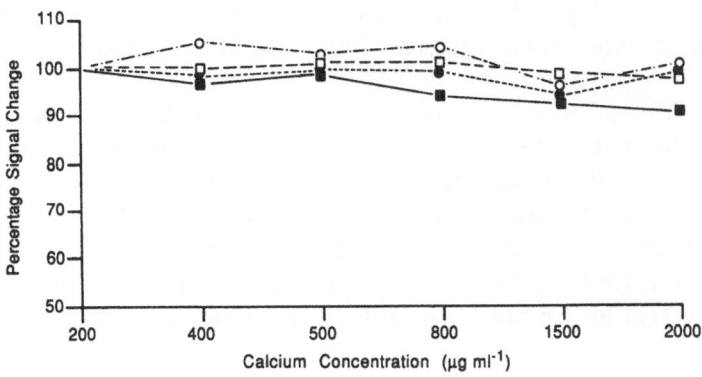

Figure 8.4 The effect of increasing calcium concentration on some analyte elements at 1 μg ml^{-1}.

Table 8.17 Some possible magnesium, phosphorus and calcium polyatomic ion interferences[a]

Element	M$^+$	Abundance(%)	MH$^+$	MO$^+$	MOH$^+$	MO$_2^+$	MAr$^+$
Mg	24	79	25	40	41(K)	56(Fe)	64
	25	10	26	41	42	57(Fe)	65(Cu)
	26	11	27(Al)	42	43	58	66(Zn)
P	31	100	32(Si)	47(Ti)	48(Ti)	63(Cu)	71(Ga)
Ca	40	96.9	41(K)	56(Fe)	57(Fe)	72(Ge)	80(Se)
	42	0.65	43	58	59(Co)	74(Ge)	82(Se)
	43	0.14	44	59(Co)	60(Ni)	75(As)	83
	44	2.08	45(Sc)	60(Ni)	61	76	84
	46	0.003	47	62(Ni)	63(Cu)	78(Se)	86(Sr)
	48(Ti)	0.19	49(Ti)	64	65(Cu)	80	88(Sr)

[a] From Date *et al.* (1987a).

Table 8.18 Some possible manganese and iron polyatomic ion interferences[a]

Element	M$^+$	Abundance (%)	MH$^+$	MO$^+$	MOH$^+$	MO$_2^+$	MO$_2$H$^+$	MAr$^+$
Mn	55	100	56(Fe)	71(Ga)	72(Ge)	87	88(Sr)	95(Mo)
Fe	54(Cr)	5.8	55(Mn)	70	71(Ga)	86(Sr)	87	94
	56(ArO)	91.7	57	72(Ge)	73	88(Sr)	89(Y)	96
	57	2.14	58	73(Ge)	74(Ge)	89(Y)	90(Zr)	97(Mo)
	58(Ni)	0.31	59(Co)	74(Ge)	75(As)	90(Zr)	91	98(Mo)

[a] From Date *et al.* (1987a).

providing that a correction is made, accurate data may be obtained (Date *et al.*, 1987a). A number of potential polyatomic ion interferences should also be borne in mind. Calcium has six isotopes which may be combined with O, H, OH, O_2 and Ar and potentially lead to interference effects up to 88 m/z (Table 8.17). The severity of the interferences should be established under the particular operating conditions used for analysis since their magnitude may vary from day to day and certainly from one instrument to another. In practice, the most serious effects are from $^{44}Ca^1H$ (^{45}Sc) and $^{48}Ca^1H$ (^{49}Ti).

8.3.5.3 *Iron-rich samples*   For the analysis of iron-rich samples such as iron ores and manganese nodules, a range of non-spectroscopic matrix effects may occur. Date *et al.* (1987a) report similar levels of analyte signal suppression and enhancements both for a Ca and Fe matrix at 500 μg ml^{-1}. Polyatomic ion interferences from Fe combined with H, O, OH, O_2, O_2H and Ar are shown in Table 8.18. Again, in practice, few interferences are likely to be serious except for Ge which has an FeO interference on several isotopes.

8.4   Environmental applications

ICP–MS has been applied to a number of environmental applications both as a multi-element technique and for single element determinations.

8.4.1   *Multi-element applications*

The application of ICP–MS to the multi-element analysis of natural waters, biological materials and marine sediments was recently reviewed by Beauchemin (1989). The work includes a review of the digestion procedures developed for multi-element analysis of biological materials and marine sediments in order to minimise spectroscopic interferences.

The measurement of trace metal concentrations in marine organisms can be important in a wide range of applications such as an indication of pollution or in studies of biochemical processes. Although the number of general

applications is increasing in many areas, there are still few which are devoted
to the analysis of biological materials for environmental studies. Trace element
concentrations in such sample types may be low and therefore minimum
dilution is ideal. However, major elements may be high in concentration
leading to severe matrix effects. Biological tissue may be brought into solution
using HNO_3 (Ridout et al., 1988) or a HNO_3/H_2O_2 mixture (Beauchemin
et al., 1988b) providing an ideal matrix for ICP–MS analysis. The absence
of Cl and S from the dissolution acids allows the determination of elements
such as V and As which may otherwise be precluded.

Beauchemin et al. (1988d) have determined 13 elements (Na, Mg, Cr, Fe,
Mn, Co, Ni, Cu, Zn, As, Cd, Hg and Pb) in two marine biological reference
materials, dogfish liver DOLT-1 and dogfish muscle DORM-1 For all
elements except Hg, sample powders were digested using HNO_3/H_2O_2
mixture. For Hg, a modified procedure using only HNO_3 was employed.
Elemental concentrations were determined by isotope dilution and/or standard
addition, the isotopic spike for the former being added prior to digestion.
Procedural blanks were prepared for each preparation scheme. For most
elements determined, the data compared well with the accepted values
whether calibration was by standard addition or isotope dilution. However,
it was noted that the setting of the instrument resolution was critical for the
determination of Ni in both of the reference materials. The ratio of Cu:Ni
in the two reference materials was 5 (DORM-1) and 80 (DOLT-1). For the
initial elemental determinations, the resolution was set to 'low' resolution
mode (Sciex Elan) which corresponds to a peak width of 1.1 m/z at 10% of
the peak height. At this setting, ^{63}Cu was not resolved from the adjacent
^{62}Ni peak. It was therefore necessary to increase the resolution to adequately
resolve these peaks. The determination of Cr in DOLT-1 was hampered by
an interference at 52 m/z resulting from either $^{40}Ar^{12}C$ or $^{35}Cl^{16}O^1H$.
Unfortunately the authors were unable to establish which of these was the
most likely.

Ridout et al. (1988) determined 16 elements in reference material lobster
hepatopancreas TORT-1. Suppression of the analyte signal occurred for most
elements (51–208 m/z) in samples containing high levels of dissolved solids.
Dilution of the samples resulted in data which were in good agreement with
reference values.

8.4.1.1 *Halogens* One elemental group which is not easily determined by
other instrumental techniques, is the halogens; Cl, Br and I. The sensitivity
of these elements is poor by ICP–MS due to their low degree of ionisation
in the argon plasma, Cl = 0.9%, Br = 4.5% and I = 33.9%. However, all three
have been successfully determined in natural waters and environmental
samples (Date and Stuart, 1988). Waters were analysed directly, while solid
samples (urban particulates) were prepared using a Na_2CO_3 (with ZnO)
fusion. Both sample types were analysed under the same operating conditions

Table 8.19 Instrumental operating conditions for the determination of halogens[a]

ICP–MS instrument	VG PlasmaQuad
Forward power	1.3 kW
Reflected power	< 10 W
Coolant gas flow rate	$14.0 \, l \, min^{-1}$
Nebuliser gas flow rate	$0.75 \, l \, min^{-1}$
Auxiliary gas flow rate	$0.01 \, l \, min^{-1}$
Sampling distance	10 mm
Signal optimisation	^{59}Co for maximum signal
Measurement mode	Peak hopping
Measurement time per peak	^{35}Cl = 10 s
	^{79}Br = 5 s
	^{81}Br = 5 s
	^{127}I = 10 s

[a] After Date and Stuart (1988).

shown in Table 8.19. The peak hopping mode of measurement was chosen since only four, widely-spread isotopes were measured. External calibration was carried out using two standard solutions, in a fusion blank matrix, containing either $100 \, \mu g \, ml^{-1}$ Cl, $10 \, \mu g \, ml^{-1}$ Br, $1 \mu g \, ml^{-1}$ I or $10 \, \mu g \, ml^{-1}$ Cl, $1 \, \mu g \, ml^{-1}$ Br, $0.1 \, \mu g \, ml^{-1}$ I. The presence of matrix matched standards with respect to the flux content was found to be critical. All of the halogens determined against a non-matrix matched standard gave low and variable recoveries from -23% (Cl) to -39% (I), indicating suppression of the analyte signal in the presence of Na ($2170 \, \mu g \, ml^{-1}$ Na in the solution analysed). However, with matrix matching, there was excellent agreement between measured and reference values.

8.4.2 Single-element applications

8.4.2.1 *Mercury* Although multi-element determination is desirable for many applications, a single element may be required particularly for speciation studies. Beauchemin *et al.* (1988c) demonstrated the use of flow injection for the determination and speciation of Hg in marine biological samples. The concentration of total Hg in the samples was between 0.13 and $0.7 \, \mu g \, g^{-1}$ which necessitated separation of the Hg from the sample. Separation of the Hg may be carried out using a toluene extraction which is subsequently treated with cysteine acetate solution to back extract the organomercury into an aqueous medium. Unfortunately the cysteine extract contains very high levels of salts (4.3% Na) and low levels of Hg. Since Hg is only partially ionised (43%) in an argon plasma, the direct analysis of highly concentrated solutions is required if this method is used. Beauchemin *et al.* (1988c) overcame this problem by using a flow injection method for sample introduction. A $100 \, \mu l$ loop was used and calibration was made using isotope dilution.

Haraldsson *et al.* (1989) determined Hg without any pre-concentration step by introduction of the element as a gaseous hydride. The sample is placed in a 200 ml reaction vessel and a small volume of potassium permanganate is added. Air is removed from the vessel via a bypass valve to prevent the plasma being extinguished. Sodium borohyride is added throughout the analysis and Hg is evolved. The signal produced by this method is a transient one and data are collected for a period of about 10 min when using a 200 ml vessel. Calibration was carried out by either standard addition or isotope dilution (using ^{199}Hg spike). The precision of the method was determined by replicate analysis of a 100 pg sample of Hg and was found to be better than 3% RSD. The detection limit was restricted by variations in the blank signal. For measurement of natural samples, the optimum amount of Hg in the reaction vessel was 100–500 pg. Within this range precision was good and memory from one sample to another was minimal. The method was tested using marine reference material MESS-1 which was decomposed using a microwave digestion procedure. It was noted that the sensitivity in the reference material was lower than in a synthetic test solution. The concentration of Hg found in MESS-1 was 164 ng g^{-1} with a 5.6% RSD compared with a reference value of 171 ± 14 ng g^{-1}.

8.4.2.2 *Platinum*   There have been some recent concerns about the discharge of Pt into the environment from automobile exhaust catalysts. Mukai *et al.* (1990) have developed an on-line separation procedure to allow the determination of trace levels of Pt in environmental samples such as airborne particulates. Refractory oxide interferences resulting from the formation of HfO, and non-spectroscopic matrix effects are eliminated by on-line separation of the matrix elements using a cation-exchange column inserted into a flow injection apparatus. Recoveries of Pt could not be assessed in the airborne particulate reference samples since no Pt data are available. However, two ore reference materials were also processed (PTC-1 and SU-1a) and the results are given in Table 8.20. The recoveries for both samples are a little low compared with the certified values but are within the range of other published data.

Table 8.20 Determination of platinum in certified reference materials using flow injection ICP–MS (for ICP–MS data $n = 4$)[a]

Sample	ICP–MS	Literature value	Certified value
PTC-1	2.85 ± 0.73	2.44 ± 0.38[b] 2.48[c]	3.0 ± 0.2
SU-1a	0.208 ± 0.03	0.32 ± 0.05[b]	0.41 ± 0.06

[a] After Mukai *et al.* (1990).
[b] Grote and Kettrup (1987).
[c] Date *et al.* (1987b).

8.4.2.3 *Uranium* Trace level U determination in most environmental sample types by ICP–MS is relatively straightforward. Calibration may be carried out using external calibration standards, isotope dilution or standard addition. High levels of UO^+ have been reported on some instruments (Boomer and Powell, 1987) hampering accurate, low level determination.

8.4.2.4 *Lead* The release of Pb into the environment from automotive fuel and its effect on human health, is of wide concern. The determination of Pb isotope ratios by ICP–MS has been carried out in a wide range of sample types (chapter 11). However, the determination of total Pb is also an important facet of many environmental studies and its determination at low concentrations has received some attention (e.g. Lasztity *et al.*, 1989). Lead determination may be carried out using external calibration, giving accurate and precise data (Ridout *et al.*, 1988) although some authors have preferred the use of isotope dilution. The accurate determination of total Pb by isotope dilution is limited by the same factors as those affecting any accurate isotope ratio measurement and the particular problems associated with Pb, such as Hg isobaric overlap, should be taken into account if this method is employed (see section 11.3.7). Lasztity *et al.* (1989) have developed an on-line isotope dilution technique by coupling a commercial flow injection instrument to the ICP–MS. A range of different sample types including environmental, geological, biological and food samples was prepared by either fusion, acid dissolution in a microwave oven, high pressure digestion, dry ashing or open acid dissolution. Acid dissolution in sealed containers heated in a microwave oven gave the best recoveries for lead in biological and geological samples. The precision of the method, using optimum sample to spike ratios and a steady-state merging stream for flow injection, is better than 1%.

8.5 Nuclear applications

8.5.1 *Uranium matrices*

The analyses of sample types associated with the nuclear industry are in general, poorly documented, partly as a result of the secrecy which often surrounds this industry. However, the determination of trace impurities in uranium is one area of particular interest (e.g. Palmieri *et al.*, 1986; Blair, 1986; Allenby, 1987). Uranic analysis by ICP–AES and emission spectrography is hampered by formation of complex optical spectra and serious interference effects. In ICP–MS, by contrast, there are few spectroscopic interferences from a U matrix. These are restricted to the formation of U^{2+} (117, 117.5, 119 m/z) which causes an interference on some of the Sn isotopes (although ^{118}Sn remains free from interference), and to UO^+ (250, 251, 254 m/z) which gives no problems. More serious are the non-spectroscopic matrix effects which may be severe. A $1000\,\mu g\,ml^{-1}$ U matrix causes significant analyte suppression on most elements. Two approaches may be taken to alleviate this problem. The first is by matrix matching of standards and samples, and the second is by separation of the analytes from the matrix (Palmieri *et al.*, 1986).

The dissolution of some uranic samples using HF, initial high F content or the addition of F to prevent hydrolysis, necessitates the use of a corrosion resistant spray chamber, nebuliser and torch injector (see section 3.4) and these should be fitted if F is present. Isotopes for the elements of interest should be selected from Table 8.1. If only a few elements are required, the instrument should be set in the peak hopping mode particularly if the isotopes are widely spread across the mass range. If a complete elemental coverage is required, the scanning mode should be used. For multi-element determinations, the use of internal standards may not be appropriate to correct for signal drift, which can be significant in refractory matrices. The application of internal standards to correct for non-spectroscopic matrix effects is also in question. The degree of analyte suppression may be very variable across the mass range (e.g. 80% Li, 20% Ce) and matrix matching of standards and samples will compensate better for such effects.

8.5.2 *Lithium and boron matrices*

High purity lithium and boron compounds are of considerable interest in nuclear applications. Some of the lithium compounds used are highly soluble

Table 8.21 Effect of using ^{115}In as an internal standard on precision and accuracy of a multi-element standard solution[a]

| Isotope | Concentration ($\mu g\,ml^{-1}$) | |
	Without internal standard	With internal standard
^{9}Be	8.8 ± 0.5[b]	9.7 ± 0.3
^{27}Al	6.7 ± 0.6	8.6 ± 0.5
^{48}Ti	9.0 ± 0.3	9.7 ± 0.4
^{55}Mn	9.0 ± 0.3	9.6 ± 0.2
^{74}Ge	9.0 ± 0.4	9.6 ± 0.3
^{88}Sr	9.3 ± 0.2	10.0 ± 0.1
^{98}Mo	9.1 ± 0.4	9.7 ± 0.3
^{118}Sn	9.3 ± 0.2	9.9 ± 0.2
^{133}Cs	9.3 ± 0.1	10.0 ± 0.1
^{140}Ce	9.6 ± 0.2	10.3 ± 0.2
^{184}W	9.1 ± 0.5	9.7 ± 0.3
^{197}Au	9.6 ± 0.3	10.3 ± 0.2
^{238}U	9.9 ± 0.2	10.5 ± 0.2

[a] After Stotesbury *et al.* (1989).
[b] Mean $\pm$ standard error ($n = 8$).

in an aqueous medium, e.g. lithium hydride, and can easily be brought into solution using dilute nitric acid. The solutions should be diluted such that the final concentration of lithium is $\sim 1000\,\mu g\,ml^{-1}$ in 1% HNO_3. A set of matrix matched standards should be prepared using lithium carbonate. Ideally, three standards are used containing 0, 5 and $10\,ng\,ml^{-1}$ of each analyte of interest.

Boron powder may be dissolved for analysis in nitric acid and diluted to give final concentration of $500\,\mu g\,ml^{-1}$ B. The use of a pressure vessel has been shown to aid complete dissolution of boron samples. Matrix matched aqueous standards should be prepared using boric acid with analyte concentrations at 0, 5 and $10\,ng\,ml^{-1}$.

The criteria for using the peak hopping or scanning modes of instrument operation should be considered in the light of the number and range of elements which are to be determined. Internal standards may prove useful in these types of sample matrix. The data shown in Table 8.21 show a marginal improvement in precision for low mass elements when ^{115}In is used as an internal standard. The effect on accuracy is however, significant, and most elements exhibit improved accuracy when internal standardisation is implemented.

8.5.3   *Zirconium and hafnium alloys*

The determination of ultratrace levels of U is typically required in these materials, particularly ^{235}U. The levels of total U are usually low ($< 1\,\mu g\,g^{-1}$) and this necessitates separation of U from the matrix (Beck and Farmer, 1988). Th, Gd and Sm may also be required at the $\mu g\,g^{-1}$ level in both Zr and Hf alloys.

Samples may be prepared by acid dissolution using HF and HNO_3 and ^{233}U is suitable as an internal standard. No significant matrix effects were recorded in the presence of $< 2000\,\mu g\,ml^{-1}$ Zr (Beck and Farmer, 1988) on either Sm, Gd, Th or U. The lack of any matrix effects means that calibration could be carried out using mixed aqueous standards solutions containing the range of expected analyte concentrations. The typical lower levels which are specified for these sample types are from $0.7 \pm 0.007\,\mu g\,g^{-1}$ U to $4 \pm 1.6\,\mu g\,g^{-1}$ Sm, values which are between 10 and 100 times greater than the detection limits for these elements.

8.6   Industrial applications

The field of industrial analysis covers a vast range of sample matrices. Indeed, many of the applications are extremely specialised and recommendations with respect to sample preparation and instrument optimisation are beyond the scope of this chapter. Calibration strategies, mode of sample introduction

HANDBOOK OF ICP–MS

and instrument hardware for example, should be considered in the light of individual sample types, the range of elements required, etc., paying particular attention to the hazardous properties of some industrial sample types.

8.6.1 *Metals*

One group of samples, about which general guidelines for analysis can be given, are metals, e.g. steels (Vaughan and Horlick, 1989), Fe samples (Moore *et al.*, 1990), Ni based alloys (McLeod *et al.*, 1986; Meddings and Ng, 1989) and Ag alloys (Longerich *et al.*, 1987a). In general, sample preparation is relatively straightforward and usually involves a mixture of HCl/HNO_3, with the addition of HF in some cases (see section 7.3.1.3). The major potential analytical problem with any metallurgical sample is that it has a single (or two-three) element matrix. Thus, the formation of polyatomic ion species as a result of combinations of major elements with O, H, OH, O_2, O_2H, Ar, etc. may cause significant interference problems. The spectral interferences resulting from the formation of FeO, FeOH, NiO and NiOH are shown in Table 8.22. The number of affected isotopes is large from these two elements alone, and spans from 70 to 81 m/z. In addition, if Cl or S based acids are used for dissolution, polyatomic ion interferences will be further compounded.

Nickel-based alloys are perhaps one of the most difficult sample types to

Table 8.22 Spectral interferences due to FeO, FeOH, NiO and NiOH

Species	Abundance of Fe or Ni isotope	Affected mass	Affected species[a]
FeO	5.82	70	Zn(0.62), Ge(20.51)
	91.66	72	Fe(27.4)
	2.19	73	Ge(7.76)
	0.33	74	Ge(36.56), Se(0.87)
FeOH	5.82	71	Ga(39.84)
	91.66	73	Ge(7.76)
	2.19	74	Ge(36.56), Se(0.87)
	0.33	75	As(100)
NiO	68.3	74	Ge(36.56), Se(0.87)
	26.1	76	Ge(7.8), Se(9.0)
	1.13	77	Se(7.6)
	3.59	78	Se(23.6)
	0.91	80	Se(49.7)
NiOH	68.3	75	As(100)
	26.1	77	Se(7.6)
	1.13	78	Se(23.6)
	3.59	79	Br(50.7)
	0.91	81	Br(49.3)

[a] Isotopic abundance shown in parentheses.

analyse. Although Ni is the major componant (approx. 60% w/w) other elements such as Al, B, Co, Cr, Mo, Ti, V and Zr also occur in major amounts (0.5–20% w/w). In this type of matrix, therefore, a large number of potential interferences exist both from polyatomic ion formation and from non-spectroscopic matrix effects from Ni. A number of precautions should be taken to minimise these effects. If possible, Cl and S based acids should be avoided during sample preparation although it should be remembered that S may occur as a major component in the sample itself. Calibration should be carried out using either matrix matched standards or by standard additions. In either case both spectroscopic and non-spectroscopic effects can be reduced. Isotope dilution is not recommended in these complex matrices principally because two isotopes of each element must be free from interference effects. The use of an internal standard may help to compensate for non-spectroscopic matrix effects and Rh has been successfully employed for this purpose in the analysis of steels (Vaughan and Horlick, 1989) for the determination of Co, Cu, Mo and W.

The extent of non-spectroscopic matrix effects depends on the major composition of the sample. In all cases, effects will be reduced by dilution of the sample. The effect of increasing Fe concentration on $100\,\text{ng ml}^{-1}$ Rh and Ce is shown in Figure 8.5. In both cases there is significant suppression of the analyte signal above about $10\,\mu\text{g ml}^{-1}$ Fe which increases to about 20% at $1000\,\mu\text{g ml}^{-1}$ Fe. It is clear, however, that, in this case, the extent of

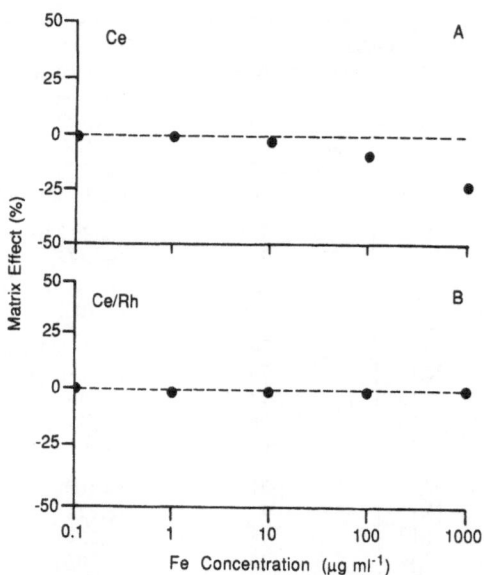

Figure 8.5 The effect of increasing Fe concentration from 0.1 to $1000\,\mu\text{g ml}^{-1}$ on Rh and Ce (after Vaughan and Horlick, 1989).

suppression on Rh and Ce is the same, and Rh could successfully be used as an internal standard for Ce in this matrix.

With any complex matrix, qualitative scans should be made prior to analysis. The extent of matrix effects should then be assessed in the particular matrix under consideration. Using the scanning mode of operation this is a relatively straightforward procedure. Providing that the above precautions are taken, fully quantitative data may be obtained for a wide range of elements in many sample types.

8.6.2 *Hydrocarbons*

The sample types analysed in this category include crude oils, lubricants, additives and organic solvents. Trace element determination in such products may be undertaken in an aqueous medium, providing that the sample has been broken down into an acid soluble form, for example by dry ashing (e.g. Al Swaidan, 1988). In this case, analysis is relatively straightforward since the original sample matrix is destroyed and the organic part of the sample is lost (as CO_2). Samples can therefore be analysed at what were initially high total solids concentrations. For example, a dilution factor of only 10 (1 ml in 10 ml) is acceptable since $> 95\%$ of the original sample material is lost during sample preparation. In this case, the actual TDS in the final solution are 200 μg ml^{-1}. The main disadvantage with this type of preparation is that volatile elements may be lost during the ashing stage.

As an alternative to ashing, samples may be dissolved in an organic solvent which is aspirated directly into the ICP (e.g. van Heuzen, 1989, 1990; Hutton, 1986; Hausler, 1987). White spirit, kerosene and xylene (the lighter solvents) are preferred, since they exhibit many of the properties of aqueous solutions. For solvents such as *n*-heptane, cyclohexane, methanol and diethyl ether for example, a significant reduction in plasma loading is necessary to prevent the plasma being extinguished. This can be achieved by use of a water cooled spray chamber operating at $< -5°C$. Although ICP–MS operation with the ligher solvents without the use of a cooled spray chamber is feasible (Hausler, 1987), better stability is obtained if cooling is applied (van Heuzen, 1989, 1990).

Instrumental operating conditions for organic analysis are somewhat different to those used for aqueous solutions. Higher forward power is required to obtain comparable sensitivity for organic solvents (1.5–1.8 kW) although nebuliser gas flow rates are comparable (Table 8.23). In addition, the introduction of O_2 into the nebuliser gas flow has considerable advantages for organic analysis. Carbon, derived largely from the solvents, condenses on the sampling cone leading to a rapid loss of sensitivity. The addition of small amounts of O_2 (a few percent) to the carrier gas, results in the oxidation of the carbon and prevents its deposition. The oxygen input is regulated by a mass flow controller. The amount of oxygen needed to prevent carbon build-up is set by visual inspection of the plasma. When organics are

Table 8.23 A comparison of ICP–MS operating for aqueous and organic solutions[a]

	Aqueous	Organics		
Characteristic	van Heuzen (1989)	van Heuzen (1989)	Hutton (1986)	Hausler (1987)
Plasma RF power (kW)	1.35	1.35	1.8	1.5
Torch	Fassel	Fassel	×[b]	×
ID of torch injector (mm)	1.5	1.2	×	×
Coolant argon flow (ml min^{-1})	14	14	15	12
Auxiliary flow (ml min^{-1})	0.5	0.7	1	0
Carrier flow (Ar) (ml min^{-1})	630	470	750	300
Carrier flow (O$_2$) (ml min^{-1})	0	40	37–52	0
O$_2$ after nebuliser (ml min^{-1})	0	0	0	5
Ar after nebuliser (ml min^{-1})	0	0	0	245
Solvent	Water	Ker/sol[c]	Wh. sp.[d]	Xylene
Sample uptake rate (ml min^{-1})	1.0	1.0	0.5	0.5
Pumped/natural	Pumped	Pumped	Pumped	Natural
Nebuliser	Conc.[e]	Conc[e]	×	Conc.[f]
Spray chamber	d.p.[g]	d.p.	d.p.	Conical
Spray chamber temp. (°C)	15	15	8–10	No cooling
Sampling cone orifice (mm)	0.75	0.75	0.75	×
Skimmer cone orifice (mm)	1.0	1.0	×	×
Lens setting optimisation	Ba(138)	V(51)	×	Pb(208)
(R) (u^{-1})[h]	1.3	1.3	×	×

[a] From van Heuzen (1989).

[b] ×, not given.

[c] 1:1 (v/v) mixture of kerosene and SHELLSOL AB. SHELLSOL is a Shell trade mark. Some data on SHELLSOL AB (or SHELL CYCLOSOL 63, trademark USA): boiling range 181–210°C, total aromatics 97.5% composition = a mixture of mainly tetramethyl benzenes with some trimethylbenzenes and naphthalenes.

[d] Wh. sp. = white spirit.

[e] Meinhard TR-230-A2.

[f] Meinhard TR-30-C2.

[g] d.p. = double pass.

[h] R = resolution quadrupole, defined as $1/\Delta(M)$, where $\Delta(M)$ is peak width at 5% of the peak height.

introduced, a green C$_2$ band emission is observed at the base, side and central injection region. Oxygen is added (slowly!) until the green emission has disappeared at the base and inner region. Excess addition of oxygen leads to the very rapid erosion of the sampling cone, and consequently should be avoided.

Polyatomic ion interferences resulting from the combination of H, N, O, S, Cl and Ar with C are a particularly serious problem for organic analysis. Table 8.24 shows the apparent concentrations corresponding to observed background levels for a number of elements. Some isotopes normally used for aqueous solution analysis, e.g. ^{44}Ca, ^{24,25,26}Mg, ^{52,53}Cr, are subject to significant interferences in an organic matrix. Furthermore, appreciable contamination in the organic solvents used is also apparent for a number of elements (Mo, Ag, Cd, Sn, Ba and Pb). It is clear that S and Cl should be

Table 8.24 Apparent concentrations corresponding to the observed background levels for some isotopes and tentative assignment of the interferent[a]

Element	Isotope (u)	Hausler (1987) (ng ml^{-1})	Hutton (1986) (ng ml^{-1})	van Heuzen (1989) (ng ml^{-1})	Possible assignment of interferent
B	10	×	×[b]	70	? B memory
B	11	30	×	n.d.[c]	? B memory
Na	23	700	×	40	Contamination[d]
Mg	24	500	920	n.d.	$^{12}C_2^+$
Mg	25	×	90	1200	$^{12}C^{13}C^+/12C_2H^+$
Mg	26	×	320	n.d.	$^{12}C^{14}N^+/^{40}Ar^{12}C^{2+}$
Al	27	300	×	70[e]	$^{13}C^{14}N^+/^{12}C^{15}N^+$
Si	28	12000	*[f]	*	$^{12}C^{16}O^+/^{14}N_2^+$
P	31	40	×	600	$^{15}N^{16}O^+$
Ca	43	×	×	680[e]	$^{12}C^{15}N^{16}O^+$
Ca	44	500000	8400	*	$^{12}C^{16}O_2^+/^{12}C^{32}S^+$
Ti	48	20	×	24	$^{14}N^{34}S^+/^{32}S^{16}O^+$
Cr	50	×	×	n.d.[g]	$^{38}Ar^{12}C^+/^{15}N^{35}Cl^+$
Cr	52	400	1200	*	$^{40}Ar^{12}C^+/^{15}N^{37}Cl^+$
Cr	53		13	700	$^{40}Ar^{13}C^+/^{40}Ar^{12}CH^+/$ $^{37}Cl^{16}O^+$
V	51	10	×	0.8	$^{38}Ar^{13}C^+/^{35}Cl^{16}O^+/$ $^{37}Cl^{14}N^+$
Mn	55	6	×	4	$^{40}Ar^{15}N^+$
Fe	56	×	20	1000	$^{40}Ar^{16}O^+$
Fe	57	40	1	1000	$^{40}Ar^{16}OH^+$
Ni	60	50	×	n.d.	Ni[h]
Cu	63	6	×	4	
Zn	66	5	×	6.5	$^{32}S^{34}S^+/^{34}S^{16}O_2^+$
Mo	98	×	×	1.4	
Ag	107	×	×	0.4	
Ag	109	2	×	0.5	
Cd	111	5	×	2.0	
Cd	114[i]	×	×	1.0	
Sn	120	3	×	3.5	
Ba	138	1	×	0.6	
Pb	208	3	×	2.6	

[a] From van Heuzen (1989).
[b] ×, not given.
[c] n.d., not determined.
[d] No special cleaning procedures to remove Na were performed before using the glassware.
[e] With improved resolution (see text).
[f] Denotes 'over linear range'.
[g] Not determined because of spectral interference by Ti and V.
[h] Nickel from sampling cone.
[i] Spectral interference by ^{114}Sn (abundance 0.65%) possible.

avoided at all stages of sample preparation or analysis and it should be remembered that crude oil samples can contain significant quantities of S (van Heuzen, 1989, 1990). The apparent presence of ^{11}B in the data from Hausler (1987) is probably a result of insufficient quadrupole resolution resulting in part of the ^{12}C peak being integrated with B. This is also the case for ^{27}Al which is adjacent to the large $^{12}C^{16}O$ peak. If these elements

Table 8.25 Quantitative determination of trace elements in NIST-1643a fuel oil by ICP–AES and ICP–MS[a]

Element	ICP–AES (Hausler, 1987) Av.[b] (μg g^{-1})	SD[c]	ICP–MS (Hausler, 1987) Av. (μg g^{-1})	CL[d]	ICP–MS (Hutton, 1986) Av. (μg g^{-1})	CL	Reference value Av. (μg g^{-1})	CL
Na	71.9	1.2	*[e]		83	4	87	4
V	58.3	2.1	53.3	0.2	54	3	56	2
Ni	30.6	2.8	23.5	0.8	29	3	29	1
Pb	< 8.3	dl[f]	2.19	0.08	2.65	0.13	2.80	0.08
Zn	6.8	1.9	2.3	0.3	2.9	0.14	2.7	0.2
Mn	3.45	0.13	*		0.21	0.02	0.19	0.02
Fe	24.6	1.8	23.1	0.4	35	5	(35)[h]	
Ca	20.4	2.7	*		×[g]		(16)	
Cr	3.8	1.9	*		1.0	0.2	(0.7)	
Co	2.6	×[i]	0.35	0.03	0.25	0.05	(0.3)	
As	< 8.4	dl[f]	< 0.07		0.15	0.02	(0.12)	
Mo	3.0	1.9	0.16	0.06	0.15	0.02	(0.12)	
Cd	3.0	1.9	< 0.05		0.04	0.004	(0.002)	
Be	×		×		0.02	0.002	(0.006)	

[a] From van Heuzen (1989).
[b] Average.
[c] Standard deviation over three measurements.
[d] Confidence limit; reported as confidence level, no data upon the number of analyses.
[e] *, obscured by contaminant peak.
[f] Detection limit less than five times noise level.
[g] ×, not given.
[h] ( ), information value only, not certified.
[i] One analysis out of three showed less than detection limit.

are to be determined, the resolution should be adjusted accordingly to completely resolve interfering peaks. Despite a number of interferences which have to be taken into account, quantitative data can be obtained for organic sample matrices (Table 8.25).

8.6.3 Other sample types

Elemental determination at ultratrace levels in high purity acids including HCl, HNO_3, $HClO_4$ and HF, has recently received some attention (Paulsen et al., 1988). Concentration of only a few tens of pg g^{-1} are reported with standardisation for most elements being carried out by isotope dilution. For the analysis of such high purity samples, contamination can be a major problem. Samples should be handled in a Class 1 clean room and a rigorous cleaning procedure adopted for all equipment with which the samples will come into contact. Analysis of HF requires replacement of glass nebuliser, spray chamber and torch injector with PTFE (nebuliser and spray chamber) or sapphire (torch injector) parts (see section 3.4). Other high purity samples which have been analysed include electronic grade quartz (Baumann and Pavel, 1989) and reactive gases (Hutton *et al.*, 1990).

8.7 Biological applications

The sample types considered in this section include many materials which are also of interest for environmental studies and applications, and the reader's attention is therefore drawn to section 8.4.

8.7.1 *Foods*

The analysis of food products for heavy and toxic metals is becoming increasingly important particularly at low levels. A review of the application of ICP–MS in food science is given by Dean *et al.* (1989). ICP–MS has been used for the analysis of a number of products, e.g. spinach, wheat, corn, lettuce, potato, soya bean (Satzger, 1988), cabbage, sprouts, lettuce and kale (Baxter *et al.*, 1989), milk (Emmett, 1988), bovine liver and oyster tissue (Munroe *et al.*, 1986) and flour (Boorn *et al.*, 1985).

In many cases, traditional digestion procedures for such samples involve the use of concentrated HCl and H_2SO_4 which are not appropriate for use with ICP–MS because of the resulting interferences. A dissolution procedure for liquid milk developed by Emmett (1988) uses a combination of HNO_3 and $HClO_4$ to ensure a complete digestion. The author notes that caution should be exercised when adding $HClO_4$ to organic samples. Of the reported values for ten elements in reference material A-11 (freeze dried milk powder), eight were in good agreement with reference values. In a study by Munro *et al.* (1986) over 40 interfering peaks were observed in the mass range $46–206 \, m/z$ during the analysis of bovine liver, oyster tissue and curly kale. A majority of these were polyatomic peaks many of which resulted from the generous use of H_2SO_4 during sample digestion, and $NH_4H_2PO_4$ which was added as a matrix modifier for a comparative ETV–AAS study to be carried out on the same samples. Unfortunately chlorine, present in some of the samples themselves, also caused a number of interference problems particularly for the determination of V. In addition, high H_2SO_4 concentrations $(2.5–10\% \, v/v)$ in the final solutions caused rapid erosion of sampling cones resulting in high blank levels for either Ni or Cu. Despite these limitations, accurate data were produced for several elements (Mo, Cd and Pb) at the sub $\mu g \, ml^{-1}$ level. Yttrium and Bi were successfully used as internal standards to improve precision for the elements reported.

The effects of major elements which are typically present in food products (e.g. Na, Mg, P, K and Ca) may be significant particularly if samples are dry ashed prior to dissolution. Analyte suppression of about 10% in the presence of $1000 \, \mu g \, ml^{-1}$ of several elements was reported in a study by Satzger (1988).

Table 8.26 A comparison of results obtained by ICP–MS with those obtained by ICP–AES and reference values for some trace elements in NIST-1570 spinach[a]

Element	ICP–MS ($\mu g\,g^{-1}$)	ICP–AES ($\mu g\,g^{-1}$)	Certified value ($\mu g\,g^{-1} \pm SD$)
Pb	1.21		1.2 ± 0.2
Cd	1.52		$(1.5)^b$
Zn	50.4	51.8	50 ± 2
	52.7[c]		
Ni	5.93		(6)
Rb	12.9		12.1 ± 0.2
Cr	4.74		4.6 ± 0.3
Mo	0.305		
Al	529	645	870 ± 50
	654[c]		

[a] From Satzger (1988).
[b] ( ), non-certified reference value.
[c] Calibrated by standard addition.

However, it should be emphasised that this figure will be critically dependent on the particular instrument and operating conditions used (see chapter 5). A comparison of ICP–MS data with reference values and that obtained by ICP–AES for SRM spinach NIST-1570 (Table 8.26) shows good agreement for eight elements at a range of concentrations.

8.7.2 *Animal tissue*

Many of the same precautions as those discussed above for accurate measurements in food products are equally applicable to the analysis of animal tissue. In order to avoid the use of H_2SO_4, Friel *et al.* (1990) developed a microwave digestion procedure using HNO_3 and H_2O_2. If total digestion of organic samples is required, precautions should be taken to ensure that the isotopes chosen for analysis are free from polyatomic ion interferences which result from the combination of C with H, N, O and Ar. Similarly, the levels of S and Cl in the samples should be monitored since both of these elements may occur at a significant concentration in animal tissue. If the major element composition of the sample is not known, it may be necessary to ascertain this prior to quantitative trace element determination. Friel *et al.* (1990) report the occurrence of CaO and CaOH species causing an interference on ^{57}Fe, ^{60}Ni and ^{65}Cu in oyster tissue, animal muscle and liver. A mathematical correction can often be made in such circumstances, providing that the concentration of Ca is not excessive. Mg, Al and Fe may also occur as major elements in such sample types. Accurate data are reported by Friel *et al.* (1990) for 24 elements across the mass range where concentrations are significantly above the detection limit, with interference corrections being made at ^{43}Ca ($^{27}Al^{16}O$), ^{51}V ($^{35}Cl^{16}O$), ^{53}Cr ($^{37}Cl^{16}O$), ^{57}Fe ($^{40}Ca^{16}O^{1}H$), ^{60}Ni ($^{44}Ca^{16}O$), ^{65}Cu ($^{48}Ti^{16}O^{1}H$) and ^{75}As ($^{40}Ar^{35}Cl$). For this work, five internal standards were used, ^{9}Be, ^{59}Co, ^{103}Rh, ^{159}Tb and ^{209}Bi. With the exception of Co, these elements are not generally considered to be biologically important and occur at very low concentrations in organic samples. At the dilution used, the contribution of Co from the organic samples was $< 1\%$ of the total Co signal. In this application the internal standards were used for two purposes, firstly to correct for analyte signal drift with time and secondly to overcome non-spectroscopic matrix effects. Be was used only to correct the Li signal, while matrix and drift corrections for elements from Mg to Fe were calculated by a linear extrapolation of the matrix effect and mass correction factor for Co and Rh. This form of correction, using multiple

Table 8.27 A comparison of measured and certified values for oyster tissue reference material NIST-1566[a]

Element	ICP–MS[b]	Certified	
		Mean	Range[c]
Li	0.293 ± 0.018	nv[c]	nv
Mg	1327 ± 33	1280	1190–1370
Al	136 ± 9	nv	nv
Ca	1602 ± 150	1500	1300–1700
V	2.4 ± 0.1	2.3	2.2–2.4
Cr	0.9 ± 0.4	0.69	0.42–0.96
Mn	17.6 ± 0.8	17.5	16.3–18.7
Fe	188 ± 10	195	161–229
Ni	1.04 ± 0.38	1.03	0.84–1.22
Cu	64 ± 2	63	59.5–66.5
Zn	863 ± 34	852	838–866
As	13.2 ± 0.5	13.4	11.5–15.3
Se	2.6 ± 0.3	2.1	1.6–2.6
Rb	4.52 ± 0.13	4.45	4.36–4.54
Sr	10.75 ± 0.31	10.36	9.8–10.92
Y	0.432 ± 0.023	nv	nv
Mo	0.32 ± 0.17	$(< 0.2)^{d}$	
Cd	3.4 ± 0.1	3.5	3.1–3.9
Sb	0.08 ± 0.11	nv	nv
Cs	0.027 ± 0.001	nv	nv
Ba	3.7 ± 0.6	nv	nv
La	0.297 ± 0.022	nv	nv
Ce	0.39 ± 0.04	nv	nv
Tl	0.007 ± 0.001	(< 0.005)	

[a] From Friel et al. (1990).
[b] ICP–MS data calculated as mean and standard deviation of four replicates.
[c] 95% confidence limits.
[d] nv, no value available.
[e] ( ), for information only.

internal standards with interpolation between pairs, has proved effective for data correction (Table 8.27) although it may be less effective at the extreme ends of the mass range (Friel *et al.* 1990). A single internal standard may be effective in some applications, for example the determination of Pt in animal tissue where In has been used to improve short term precision (Tothill *et al.*, 1990).

Although many ICP–MS applications for foods and biological products require the total elemental concentration to be determined, a significant number are directed towards the form in which an element occurs. The use of ICP–MS for speciation studies is possible, providing that a chromatographic separation is performed prior to analysis. The elements of interest in such studies, tend to be those for which both a 'safe' and 'toxic' species occur. For example As^{3+} and As^{5+} have been determined in dogfish muscle DORM-1 (Beauchemin *et al.*, 1988b, 1989), Cd species in pig kidney (Crews *et al.*, 1989) and methyl Hg, ethyl Hg and Hg^{2+} in tuna fish (Bushee, 1988). An on-line separation is carried out prior to analysis using some form of HPLC. A very wide range of chromatographic conditions have been used with varying degrees of success, and these are discussed in section 4.4.

8.7.3 *Medical applications*

The use of ICP–MS is rapidly growing in this field particularly because of the speed and versatility of the technique. The elemental analysis of blood and blood products (e.g. Delves and Campbell, 1988; Lyon *et al.*, 1988a; Matz *et al.*, 1989; Vanhoe *et al.*, 1989, 1990; Viczian *et al.*, 1990b), urine and faeces (e.g. Ting and Janghorbani, 1986; Lyon *et al.*, 1988b; Heitkemper *et al.*, 1989; Lasztity *et al.*, 1989; Mulligan *et al.*, 1990; Allain *et al.*, 1990), human liver and kidney (e.g. Yoshinaga *et al.*, 1989) and drugs and related products (e.g. Suzuki *et al.*, 1988; Bushee *et al.*, 1989) have all been carried out. A large proportion of medical applications are however directed towards the use of stable isotopes for tracer studies, and these are dealt with by element in chapter 11.

Most of the sample types listed above have a common characteristic, that is they tend to contain high salt concentrations particularly NaCl. If trace level determinations of V, Cr, As or Se are required, it is normally necessary to remove most if not all of the Cl present. A chromatographic separation technique such as gel filtration can be used to remove not only Cl but also Na (as NaCl) from samples, off line, prior to analysis. Using this technique, non-spectroscopic matrix effects due to high concentrations of Na are also eliminated. Even when elements affected by Cl polyatomic species are not required, substantial dilution of samples may be necessary, in order to reduce the level of total dissolved solids to less than about $2000\,\mu g\,ml^{-1}$. Removal of the bulk of the matrix allows solutions to be analysed with a lower dilution factor.

8.8 Summary

Although solutions are the natural form for only a relatively small number
of sample types, it is clear that they can offer a number of advantages to the
analyst. The most important of these is that sample inhomogeneity can be
reduced or even eliminated when solid materials are broken down into
solution form. Calibration is certainly more straightforward in solutions than
for direct solids analysis, since the mere process of dissolution can reduce a
wide variety of matrices of different chemical and physical forms into a simple
uniform final product. Calibration methods such as isotope dilution are easily
performed in solutions and matrix matching is usually straightforward. It is
also apparent that analytical performance, e.g. precision, accuracy and long
term, is generally better than in solids analysis. Separation and pre-
concentration procedures are practicable if materials are brought into
solution, thus improving analytical performance further.

The main disadvantage of solution analysis is that the sample preparation
required may be time consuming, expensive and can lead to the introduction
of contamination. At this time, the advantages of solution analysis generally
outweigh the disadvantages, particularly for fully quantitative analysis,
although this may change with future developments in solid sampling
techniques.

9 The analysis of natural waters by ICP–MS

J. W. McLAREN

9.1 Introduction

Elemental analysis of freshwater, seawater and wastewater samples is undertaken for a wide variety of purposes, including determination of suitability for human consumption, environmental monitoring for regulatory programs, and fundamental study of the geochemical cycles of trace elements in natural waters. The rapid development of ICP–MS as a technique for the routine determination of trace elements in water at sub-parts-per-billion levels has already had a considerable impact on the field of water analysis, and it appears that ICP–MS will revolutionise this field over the next 10 years. The recent publication by the US Environmental Protection Agency (EPA) of a detailed procedure (Method 200.8) for the elemental analysis of ground, surface and drinking waters by ICP–MS is evidence of the rapid acceptance of the technique by regulatory agencies. The dream of rapid, multi-element analysis of freshwater samples with no pre-treatment other than the usual acidification, and possibly filtration, at the time of sampling, is fast becoming a reality. For seawater, however, the relatively high dissolved salt content (typically about 3.5%), in combination with the very low concentrations of many of the elements of interest, usually precludes direct analysis. All published reports to date of ICP–MS analysis of seawater have involved a chemical separation.

A careful consideration of the end use of the data is essential in the development of any analytical procedure, and is especially critical in the development of ICP–MS procedures for the analysis of natural water samples. Water analyses can be subdivided into two categories according to the end use of the results. For many regulatory or monitoring programs, the analytical methodology is required to determine whether the concentrations of certain elements are above levels mandated by health or safety considerations or legislative requirements. In some cases, method detection limits (MDLs) are specified by the responsible regulatory agency. Examples of this type of analysis would be those required by the US EPA for the analysis of ground, surface and drinking waters, and those mandated by the Ontario Ministry of the Environment (OME) Municipal/Industrial Strategy for Abatement (MISA) for the analysis of wastewaters. A second category of analyses includes those in which values for the elemental concentrations are sought regardless of how low these might be. Examples of this type of analysis are the certification analyses for the production of natural water reference materials and the analysis of natural waters for geochemical and oceanographic research.

Some regulatory detection limits are compared in Table 9.1 with the concentrations for 13 trace elements in two freshwater certified reference materials produced by the National Research Council of Canada (NRCC), SLRS-1 and SLRS-2. SLRS-1 was collected from the St Lawrence River near Quebec, while SLRS-2 was taken from the Ottawa River, a major tributary of the St Lawrence. It can be seen that the OME detection limits specified for the MISA program are considerably higher than the concentrations observed in SLRS-1 and SLRS-2, primarily because the MISA program has been set up to monitor the concentrations in industrial wastewaters which are discharged into Ontario waterways. The MDLs published by the US EPA in proposed Method 200.8 are based on the standard deviation of replicate analyses of reagent water blanks spiked at concentrations two to five times the estimated instrumental detection limits. In general, these values are much closer to the concentrations in SLRS-1 and SLRS-2, but would be inadequate for the determination of Cd, Cr, Co, Pb and V. In many cases, regulatory detection limits are based partly on the figures of merit for a

Table 9.1 A comparison of regulatory detection limits with certified concentrations in two freshwater reference materials

Element	EPA 200.8 $(\mu g\,l^{-1})^a$	OME MISA $(\mu g\,l^{-1})^b$	SLRS-1 $(\mu g\,l^{-1})$	SLRS-2 $(\mu g\,l^{-1})$
Al	1.0	30	23.5	84.4
Be	0.3	10	–	–
Cd	0.5	2	0.02	0.03
Cr	0.9	20	0.36	0.45
Co	0.09	20	0.04	0.06
Cu	0.5	10	3.6	2.8
Pb	0.6	30	0.11	0.13
Mo	0.3	20	0.78	0.16
Ni	0.5	20	1.1	1.0
Ag	0.1	30	–	–
Tl	0.3	30	–	–
V	2.5	30	0.66	0.25
Zn	1.8	10	1.3	3.3

ᵃ EPA 200.8, detection limits for total recoverable elements in waters by proposed Method 200.8.
ᵇ OME MISA, detection limits required by the Ontario Ministry of the Environment for monitoring of waste waters under the Municipal/Industrial Strategy for Abatement Program.

particular instrumental method such as ICP–AES or ICP–MS and partly on the expected concentrations of these elements in the samples to be analysed.

9.2 Water sampling procedures for ICP–MS

Thompson and Walsh (1989) have reviewed general aspects of sampling, filtration, stabilisation and storage of water samples destined for analysis by inductively coupled plasma-atomic emission spectrometry (ICP–AES). Much of this information is equally applicable for use with ICP–MS. In view of the fact that ICP–MS detection limits for many elements are often two or even three orders of magnitude lower than corresponding ICP–AES detection limits, avoidance of accidental contamination during the sampling process is even more difficult than indicated by Thompson and Walsh. This has been illustrated in a report by Henshaw et al. (1989), who evaluated the performance of ICP–MS in the multi-element analysis of a large number of lake water samples. A comparison of instrumental detection limits (IDLs) with 'system detection limits' (SDLs) based on 22 field blanks indicated that for a number of elements significant contamination occurred during the sampling process. Aluminium and zinc were identified in particular as elements for which significant contamination occurred during the sampling or sample treatment process; other elements for which the SDLs were considerably higher than the IDLs were Cu, Ni and Pb. The authors explained that sampling protocols for this large survey had already been established when ICP–MS was incorporated into the program, which illustrates that exploitation of the full potential of ICP–MS places even more stringent requirements on the sampling procedures than are currently in place.

9.2.1 Filtration, acidification and storage

Sampling and storage of natural waters for trace metals have been extensively reviewed by Sturgeon and Berman (1987). Included in this review is a summary of available data on the trace metal contents of materials often used in the construction of samplers, sample handling equipment and storage containers; materials to be avoided include all metals or plastic coated metals, rubber and soft glass. With few exceptions, borosilicate glass should also be avoided, although it is useful for the collection and storage of samples for mercury determinations. Plastics have found widespread application in trace metal work. Among these materials, the most desirable are fluorinated ethylene polymers (e.g. PTFE, FEP, PFA), polypropylene and high-density polyethylene; polyvinylchloride (PVC) and structural nylon should be avoided. Even the best plastic materials must be rigorously cleaned before use. Many of the recommended procedures involve leaching for several days with dilute nitric and/or hydrochloric acid solutions of gradually decreasing concentration, such that the pH of the final solution is identical to the pH of the samples after acidification.

Sampling procedures for natural waters destined for trace element analysis normally require passage of the sample through a plastic membrane filter of nominal 0.45 μm pore size, often on site. 'Dissolved' metals are operationally defined as those which pass through the filter, while an estimate of 'particulate' metals can be obtained, if desired, by analysis of the filter. The filtered sample is normally acidified with about 1% w/v mineral acid to prevent the formation of precipitates or the adsorption of trace elements onto sample container walls. Samples are sometimes also cooled or frozen to stabilise them during shipment to the laboratory. The addition of hydrochloric acid to samples for ICP–MS analysis should be avoided, as chlorine-containing polyatomic species can interfere with the determination of V, Cr and As (see section 5.5.2).

Some of the data of Henshaw et al. (1989) are compared in Table 9.2 with the values for 17 trace elements in two NRCC natural freshwater certified reference materials SLRS-1 and SLRS-2. Both of these samples were passed through 0.45 μm acrylic co-polymer capsule filters and acidified on site with 1.5 ml of high purity nitric acid per litre of water with a custom apparatus designed for field use.

Table 9.2 A comparison of instrumental and system detection limits with the certified values for the freshwater reference materials SLRS-1 and SLRS-2

Element	IDL[a] ($\mu g l^{-1}$)	SDL[a] ($\mu g l^{-1}$)	SLRS-1 ($\mu g l^{-1}$)	SLRS-2 ($\mu g l^{-1}$)
Al	0.2	4.0	23.5	84.4
V	0.1	0.1	0.66	0.25
Cr	0.4	0.4	0.36	0.45
Fe	7.0	9.0	31.5	129.0
Mn	0.06	0.1	1.8	10.1
Co	0.03	0.03	0.04	0.06
Ni	0.2	0.5	1.1	1.0
Cu	0.1	0.5	3.6	2.8
Zn	0.3	8.0	1.3	3.3
As	0.2	0.3	0.55	0.77
Sr	0.03	0.04	136.0	27.3
Mo	0.1	0.2	0.78	0.16
Cd	0.2	0.2	0.02	0.03
Sb	0.1	0.09	0.63	0.26
Ba	0.05	0.09	22.2	13.8
Pb	0.3	0.6	0.11	0.13
U	0.1	0.09	0.28	0.05

[a] IDL, instrument detection limit (3σ criterion); SDL, system detection limit based on 22 field blanks. IDL and SDL data from Henshaw et al. (1989).

9.3   Direct water analysis by ICP–MS

9.3.1   *Pneumatic nebulisation*

The vast majority of ICP–MS water analyses reported to date have been accomplished with pneumatic nebulisation. The data in Table 9.2 indicate that, if no significant contamination occurs during the sampling process, the direct determination of many elements in natural fresh waters is possible. Concentrations of most of the certified elements in SLRS-1 and SLRS-2 are at or above the IDLs reported by Henshaw *et al.* (1989). Whilst the direct determination of Cd and Pb would appear to be out of the question, and concentrations of a few other elements (Cr, Co, Fe and As) are uncomfortably close to the IDLs, more recent compilations of estimated IDLs for new instruments present a somewhat more optimistic picture, with values for many elements, including Pb and Cd, in the range $0.001–0.01$ $\mu g l^{-1}$. To the extent that procedural blanks do not limit the MDLs, these improved IDLs are likely to increase the number of elements which can be directly determined. Beauchemin *et al.* (1987b) reported the direct determination of Cr in SLRS-1, but Co, Cd and Pb could be satisfactorily determined only after a chemical pre-concentration, while As was determined after a 20-fold concentration by evaporation. (Lam and McLaren (1990a) made more extensive use of evaporative concentration during the certification of SLRS-2, as described in section 9.4.2.) Iron was not determined because of the severity of isobaric interferences by ArO and ArN polyatomic species on the major isotopes of iron. The direct determination of Fe is now possible with the use of a nitrogen/argon mixed gas plasma (Lam and McLaren, 1990b) in which the level of ArO is substantially reduced, thereby permitting the use of the major isotope of iron, ^{56}Fe.

The experience of Beauchemin *et al.* (1987b) and Henshaw *et al.* (1989) is generally in agreement with that of Sansoni *et al.* (1989), who compared ICP–MS with ICP–AES, GFAAS and ICP–AFS for the multi-element analysis of nearly 100 unpolluted ground and drinking waters from a granitic region. Of 36 major and trace elements considered, ICP–MS was considered unsuitable for only two (S and Si), because of severe isobaric interferences by polyatomic species present in the background spectrum. No particular difficulties were encountered in the determination of Ca, Mg, Na and K at the same time as 30 trace elements, although the median concentrations of 10 of the trace elements (Cu, Cd, Ni, Se, Pb, As, Hg, Mo, Tl and Sn) were below the detection limits. It should be noted that the determination of K requires some care to ensure that the ^{39}K peak is adequately resolved from the much larger ^{40}Ar peak.

9.3.2   *Electrothermal vaporisation and direct sample insertion*

The use of electrothermal vaporisation (ETV) and direct sample insertion

(DSI) devices for the direct analysis of natural waters has been evaluated in a collaborative study conducted by the Geological Survey of Canada and the Ontario Ministry of the Environment (Hall *et al.*, 1988b). A significant advantage offered by both types of devices over nebulisers is the opportunity to remove the water from the sample prior to the vaporisation of the analytes. Since the water in aqueous samples is the major source of oxygen for polyatomic species such as ArO, CaO and CaOH, the severity of certain isobaric interferences can be greatly reduced. An improvement in detection limits is normally also achieved because of the much greater sample transfer efficiency. Hall *et al.* demonstrated the use of ETV–ICP–MS for the direct determination of Fe and Ni in SLRS-1, made possible by the elimination of isobaric interferences by ArO and CaO, and the direct determination of As in SLRS-1 by both ETV and DSI sample introduction.

The same study also revealed some disadvantages of ETV and DSI sample introduction. The only DSI device evaluated was a tungsten wire loop, which was found to have levels of Fe, Ni and Mo impurities which precluded the determination of these elements in SLRS-1. A variety of materials (graphite platforms and W, Ta and Re filaments) was available for the ETV device; the best choice was found to depend on a number of factors, including volatility of the analyte, chemical interaction between the analyte and filament, impurities in the filament material and the possibility of isobaric interference by polyatomic species containing the filament material. Limitations in the speed of the data acquisition system also restricted the number of elements which could be determined simultaneously.

9.3.3 *Gas phase injection*

Although there have been several reports concerning the use of hydride generation in combination with ICP–MS, the use of this technique for the analysis of natural waters has not yet been reported. This is somewhat surprising in view of the fact that the direct determination of many of the gaseous hydride-forming elements in natural waters is usually impossible because of their low concentrations. Beauchemin *et al.* (1987b) were able to directly determine Sb in SLRS-1, but an evaporative pre-concentration was necessary for As, and the determination of Se was not attempted. The relatively poor ICP–MS detection limit for Se is a result of its high ionisation potential and isobaric interferences by argon dimeric species on the three most abundant isotopes. Sansoni *et al.* (1988) found that the median concentration of Sb in about 100 unpolluted ground and drinking waters was only slightly above their detection limit of $0.01 \mu g \, l^{-1}$, while the median values for As, Se and Sn were below this level.

The determination of mercury at $ng \, l^{-1}$ levels in Swedish lake water samples was accomplished by Haraldsson *et al.* (1989) by the direct introduction of Hg vapour generated from the sample by reduction with sodium borohydride.

The more commonly used reducing agent, tin(II) chloride, was found to contaminate the instrument and to seriously degrade the detection limit for tin in other work. The absolute detection limit, defined as three times the standard deviation of the blank, was 8 pg. Thus, to achieve a detection limit of $0.1 \, \text{ng} \, l^{-1}$ for water analysis, a sample volume of about 80 ml would be required.

9.4 Water analysis with chemical separation and/or pre-concentration

Chemical separation and pre-concentration techniques have been used to extend the applicability of ICP–MS to natural waters with salt contents too high for direct introduction, and to the analysis of waters for elements at concentrations too low for direct determination. In some cases, for example the determination of many of the trace elements in seawater, both a separation from the sea salt matrix and a pre-concentration are required. In the case of determination of trace elements in surface fresh waters, only a concentration is required. There has been considerable recent interest in the development of new methods, or the adaptation of existing methods, for use 'on-line' with ICP–MS instrumentation. On-line procedures, when used in combination with flow injection techniques, offer advantages in the reduction of both sample preparation time and required volumes.

9.4.1 *Seawater*

The relatively high salt content of seawater prohibits its direct analysis by ICP–MS. Even if the obvious problems of salt deposition on the torch, sampling interface, or ion lenses can be avoided by the use of flow injection techniques instead of continuous nebulisation, at least a 10-fold dilution is necessary in order to alleviate severe suppression of sensitivity by the high salt concentration. This unavoidable dilution will place the concentrations of all but a few of the elements of environmental interest too low for accurate determination, as can be seen by an examination of Table 9.3, in which the certified concentrations of 10 trace elements in the NRCC reference materials SLEW-1 (estuarine water), CASS-2 (coastal seawater) and NASS-2 (open-ocean seawater) are compared.

Two procedures developed at NRCC for the separation and pre-concentration of trace elements from fresh and saline natural water samples have been adapted for use in the ICP–MS analysis of seawater. The first method accomplishes the separation by passage of the sample over a small column of silica-immobilised 8-hydroxyquinoline (I-8-HOQ); the adsorbed trace elements are then eluted with 10 ml of a mixture of 1 M HCl/0.1 M HNO$_3$. If a sample volume of 500 ml is processed, the method is suitable for the simultaneous determination of Cr, Mn, Co, Ni, Cu, Zn, Cd, Pb and U in

Table 9.3 A comparison of the concentrations ($\mu g\,l^{-1}$) of 11 trace elements in the NRCC natural water reference materials SLEW-1, CASS-2 and NASS-2

Element	SLEW-1[a] Estuarine water	CASS-2[b] Coastal seawater	NASS-2[c] Open-ocean seawater
As	0.765	1.01	1.65
Cd	0.018	0.019	0.029
Cr	0.139	0.121	0.175
Co	0.046	0.025	0.004
Cu	1.76	0.675	0.109
Fe	2.08	1.20	0.224
Pb	0.028	0.019	0.039
Mn	13.1	1.99	0.022
Mo	–	9.01	11.5
Ni	0.743	0.298	0.257
Zn	0.86	1.97	0.178

[a] Salinity 11.6 parts per thousand.
[b] Salinity 29.2 parts per thousand.
[c] Salinity 35.1 parts per thousand.

most samples. It has been successfully applied in the certification of SLEW-1, CASS-1 (McLaren *et al.*, 1985), CASS-2 (McLaren *et al.*, 1990) and NASS-2 (Beauchemin *et al.*, 1988c).

A more recently adapted method, originally reported by Nakashima *et al.* (1988a), involves the co-precipitation of trace elements with iron and palladium by reduction with sodium borohydride. The precipitate is separated by filtration, re-dissolved in nitric acid, and diluted to a final volume of 15 ml. For a 600 ml seawater sample, a 40-fold pre-concentration is achieved, and the simultaneous determination of Cr, Mn, Ni, Co, Cu, Zn, As, Mo, Cd, Pb and U is possible. The method was applied during the certification of CASS-2 (McLaren *et al.*, 1990).

Falkner and Edmond (1990) developed a method for the determination of gold at femtomolar levels in seawater, based on anion exchange pre-concentration of $Au(CN)_4^-$ formed by the addition of KCN to seawater at pH 8. Seawater samples of 4–8 l were passed over a small column containing about 2 ml of anion exchange resin. The gold was eluted with 40 ml of hot nitric acid. After evaporation to reduce the volume and further chemical treatment to ensure solubilisation of the gold, a final solution volume of only 1 ml was reached. Flow injection (of $120\,\mu l$ aliquots) was used to allow multiple determinations and standard additions.

Very recently, a method for the determination of the rare earth elements (REEs) and yttrium in seawater has been reported by Shabani *et al.* (1990). A pre-concentration factor of up to 200 is achieved by a 2-step solvent extraction procedure. Results for all 14 REEs in two seawater samples were presented and shown to be in good agreement with values obtained by isotope dilution thermal ionisation mass spectrometry. Concentrations of some of the heavier REEs were less than $1 \, \text{ng} \, l^{-1}$.

9.4.2 Freshwater

Pre-concentration of trace elements in freshwater samples can be achieved either by evaporation or by chemical procedures like those described in the previous section. Evaporation is certainly the simplest method to achieve a 10- to 20-fold concentration, although care must be taken to avoid contamination of the sample during the process. For example, Beauchemin et al. (1987b) found evaporation a convenient method to concentrate a sample of the river water SLRS-1 for As determination. Lam and McLaren (1990a) used a 10-fold concentration by evaporation to increase the number of elements which could be determined in SLRS-2 without a chemical pre-concentration. As shown in Table 9.4, accurate results were obtained for all but 3 (Co, Ni and Pb) of 17 trace elements certified in SLRS-2. It was still necessary to employ a chemical separation and pre-concentration (by adsorption of silica-immobilised oxine) in order to obtain accurate results for Co and Ni, both of which are subject to isobaric interferences by CaO and CaOH species. The probable explanation for the high Ni result is isobaric interference on ^{60}Ni by $^{44}\text{Ca} \, ^{16}\text{O}$, a problem also reported by Beauchemin et al. (1987b) and Sansoni et al. (1988) for which the solution is either arithmetic correction or a chemical separation of the

Table 9.4 ICP–MS certification data for the river water reference material SLRS-2[a]

Element	Direct determination ($\mu g \, l^{-1} \pm SD$)	10 × evaporation ($\mu g \, l^{-1} \pm SD$)	50 × 1-8-HOQ ($\mu g \, l^{-1} \pm SD$)	Certified value ($\mu g \, l^{-1} \pm SD$)
Al	80.5 ± 3.1	–	–	84.4 ± 3.4
As	–	0.78 ± 0.01	–	0.77 ± 0.09
Ba	–	13.6 ± 0.1	13.8 ± 0.1	13.8 ± 0.3
Cd	–	0.028 ± 0.003	0.030 ± 0.002	0.028 ± 0.004
Co	–	–	0.069 ± 0.009	0.063 ± 0.012
Cr	–	0.50 ± 0.03	0.31 ± 0.01	0.45 ± 0.07
Cu	–	2.83 ± 0.05	2.72 ± 0.04	2.76 ± 0.17
Fe	125.0 ± 2	–	–	129.0 ± 7.0
Mn	10.2 ± 0.2	–	–	10.1 ± 0.3
Mo	–	0.159 ± 0.005	–	0.16 ± 0.02
Ni	–	1.45 ± 0.04	1.07 ± 0.05	1.03 ± 0.10
Pb	–	0.17 ± 0.01	0.131 ± 0.006	0.129 ± 0.011
Sb	–	0.31 ± 0.05	–	0.26 ± 0.05
Sr	–	27.1 ± 0.1	27.4 ± 0.2	27.3 ± 0.4
U	–	0.049 ± 0.001	0.049 ± 0.001	0.049 ± 0.002
V	–	0.26 ± 0.01	–	0.25 ± 0.06
Zn	–	3.4 ± 0.1	3.3 ± 0.1	3.33 ± 0.15

[a] Data from Lam and McLaren (1990a).

Ni from the bulk of the Ca. Garbarino and Taylor (1987) found arithmetic correction to be unreliable, especially at calcium concentrations above 50 mg/l. Chemical separation is the preferred route for Co, not only because of its rather low concentration, but also because of the possibility of isobaric interference by $^{43}Ca^{16}O$ and $^{42}Ca^{16}OH$. A more accurate and precise result was also obtained for Pb after chemical pre-concentration. The high Pb result may have been caused by contamination during the evaporation, in spite of the fact that the operation was conducted in a Class 10 clean laboratory. In contrast, the result for Cr obtained after chemical pre-concentration is in poor agreement with the certified value, despite the use of isotope dilution methodology. This surprising result was observed previously during the certification of SLRS-1, and is believed to be due to a failure to obtain chemical equilibration of the ^{53}Cr isotopic spike with the Cr present in the water. It would appear that the degree of retention of the ^{53}Cr spike on the silica-immobilised oxine (I-8-HOQ) differs from that for the Cr present in the water. The failure to achieve chemical equilibration is irrelevant in the case of direct determination, as is indicated by the better agreement of the result with the certified value.

It should be noted that the accurate determination of As in some freshwaters may be complicated by the presence of concentrations of chloride high enough to give rise to significant isobaric interference by $^{40}Ar^{35}Cl$ on ^{75}As. This was apparently not a problem for the two NRCC freshwater standards. At higher chloride concentrations, arithmetic correction may be required unless a chemical separation is performed (e.g. by means of the reductive co-precipitation procedure described in the previous section).

9.4.3 On-line separation and pre-concentration

On-line methods for the separation and pre-concentration of trace elements from natural waters offer savings both in time and sample consumption. In addition, the risk of contamination from airborne sources in a closed on-line system is less than in an open laboratory. The chemical separation and/or concentration is often accomplished by a miniaturised version of a well characterised off-line technique (e.g. Nakashima et al., 1988b). Flow injection methodology is usually incorporated to facilitate both the delivery of a relatively small sample volume to a miniature column which effects the separation, and the subsequent transfer of the analytes to the ICP in a volume normally less than 1 ml. If concentration of the analytes is required, pre-concentration factors of 100 or more can be achieved with sample volumes as small as 100 ml. The success of this approach depends on a number of factors, perhaps the most important of which is the control of blanks.

Plantz et al. (1989) described an on-line method for the separation and pre-concentration of trace elements from high-salt matrices such as seawater and urine based on adsorption of their neutral bis(carboxymethyl)dithiocarbamate complexes onto a styrene-divinylbenzene resin (XAD-4) at pH 3.5, followed by elution with 0.1 M ammonium hydroxide. The complexing agent is first added to the sample, which is mixed with, and carried onto the column by a 0.05 M formate buffer at pH 3.5. The neutral complexes are adsorbed. The pH change (from 3.5 to 11) caused by the strongly basic eluent results in the dissociation of the carboxylic acid groups of the complexing agent and release of the now anionic complexes from the column. Pre-concentration factors of about 20 were achieved by using a 5.0 ml injection loop for sample loading, and a 0.5 ml loop for the elution. The feasibility of the method was illustrated by its application to the separation of Cr, Co and Ni from the NRCC coastal seawater reference material CASS-1.

Beauchemin and Berman (1989) adapted the off-line I-8-HOQ method described in section 9.4.1, and miniaturised by Nakashima et al. (1988b), for use on-line with ICP–MS. A small column containing about 80 mg of I-8-HOQ was placed in the sample stream; a Teflon sample injection valve immediately downstream of the column allowed passage of the stream either to the nebuliser or to waste. Sample volumes of up to 10 ml were passed though the column prior to elution with 1 ml of a 2 M HCl/0.1 M HNO_3 mixture. Surprisingly, it was found to be necessary to add an ammonium acetate buffer to the samples in order to obtain reproducible elutions. Although a pH adjustment of acidified samples to approximately 8.0 is necessary in the off-line procedure, buffering is not required, nor was it found to be necessary by Nakashima et al. (1988b). Sample volumes of only 1 ml sufficed for the determination of Mn and Cu in the river water reference material SLRS-1; 10 ml were sufficient for the determination of Co, Pb and U. Co-elution with Ni of some Ca retained on the column degraded accuracy for Ni because of isobaric interference by $^{44}Ca^{16}O$ on ^{60}Ni. The method was also applied to the determination of Mn, Mo, Cd and U in the seawater reference material NASS-2. High blanks for Fe and Zn precluded their determination in either SLRS-1 or NASS-2.

Heithmar et al. (1990) employed a more elaborate device in which metals were adsorbed onto a prototype macroporous iminodiacetate resin column developed by Dionex Corporation. Since water samples were normally acidified, they were first mixed on-line with an ammonium acetate buffer solution to adjust the pH to 5.5. After passage of up to 10 ml of the sample, the elution was performed with 1 M HNO_3. Effective separation of nine elements (Ti, V, Mn, Fe, Co, Ni, Cu, Cd and Pb) from potentially interfering elements such as Na, Ca and Mg, and good recoveries from spiked samples of synthetic seawater and digested wastewaters were demonstrated. High blanks for Zn prevented its inclusion in the scheme.

Boomer et al. (1990) also used a commercially available column for on-line pre-concentration. A Dionex Ionpac cation guard cartridge was shown to be effective for the removal of Cu, Cd and Pb from aqueous solutions over the pH range 3–9. Solutions were loaded at a flow rate of 2.0 ml min^{-1}; the eluent was 5% HNO_3 at 3.5 ml min^{-1}. Detection limits were improved by a factor of 10–20 when 20 ml samples were passed over the column.

The desirability of on-line versus off-line methods for the separation and pre-concentration of trace elements from natural waters prior to ICP–MS determination hinges on a number of factors. If too large a concentration factor is required, the efficiency of instrument utilisation may be unacceptably low. It would also appear that, although the risk of airborne contamination is clearly reduced with closed, on-line systems, the pumps, valves and other plumbing may be significant sources of contamination.

9.5 Calibration strategies

All of the calibration strategies for ICP–MS, described in chapter 6, can be put to good use in the analysis of natural waters (Taylor, 1989). The large numbers of samples generated in survey and monitoring programs will usually dictate the use of external calibration, or even semiquantitative analysis, in order to maximise throughput. In contrast, isotope dilution, the most time-consuming calibration strategy, is the method of choice in the certification of natural water reference materials. The method of standard additions, which occupies an intermediate position as far as throughput is concerned, has seen limited use as an adjunct to isotope dilution for certification of monoisotopic elements in reference materials, and may also find applications in the analysis of samples containing high levels of dissolved salts, e.g., seawater and some groundwaters.

9.5.1 External calibration

External calibration is the method of choice for ICP–MS analysis in large survey and monitoring programs. A common feature of many of the reported applications (Boomer and Powell, 1986; Garbarino and Taylor, 1987; Henshaw et al., 1989) is the use of one or more internal standard elements not only to improve precision, but to compensate for multiplicative interferences. The use of multiple internal standards is based on the observation of pronounced mass dependence of the drift characteristics of many instruments, regardless of whether this is induced by the samples or arises internally. Mass dependence of interference effects has also been commonly observed; the magnitude of suppression by a high concentration of a concomitant element (e.g. Na) is often inversely proportional to analyte mass. The evolution of ICP–MS instrumentation has in general reduced, but not eliminated, these effects. Based on the performance of an early instrument, Taylor (1989) recommended that internal standards be chosen to match the analytes fairly closely with respect to mass, and if possible, ionisation potential; of course, the chosen elements must be present at only negligible concentrations in the samples and will preferably be mono-isotopic. Based on these criteria, the following isotopes were suggested: ^{45}Sc, ^{103}Rh, ^{115}In, ^{159}Tb and ^{181}Ta. In preliminary analyses by external calibration prior to isotope dilution determination of seven elements in standard reference water samples, ^{72}Ge was used for Ni, Cu, Sr and Cd, ^{141}Pr for Ba, and ^{187}Re for Tl and Pb (Garbarino and Taylor, 1987). Method 200.8, developed by the US EPA, requires the use of internal standardisation in all analyses to correct for instrumental drift and physical interferences. For analyses covering the full mass range, a minimum of three internal standards, chosen from a list of nine isotopes (^{6}Li, ^{45}Sc, ^{89}Y, ^{103}Rh, ^{115}In, ^{159}Tb, ^{165}Ho, ^{175}Lu and ^{209}Bi) must be used.

Henshaw et al. (1989) developed a method for the analysis of 250 lake water samples for 49 elements. An interesting feature of this work is that external calibration was performed for only 21 of the elements, while semi-quantitative results for the other 28 were based on 'surrogate' standards chosen from the first 21. Accuracy of these latter data was found to be $\pm 25\%$ provided that the ionisation potentials of the analyte and its surrogate were similar; this requirement posed some difficulties for the determination of elements with high ionisation potentials, e.g. Zn. A close match in mass is not required because both the analyte and surrogate raw signals are ratioed to whichever of four internal standard elements (^{49}Ti, ^{115}In, ^{159}Tb or ^{209}Bi) matches it most closely with respect to mass. Equally important as the accuracy of the data was the fact that long term precisions were better than $\pm 10\%$, greatly facilitating the comparison of data obtained over several months.

9.5.2   Standard additions

Rather limited use has been made of the method of standard additions, a potentially useful strategy for the analysis of samples with high dissolved solids. Analysis by external calibration may give inaccurate results if concomitant elements present in the samples, but not in the calibration solutions, are at concentrations high enough to cause multiplicative interferences (usually suppression). The major reason for the infrequent use of standard additions is probably the success that has been achieved in the use of internal standards in analyses by external calibration to compensate for these effects. This seems to be a more efficient solution to the problem. The method of standard additions has been used at the NRCC for the certification of the mono-isotopic elements Mn, Co and As in fresh and saline natural water reference materials (McLaren et al., 1985; Beauchemin et al., 1987b; McLaren et al., 1990).

9.5.3   Isotope dilution

Isotope dilution is the preferred method for the certification of elemental concentrations in natural water reference materials, both at the NRCC

(McLaren *et al.*, 1985, 1990; Beauchemin *et al.*, 1987b, 1988c) and at the United States Geological Survey (Garbarino and Taylor, 1987; Taylor, 1989). In practice, the precision and accuracy of results obtained by other strategies will only occasionally match those of isotope dilution. The NRCC has employed isotope dilution ICP–MS both for the direct determination of trace elements in natural fresh waters and for their determination in fresh and saline waters after separation and pre-concentration. The US Geological Survey has developed a method for the direct determination of seven elements (Ni, Cu, Sr, Cd, Ba, Tl and Pb) in a series of 48 standard reference water samples in which the concentrations of some trace elements have been augmented by spiking.

10 Analysis of solid samples

10.1 Introduction

Since the appearance of the first commercial ICP systems in analytical laboratories in the late 1970s, sample dissolution has been a prerequisite for most applications. Indeed, the analysis of samples in solution remains the usual method in both IPC–AES and ICP–MS. However, many sample types occur in a solid form, e.g. rocks, minerals, metals, plants, chemical compounds and their direct analysis would, at least, decrease sample handling and could provide advantages over the solution technique.

A number of real advantages can be identified with respect to the analysis of solids. Dilution of a sample normally results during a dissolution procedure, reducing the actual determination limits available. Some solid sample introduction techniques, e.g. laser ablation, allow the sample to be analysed directly and therefore introduce little dilution of the sample. The time required for some solution preparation schemes can be extensive, typically several hours to one day. This preparation time may be considerably reduced for solid sample introduction where samples are typically ground to a fine powder and suspended in a fluid medium, mixed with a binder and pelletised or analysed directly if required. Contamination from reagents is also a factor which may control determination limits in solution analysis. In general, contamination can be minimised for solids analysis since sample handling is minimal, and the agents added as binders, for example, tend to have an organic matrix. One of the major advantages of solids analysis for some sample types such as ceramics, refractory minerals and catalysts is simply that dissolution is not required. Such samples are difficult to bring into solution, and usually require the addition of an alkali flux in order to produce a low temperature melt, which is soluble in dilute acid. The addition of a flux severely limits determination limits by ICP–MS (see chapter 7) due to the enhanced levels of total dissolved solids (TDS). In addition, some elements, e.g. B, Cr, Sn, may be lost as volatile species from the sample during the dissolution procedures commonly used. A reduction or elimination of the levels of H_2O entering the plasma is a highly desirable feature, since it results in a reduction in the levels of some polyatomic ions and this is possible with some methods of solid sample introduction (see chapter 5). One final advantage, specifically using laser ablation, is that both bulk sampling and discrete profiling can be carried out. In this way detailed elemental or isotopic distributions through a sample can be investigated, e.g. mineral zoning.

There are, however, a number of limitations which currently exist with most solid sampling techniques. Sample inhomogeneity is a serious consideration which can to some extent be overcome when a sample is brought into solution. In addition, particles of solid material can be fractionated during transport of the sample into the ICP particularly when the vapour generation process is remote from the ICP. The most serious problem is, however, calibration and quantitative measurement. This will be discussed below for each of the introduction techniques but usually requires that the sample and standard are of similar chemical matrix and physical form. Even when this is achieved, a mis-match of elemental response between sample and standard may occur.

Although a number of different methods of solid sample introduction have been used with ICP–AES (the reader's attention is drawn to Routh and Tikkanen, 1987, for a full review of these methods), only a limited number have been evaluated with ICP–MS, i.e. laser ablation, slurry nebulisation, direct sample insertion, powdered solids introduction and arc nebulisation. Slurry nebulisation and laser ablation have, to date, received the most attention. The former requires no special additional equipment except for an appropriate nebuliser and can therefore be used on any ICP–MS instrument. The latter is available at considerable extra cost to the basic instrument but offers the possibility of both direct bulk analysis and discrete profiling.

There are a number of general requirements for any method of solid introduction. The technique should be capable of introducing a representative sample aliquot. In addition, the rate of sample introduction should be constant and reproducible if continuous introduction techniques are used, e.g. slurry nebulisation, or of sufficient duration if introduction is by a fixed volume technique, e.g. direct sample insertion. It is important to note that limitations on plasma loading apply equally to solid and solution sample introduction. At any given set of operating conditions, the plasma energy is only sufficient to dissociate and ionise a finite amount of material without unacceptable disturbance to the equilibrium. If this load is exceeded, undissociated material passes through the plasma and will deposit on the sampling cone orifice, resulting in a loss of sensitivity. In addition, if major particles pass through, some elements may not be released into the plasma and therefore not determined. Deposition on the sampling cone can also occur if the load is unacceptable or particles of solid are too large for the dwell time provided by sampling extraction position and gas flow rate.

10.1.1 Calibration

Quantitation can be achieved in a number of ways using different basic methods of calibration. In slurry nebulisation for example, calibration is achieved using synthetic aqueous based standards which do not necessarily have the same matrix as the sample itself. Standards can be prepared covering the range of expected concentrations for a large number of elements. In laser ablation, by contrast, the standard is normally in a solid form and the use of 'gel' standards for laser ablation work was explored by Gray (1985b). The 'gels' are elemental mixtures, added to silicon dioxide, co-precipitated and dried to form a stable calibration standard (Date, 1978). However, data suggest that these matrices are too 'ideal' to calibrate most sample types. An alternative method is the use of certified reference materials which are available in many different matrices. This approach is not ideal, many reference materials are poorly characterised, having good values for only a few elements and are specifically designed as reference points, not as primary standards. Many materials are also in short supply and their use with destructive solid sampling techniques soon results in depletion of stocks. However, with a linear response of over 5 orders of magnitude, a single point calibration can be used. A viable alternative for some methods of solid sample calibration is the use of synthetic glasses. These can be produced with a range of elemental concentrations, of homogeneous composition and in large volume.

10.2 Slurry nebulisation

A slurry may be defined as a uniform suspension of small particles. To be of practical use as a method of introducing samples into an ICP spectrometer, the slurry must be of low viscosity. In an ideal situation, the transport properties of a slurry should be similar to those of an aqueous solution, since the behaviour of the slurry in the spray chamber, torch and plasma will then be similar to that of a solution. If these criteria are fulfilled, it should then be possible to calibrate the system using aqueous standards. In an ideal slurry, all particles would be of uniform small size ($< 5 \, \mu m$) so that they will remain suspended in a fluid medium during an analytical run. Some, if not all, of these aims have been achieved with some degree of success in a range of sample types including catalysts, ceramics, coals, soils and rocks.

10.2.1 *Grinding techniques*

A wide range of grinding techniques and materials have been used in slurry preparation. The most suitable grinding method and the times taken to reduce the particle size of a sample are highly dependent on the sample type, particularly its mechanical strength. In general, brittle hard materials can be ground with the most success. However, some geological samples may present a particular problem since they are generally composed of a range of minerals of varying grain size, hardness and friability. In principle, if the particle size is reduced to less than about $10 \, \mu m$, the slurry may be expected to behave in a similar manner to a solution during the nebulisation process.

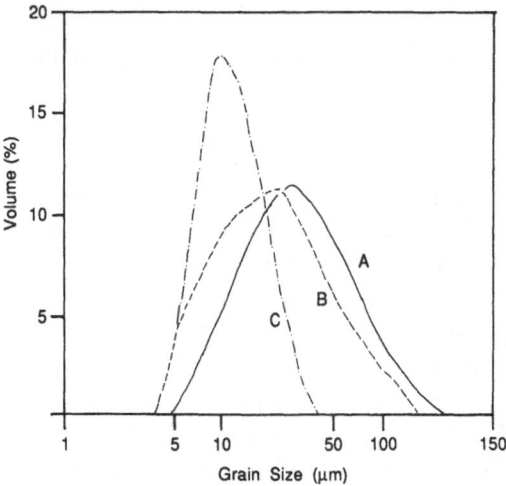

Figure 10.1 Particle size distribution of the USGS glass SRM GSE: (A) original, (B) after 20 min regrinding and (C) after an additional 40 min grinding (after Verbeek and Brenner, 1989).

Unfortunately, the classic grinding techniques used in many laboratories do not generally fulfil this requirement and some can also lead to contamination of the sample. Sample throughput may also be limited.

The grain size distribution of a USGS SRM Glass is shown in Figure 10.1 after different grinding periods in a rotary grinder using tungsten carbide rings. The original distribution pattern (A) is shifted by only a small amount after 20 min grinding. After a further 40 min, the mean grain size is around 10 μm but with about 50% of the sample still remaining over this grain size. An alternative grinding method (e.g. Jarvis and Williams, 1989) is the use of small plastic bottles containing grinding beads composed of either zirconium oxide or agate (see below for details). The sample is added to the bottle and beads and is shaken vigorously for a number of hours. The grain size distribution which can be achieved, varies according to the shaking time. Jarvis and Williams (1989) examined a range of geological sample types (granite, diorite, limestone and sandstone) which were ground for 2, 6 and 15 h using this method. It is clear from this study that the physical properties of a sample are extremely important in limiting the effectiveness of the grinding process. Relatively homogeneous materials containing minerals of low hardness, such as the limestone, can be reduced in grain size to less than 6 μm in only 2 h. Harder samples may not require longer grinding times provided the sample is of relatively uniform mineralogy (e.g. sandstone which is composed dominantly of quartz). The diorite and granite, however, represent samples which are of variable mineralogy. The grain-size distribution plots of these samples are more complex, indicating that a prior knowledge of sample mineralogy may be helpful to achieve optimum grinding times and performance.

10.2.2 Dispersing agents

In addition to a range of grinding methods, a number of different 'suspending' or 'dispersing' agents have been employed. These agents are required to keep the ground particles suspended and to prevent flocculation. Nearly every published application has involved the use of a different dispersing agent, including aqueous ammonia (Ebdon and Collier, 1988b: ICP–AES), sodium hexametaphosphate (Ebdon et al., 1989: ICP–AES) and tetrasodium pyrophosphate (Jarvis and Williams, 1989: ICP–MS). In addition, a number of commercially available products have been used including Triton X-100 (Ebdon and Wilkinson, 1987b: ICP–AES), Calgon (sodium polyphosphate and sodium carbonate) and Dispex (sodium polyacrylate) (Ebdon and Collier, 1988b: ICP–AES), Viscalex HV30 and NOPCO NPZ (Fuller et al., 1981: ICP–AES) and Aerosol OT (Ebdon et al., 1988: ICP–MS). To date there has been no systematic study of these dispersing agents, and their choice seems to be very much a matter of personal preference.

10.2.3 Particle size distributions

At the time of writing, relatively little data have been published for slurry nebulisation with ICP–MS. A review by Jarvis (1991) documents the current state of the technique with respect to geological samples by both ICP–AES and ICP–MS. A number of studies have, however, examined transport effects (e.g. Ebdon and Collier, 1988a, b) and the temperature distribution in the plasma during the introduction of slurried samples (Verbeek and Brenner, 1989) using ICP–AES. A number of general conclusions can be drawn. In general, atomisation efficiencies and therefore sensitivity are biased in favour of slurries containing the highest proportion of small particles (Halicz and Brenner, 1987; Ebdon and Wilkinson, 1987a). Grinding problems led to a series of experiments by Ebdon and Collier (1988a, b) who used a natural kaolin clay sample (which required no grinding) to monitor fractionation of the sample during its passage through the nebuliser, spray chamber and torch. They concluded that spray chamber design was an important factor (see chapter 3) and that the Scott double pass design (supplied by most ICP–MS instrument manufacturers) was unsuitable for slurry nebulisation since it led to excessive fractionation of the sample. In the authors' laboratory, a single pass design is preferred.

Beyond the spray chamber, the torch injector can also limit the size of particles which are able to enter the plasma. Ebdon and Collier (1988a) observed a cut-off point of 2–2.5 μm for a 2.2 mm injector tube while a 3 mm i.d. injector apparently limited the particle size passing through it to 6–7 μm diameter. However, while a 4 mm injector, for example, will allow droplets or particles < 16 μm to pass through freely, it also produces an unstable plasma (Ebdon and Collier, 1988a) which must lead to both poor precision and accuracy.

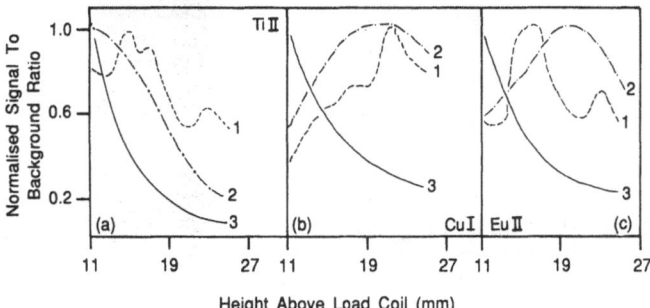

Figure 10.2 Variation in normalised signal to background ratio with height in an argon-nitrogen ICP for (1) glass slurry, (2) aqueous solution and (3) slurry dispersant blank (after Verbeek and Brenner, 1989).

Differential fractionation of the slurry by grain size anywhere in the system can lead to poor analytical data if different grain-size fractions are of different compositions as might be expected in geological samples which are generally polymineralic. In an attempt to overcome such sample inhomogeneity problems, Verbeek and Brenner (1989) prepared some fused silicate glasses which were subsequently re-ground to reduce their mean particle size to about 12 μm. These authors reported the signal to background ratios (SBR) across the plasma profile of an ICP–AES for the slurried glasses and for comparable aqueous solutions. Three elements were studied, Ti (334.941 nm) and Eu (381.967 nm) using ion lines while Cu was monitored using the 324.754 nm atomic line. Although the profile for the slurry and aqueous solution are broadly similar, it is interesting to note that the changes in SBR for aqueous solutions are smooth while those for the slurries are rather erratic (Figure 10.2). The authors concluded that the differences in profiles observed in the plasma were due to differential decomposition and volatilisation of the glass samples, probably due to the variable temperature across the plasma profile and sample particle location in the central channel. The effects were reduced by use of a mixed-gas argon-nitrogen plasma compared with an all argon plasma probably because the former provides a hotter and more compact central channel.

10.2.4 Applications of slurry nebulisation

In ICP–AES applications slurries are typically analysed at concentrations of between 1 and 30% m/v. However, for analysis by ICP–MS it is necessary to limit the levels of total solids (or total dissolved solids) to less than 2000 μg ml^{-1} in order to prevent blocking of the sampling cone orifice (Williams and Gray, 1988). A feasibility study was undertaken by Williams *et al.* (1987) for the analysis of some soils and industrial catalysts while a further study by Ebdon *et al.* (1989) examined the analysis of coal by ICP–MS. In the work by Williams *et al.* (1987), a number of trace and major elements

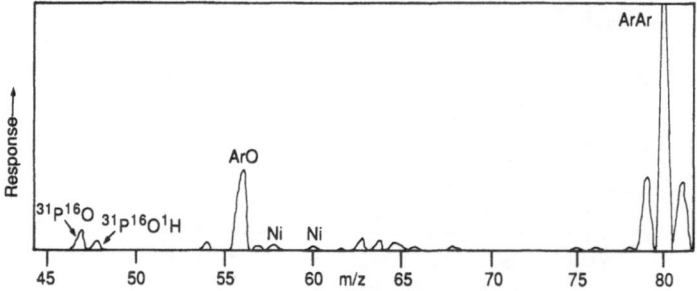

Figure 10.3 Blank spectrum for a 0.05 g per 100 ml solution of $Na_4P_2O_7$ showing phosphorus polyatomic peaks at 47 and 48 m/z and Ni contamination from the sampling cone (from Williams *et al.*, 1987).

(principally transition metals) were determined at a dilution of 0.05 g in 100 ml, while major elements were measured in a preparation diluted a further 200 times. Samples were ground using a bottle and bead method (see below) and dispersed in a 0.05% m/v solution of tetrasodium pyrophosphate ($Na_4P_2O_7$). As a compromise, a 3 mm diameter torch injector tube was used and gas flow rates were optimised to give maximum signal whilst maintaining polyatomic and oxide ion interferences at a minimum. The blank spectrum for a 0.05% m/v solution of $Na_4P_2O_7$ (Figure 10.3) clearly shows small amounts of Cu (63, 65 m/z), Zn (66, 68 m/z) and Br (79, 81 m/z) contamination. Some Ni is also present, which is thought to be derived from the sampling cone. In addition to these elemental peaks, two polyatomic ion peaks are seen at 47 and 48 m/z resulting from the formation of $^{31}P^{16}O$ and $^{31}P^{16}O^1H$, respectively. These interferences which coincide with two of the isotopes of Ti, however, are not serious in practical analysis, since other isotopes are available for the Ti determination. Low level of contamination and minimum polyatomic ion formation, therefore make $Na_4P_2O_7$ a useful dispersing agent. In addition, the viscosity of a 0.05% m/v solution is close to that of water, making it well suited for use in ICP–MS analysis. Some major and trace element data obtained by Williams *et al.* (1987) for a mixed and chromium oxide catalyst are shown in Table 10.1. There is reasonably good agreement with the reference values.

The determination of a wide range of elements in some geological SRMs was recently undertaken by slurry nebulisation ICP–MS (Jarvis and Williams, 1989). The samples were ground using the bottle and bead method (see Jarvis and Williams, 1989) and analysed at a total solids concentration of 0.2%.

Method (from Jarvis and Williams, 1989)
Equipment
  (i) 30 ml high density plastic bottles
 (ii) zirconia or agate grinding beads
(iii) Laboratory flask shaker

Table 10.1 A comparison of results for two catalyst samples by slurry nebulisation[a]

	Slurry ICP–MS	Slurry ICP–AES	Solution ICP–AES
Mixed oxide catalyst			
Major element (wt%)			
Al_2O_3	53	49	54
SiO_2	1.2	1.3	1.3
MgO	7.9	7.8	7.9
CuO	–	14	19
Trace element ($\mu g\,g^{-1}$)			
Mn	100	70	88
Fe	700	810	1100
V	16	16	< 40
Cr	16	23	< 50
Co	15	11	< 40
Ni	41	42	37
Chromium (III) oxide catalyst			
Major element (wt%)			
SiO_2	1.2	1.3	1.3
CrO	–	83	91
Trace element ($\mu g\,g^{-1}$)			
Al	870	1000	900
Mg	250	215	450
Mn	20	20	17
Fe	1030	900	1200
V	179	155	162
Co	17	18	24
Ni	< 16	26	24
Cu	52	57	112

[a] From Williams *et al.* (1987).

Reagents
(i) Tetrasodium pyrophosphate ($Na_4P_2O_7$) solution 0.05% w/v

Procedure
0.1 g of powdered sample is weighed into a new 30 ml plastic bottle and 2 ml of tetrasodium pyrophosphate solution are added. To this, 10 g of either zirconia or agate beads (see text) are added. The bottles are sealed with plastic tape to prevent leakage. The bottles are placed on a laboratory flask shaker and shaken vigerously for 12–15 h or overnight. Transfer to 100 ml volumetric flasks and make up to volume with 0.05% w/v tetrasodium pyrophosphate solution. Analyse immediately.

The modified operating conditions used for slurry nebulisation ICP–MS are given in Table 10.2. The RF forward power is increased to 1.5 kW and the standard 1.5 mm injector is replaced with a 3 mm diameter injector following the work of Williams *et al.* (1987) and Ebdon and Collier (1988a). The De Galan nebuliser used is a commercially available Babington-type

Table 10.2   ICP–MS operating conditions used for slurry nebulisation[a]

Forward power	1.5 kW
Reflected power	< 5 W
Plasma gas	Ar
Coolant gas flow	13 l min^{-1}
Nebuliser gas flow	1.1 l min^{-1}
Auxiliary gas flow	0.5 l min^{-1}
Sample uptake	2.0 ml min^{-1}
Torch injector diameter	3 mm
Nebuliser	De Galan, V-groove, high dissolved solids
Spray chamber	Water cooled single pass spray chamber maintained at 13°C
Ion lenses	Optimised on ^{59}Co and ^{238}U

[a] After Jarvis and Williams (1989).

design which is tolerant to high levels of dissolved solids (see section 3.2.4). Calibration was made using two standards prepared in 0.05% $Na_4P_2O_7$, one for those elements from nitric acid stocks and one for those from an HCl matrix.

To help alleviate initial signal loss due to deposition of material on the sampling cone orifice, a slurry sample should be continuously aspirated for 10 min prior to calibration and analysis, and using this procedure an equilibrium can be established between condensation and evaporation (Williams and Gray, 1988). If this 'priming' process is adopted, signal drift due to further cone blockage is minimised and can easily be corrected (see section 6.9.2). To ensure that a uniform suspension is maintained a magnetic stirrer has been used (Ebdon and Collier, 1988b), however, this was found to be impractical for some geological sample types since it results in the removal of magnetic mineral phases from the sample. Instead, a small ultrasonic tank is recommended to agitate the sample during analysis.

10.2.4.1 *Detection limits, precision and accuracy*   An assessment of ICP–MS instrumental performance was made by Jarvis and Williams (1989) under the modified operating conditions used for slurry nebulisation analysis (Table 10.2). Instrumental detection limits were determined in a matrix of $Na_4P_2O_7$ for a number of elements across the mass range (Table 10.3). Clearly neither the use of this particular dispersing agent nor the modified operating conditions appear to compromise analytical performance.

An important factor in the assessment of any analytical method is the reproducibility of the analytical data, both for replicate sample preparations and repeated analysis of a single sample. An estimate of intrasample precision for some volatile elements in SRM marine sediment MESS-1 is shown in Table 10.4 and is better than 10% RSD for most elements. Intersample precision is a little poorer, although also typically better than 10% RSD (Table 10.5). In general, those elements which display poorest precision are

Table 10.3 A comparison of ICP–MS three sigma detection limits[a]

Element	Isotope	Detection limit (ng ml^{-1})	
		1% HNO$_3$ 1.5 mm injector	Na$_4$P$_2$O$_7$ 3 mm injector
Be	9	1.16	0.72
B	11	3.78	11.1
Cr	52	0.11	1.52
Ge	74	0.96	0.32
As	75	2.84	0.20
Nb	93	0.11	0.92
Mo	95	0.24	0.15
Sn	120	0.21	0.12
Sb	121	0.19	0.08
Ta	181	0.09	0.03
W	184	0.11	0.07
Bi	209	0.11	0.96
Th	232	0.04	0.02
U	238	0.02	0.04

[a] From Jarvis and Williams (1989).

Table 10.4 Intrasample precision for some volatile elements in marine sediment SRM MESS-1

Element	Determination		
	Mean[b] (μg g^{-1})	SD	RSD (%)
Cr	62.5	1.61	3
As	8.90	0.31	3
Cd	1.26	0.10	8
Sn	12.2	1.21	10
Sb	0.78	0.12	15

[a] From Jarvis (1991).
[b] Mean values calculated from $n = 3$ (triplicate analysis of a single slurry).

present at the lowest concentrations (e.g. Sb). The accuracy of the data for some geological samples was assessed by comparison with reference values (Jarvis and Williams, 1989). Accuracy was found to be satisfactory for a range of different elements including U and Th. Data for some geologically important elements for the USGS SRM AGV-1 are given in Table 10.6.

10.2.4.2 *Determination of chlorine* There are a number of other elements which can be usefully determined using slurry nebulisation ICP–MS. Traditionally, elements such as the halogens are determined in solution using an ion selective electrode. The process of sample decomposition may lead to

Table 10.5 Intersample precision for some volatile elements in syenite SRM SY-2[a]

Element	Mean[a] ($\mu g\,g^{-1}$)	SD	RSD (%)
		Determination	
Cr	9.19	0.19	2
As	15.2	0.08	< 1
Sn	4.92	0.09	2
Sb	0.32	0.07	21

[a] From Jarvis and Williams (1989).
[b] Mean values calculated from $n = 3$ (single analysis of three sample preparations).

Table 10.6 Trace element concentration in SRM AGV-1 determined by slurry nebulisation ICP–MS[a]

Element	Slurry ICP–MS ($\mu g\,g^{-1}$)	Reference value ($\mu g\,g^{-1}$)
Be	2.35	2
B	12.8	7
Cr	11.4	12
Ge	1.3	1.25
As	0.20	0.84
Nb	14.4	15
Mo	1.60	3
Ag	0.75	0.104
Cd	0.86	0.061
Sn	4.06	4.2
Sb	3.88	4.4
Ta	0.90	0.92
W	0.64	0.53
Bi	0.54	0.054
Th	6.56	6.5
U	1.98	1.89

[a] ICP–MS data from Jarvis and Williams (1989), reference values from Govindaraju (1989).

loss of these elements or may cause them to be complexed. The determination of chlorine using slurry nebulisation has been investigated (Jarvis, 1991). Chlorine is only partially ionised (0.9%) in an argon plasma and its sensitivity is rather poor with a detection limit of around $13\,\mathrm{ng\,ml^{-1}}$. Some geological reference materials were prepared for analysis using the methods of Jarvis and Williams (1989) with calibration using a standard solution containing $50\,\mu g\,ml^{-1}$ Cl. A comparison of measured with reference values is shown in Table 10.7. At these relatively high concentrations, 0.1–3% (w/w), the data agree well and the technique would clearly be useful for the determination of Cl in these sample types.

Table 10.7 Determination of chlorine in some geological standard reference materials by slurry nebulisation ICP–MS[a]

Sample	Lithology	Measured (wt%)	Reference (wt%)
MAG-1	Marine mud	3.46	3.09
BCSS-1	Marine sediment	1.20	1.12
MESS-1	Marine sediment	0.82	0.82
NIM-L	Lujavrite	0.11	0.12
Mica-Mg	Biotite mica	0.14	0.08

[a] From Jarvis (1991). Reference values from Govindaraju (1989).

10.2.4.3 *REE determination* Although the REE are normally determined by ICP–MS in solutions prepared for trace element analysis, successful determination using slurry nebulisation is also possible (Mochizuki *et al.*, 1989; Jarvis, 1991). In the work by Jarvis (1991) samples were ground by the method described above but using agate grinding beads. Although zirconia beads are preferred because they are of low cost and are disposable, they lead to significant contamination of the sample with heavy REE. Agate is an excellent alternative, but the cost of each bead is high and they must therefore be cleaned and re-used many times to be cost effective.

In summary, slurry nebulisation is a useful alternative method of sample introduction for the direct analysis of solids. Providing that the sample can be reduced to a sufficiently small and uniform particle size, measured data should be of comparable quality to that obtained by solution nebulisation. Certainly instrument performance is not degraded by the modified operating conditions required for slurry nebulisation. The technique is best applied to materials for which alternative methods of sample preparation are unsuitable and to the determination of elements which are lost as volatile species during other methods of sample preparation. Problems of contamination and particle size reduction currently limit the widespread application of slurry nebulisation.

10.3 Laser ablation

The use of a laser to extract material from a sample for analysis, can be traced back to the early 1960s. Since this time, lasers have been used not only to extract a sample for transport to an ionising or atomisation source, but also as a form of ion source in their own right, e.g. laser ionisation mass spectrometry (LIMS) and laser microprobe mass spectrometry (LMMS). Although LMMS is an established technique its primary advantage of $< 25\,\mu m$ spacial resolution is offset by the major disadvantage that it is a qualitative technique. The interaction of laser light with a solid, the effects of wavelength, influence of repetition rate, etc. will not be dealt with here. The reader's attention is directed to the excellent volume on laser microanalysis

(Moenke-Blankenburg, 1989) and review of laser vaporisation in atomic spectroscopy by Dittrich and Wennrich (1984).

Laser ablation for micro-sampling has been used as a sample introduction technique with ICP–AES for a number of years (e.g. Thompson et al., 1981, 1989, 1990). The early work by Thompson et al. (1981) showed good signal reproducibility on SRM steels. In addition, calibration curves were shown to be linear over a wide range of concentrations while detection limits on the ICP–AES used were in the range $20–50 \mu g\, g^{-1}$ in a steel sample. Thus a number of desirable features of the technique were demonstrated.

The first application of laser ablation for solid sample introduction into ICP–MS was published by Gray (1985b) using a ruby laser with the Surrey research ICP–MS system. Further work by Arrowsmith (1987a, b), Tye et al. (1987), Arrowsmith and Hughes (1988) and more recently Hager (1989) and Mochizuki et al. (1988, 1990), have used both a ruby (wavelength = 694 nm) and Nd:YAG (wavelength = 1064 nm) laser with commercial ICP–MS instruments.

10.3.1 What is a laser?

The principle of the action of lasers is based on the induced, or stimulated, emission of radiation. The term laser is an acronym for 'light amplification by stimulated emission of radiation'. The beam produced is of coherent monochromatic light of very great intensity. When an atom or molecule in its ground state absorbs energy (photons) it is excited to a higher energy level. The energy absorbed may then be spontaneously released to yield a ground state atom once more. In a laser, the excited atom is struck by a photon of exactly the same energy as the one which would be emitted. The excited atom is stimulated to emit a photon and returns to ground state with the result that two photons of precisely the same wavelength are produced. This process is then repeated many times. In a ruby laser, this is achieved using Cr^{3+} ions in the ruby structure while in a YAG (yttrium aluminium garnet) system Nd^{3+} ions are used.

10.3.2 Modes of operation

For most laser ablation techniques, solid-state lasers operated in a pulsed mode are generally used. The optical medium (such as a Cr doped ruby rod or Nd doped YAG) is contained in a resonator composed of two reflectors at either end of the rod. In parallel with the resonator, there is a flash tube which acts as an optical pump. A reflector surrounds both the resonator and flash tube. If the laser is simply pumped by the pulsed flash tube, and the radiation allowed to emerge when the threshold conditions for laser operation are reached, a relatively long train of pulses is emitted (normal mode). As an alternative, the laser may be operated in the Q switching mode by changing

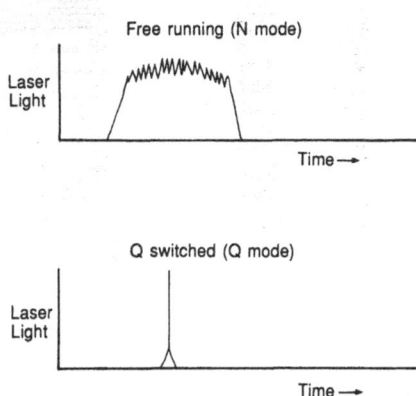

Figure 10.4   Diagram to show the duration of laser light emitted in the normal and Q-switched modes.

the quality of the laser resonant cavity. This can be done in a variety of ways. In the Q switched mode, the lasing operating is delayed, and a pulse is emitted of high power and short duration. 'Passive Q switching' can be achieved by inserting bleachable dyes between the laser rod and a total reflecting prism. In the system described by Moenke-Blankenburg (1989) and Moenke-Blankenburg *et al.* (1990) six cells of a glass chamber containing vanadyl-phthalocyanine in nitrobenzene are inserted. With increasing thickness of this absorbing layer, the number of laser pulses falls off until finally one pulse arises. The most common method of Q switching is the active one in which the light path entering the resonator is obstructed by an electro-optic switch such as a Pockels cell. With normal mode, laser pulse widths in the range $100\,\mu s$ to 1 ms are typical while the duration of semi-Q switched pulses are between 100 ns and $10\,\mu s$. The light intensity patterns for the two modes of operation are shown in Figure 10.4.

10.3.3   *System configuration*

Whichever type of laser is used, the configuration of the cell and sample introduction system is broadly similar. The light beam usually exits from the laser in a horizontal plane and is deflected by 45° or 90°, through the focusing optics onto the sample surface in a fine spot. Viewing of the sample may be via a binocular microscope or video camera connected to a monitor display. Some systems allow continuous viewing of the sample even during the ablation process (with a filter to absorb reflected radiation). In all systems, the sample is located in a quartz or optical glass cell containing an optical window, which allows the laser beam to pass through to the sample unperturbed (Figure 10.5). The dimensions of the glass cell are variable but typically are of small volume (30–$130\,cm^2$) to minimise dilution of the sample vapour. The cell contains a stage so that the sample can be rotated or moved

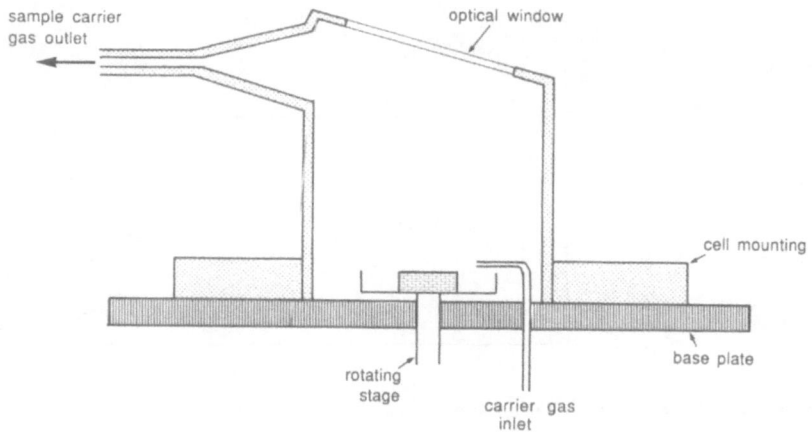

Figure 10.5   Laser ablation cell containing sample mount stage.

in an X–Y motion (rastered) without opening the cell to air. Experimental data for three novel cell designs is given by Arrowsmith and Hughes (1988), who examined a number of parameters including transport efficiency, particle fractionation and gravitational settling.

Prior to ablation and analysis, the cell is purged with a stream of argon which enters either tangentially at the base of the cell or close to the top. This produces a flow around the sample, the vapour then exits from either the top or side of the cell. This carrier gas flow is directed towards the base of the ICP torch (nebuliser and spray chamber are removed) and replaces the normal nebuliser gas flow. The cell may either be located directly under the torch following the design of Carr and Horlick (1982) or connected via a length of tygon tubing (Gray, 1985b). In the latter case the tube is usually relatively short, typically 1 m or so, although for applications where the sample is required to be ablated some distance from the ICP (i.e. for radioactive samples) the carrier tube may be several metres in length. The flexible tube acts as a buffer to the pulse from the laser and may in fact result in a less noisy signal than when the cell is located directly below the torch. Both the carrier tube and the laser cell are potential areas of contamination. Ablated material (particularly from metals) adheres to the walls and top of the cell, and to the walls of the carrier tube, particularly within the first metre. The cell walls may be lined with a plastic film for ease of cleaning and, in an ideal design, the optical window can be removed and cleaned. The tygon carrier tube is best replaced when contamination is suspected.

In the system used in our laboratory, the carrier gas flow to the torch is maintained at all times using a by-pass valve, even when the cell is opened. In this way the plasma is always punched and disturbance is minimised during sample change-over. The sample stage within the cell can accommodate samples up to about 35 mm in diameter, although if small samples are used

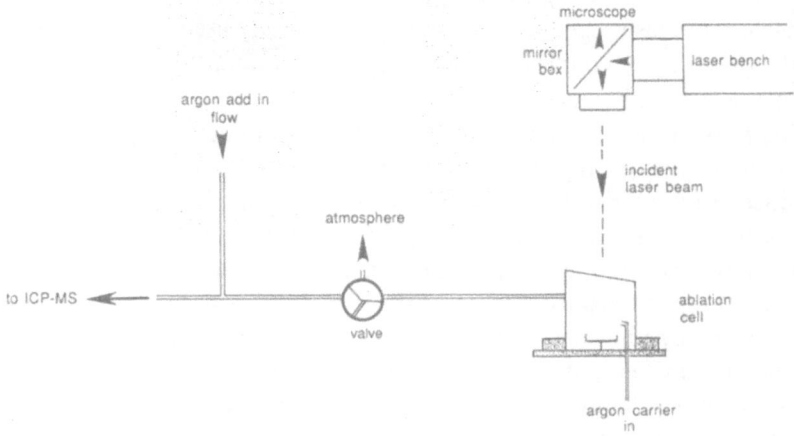

Figure 10.6 General arrangement of laser, ablation cell and gas flows (after Gray, 1985a).

several can be loaded simultaneously (e.g. standard and several samples). Data from our laboratory suggest that providing samples are of similar composition, contamination effects are negligible. Alternatively a glass window (constructed from microscope cover slips) can be placed between the samples to prevent sample cross-contamination. The ruby system used by Gray (1985b) is shown in Figure 10.6. This setup has a cell of $\sim 130\,cm^3$ which is connected to the ICP torch with a 150 cm length of 4 mm i.d. tygon tubing. It is important that the gas path is as straight as possible to avoid solids deposition at discontinuities which can contribute memory to successive analyses. Due to the shorter pulses and higher gas temperatures reached, Q switching tends to produce a bigger acoustic surge in the gas line when the laser is fired and this can disturb deposited material which has become entrained in the sample flow. This particular system has now been automated since the original work in 1985, such that the firing of the laser is synchronised with the operation of the mass spectrometer and the number of shots and pulse duration can be preset. These facilities are available on the dedicated commercial laser systems sold with ICP–MS instruments.

10.3.4 *Laser operation*

Both the ruby and Nd:YAG lasers can be operated in either the normal (Fixed Q) or Q switched mode. Although in principle, the use of Q switched pulses should give the highest proportion of vaporised material, Gray (1985b) found that in practice, this was not the case for pressed powder pellets. Recent work in our laboratory suggests that if the power input is low, the pellet surface is not efficiently ablated, while if the power input is too high, the pellet surface is fractured and cratered leading to non-reproducible sampling. For glasses however, which have a greater mechanical strength but which are almost transparent to laser light of the wavelengths typically used, Q switching may be the only way to couple the laser energy to the sample and thereby bring about effective ablation.

The normal mode of operation, by contrast, produces deep regular pits. Arrowsmith (1987a) reports that only two or three pulses are required to produce a cavity of depth approximately equal to diameter in alumina ceramic compacts. Thus a layer approximately 50 μm thick, corresponding to 1 μg of material is removed by each laser pulse. In normal mode operation, the signal is closely proportional to energy and the recorded integrals are proportional to pulse energy times the number of shots. In Q switching mode, the signal is generally smaller and the species distribution in the spectrum is different. Ablation of a pelletised powder is by explosion of contained gas such as water vapour, as much as by vaporisation (see Thompson et al., 1990 for further discussion).

The sampling strategy may be similar, for either laser type, in either mode of operation. The laser may be operated such that in normal mode successive shots are fired either in the same, or on a new spot. If the former is chosen, the drilled hole becomes successively deeper such that the focal point of the laser beam is eventually located above the sample surface. Successive shots therefore remove less material until the ablation process ceases. If the sample stage is rotated or rastered, the same volume of material can be removed with each shot. In either mode of operation, the number of successive shots is usually fixed, e.g. 10, before data acquisition begins. The transient signal produced is then collected over a relatively short time period, e.g. 30 s. If Q switching is used, again either the sample can be ablated continuously on a single spot or it may be rotated such that a large area of the sample surface is eventually removed. The single spot method is often used with a 'burn in' period to remove the surface of a sample and then for many shots producing a level signal. Vaporisation under these conditions is efficient. Data are typically acquired during the whole ablation period since the amount of material removed in, for example, 1 s is relatively small compared to the normal mode of operation. The generation of a continuous signal is preferred for systems which do not have a multi-channel analyser for data transfer, since the peak hopping mode is generally employed.

10.3.5 Sample preparation

It is sometimes desirable to analyse a solid sample directly in the form in which it originally occurs, particularly if elemental profiling is to be carried out. Surface quality is considered by Arrowsmith (1987a) to be unimportant since it is modified during the first few laser pulses (in the case of ablation on the same area). In addition, samples are not required to be flat or parallel to better than $\pm$ 250 μm across the area to be ablated. Samples suitable for this type of preparation include, for example, steels, alloys, minerals and ceramics. Mochizuki *et al.* (1988) prepared some NBS reference steel blocks by polishing with 100 mesh alumina paper followed by washing with alcohol prior to analysis. Alternatively, samples can be ground to a fine powder and pressed to form pellets. Some sample types can be pressed directly as either free standing pellets or into rings or cups. Mochizuki *et al.* (1988) prepared a number of geological reference materials using 0.5 g of sample powder placed in copper rings and pelletised to 20 tons. The resulting discs were of 2.5 mm thickness and 15 mm diameter.

Most sample types however require the addition of a binder to give the pellet some mechanical strength. A number of different substances have been used for this purpose including polyvinyl alcohol, Elvasite 2013 and Mowiol. Gray (1985b) used a solid binder (Elvasite 2013) giving a sample to binder ratio of 4:1 by weight. The mixture was shaken in a ball mill for 5 min to ensure complete mixing, prior to pressing into aluminium cups. Approximately 5 g of sample/binder mixture is required to produce a disc 3 mm thick by 32 mm diameter. The addition of a solid binder does, however, result in a significant dilution of the sample. Alternatively, a liquid binder such as 5% Mowiol can be used. The liquid is added to the powdered sample (approximately 10% by weight of the sample) which can be simply mixed in a plastic dish, and then pressed at 15 tons to form free-standing pellets. The pellets are dried in an oven overnight prior to use. Note: pellets should be stored in a drying cabinet to prevent them absorbing water. Free-standing pellets have the advantage that, providing the pellet is of sufficient thickness, both sides can be used for analysis. Care must be taken to prevent contamination during storage.

As an alternative to the preparation of pellets, powdered samples have also been mounted directly onto sticky tape. Although preparation time is minimal, reproducibility of the data is typically poor.

10.3.6 *Calibration*

Calibration remains the most important aspect of laser ablation ICP–MS and is still the limiting factor for quantitative analysis. Calibration curves are typically linear over several orders of magnitude as illustrated in Figure 10.7 for a series of gel standards containing a wide range of elements. The fit of the curves is well constrained in contrast to those illustrated in Figure 10.8 for a series of copper matrices. The scatter may reflect either a matrix effect, sample inhomogeneity or simply rather poor reference data for the SRMs. In this case the data points tend to be distributed about a linear curve, and for some elements the scatter is rather large. The precision may be partly a reflection of the poor reproducibility of successive laser shots, but more importantly, of a non-reproducible sampling process. If standard reference materials are available of similar physical and

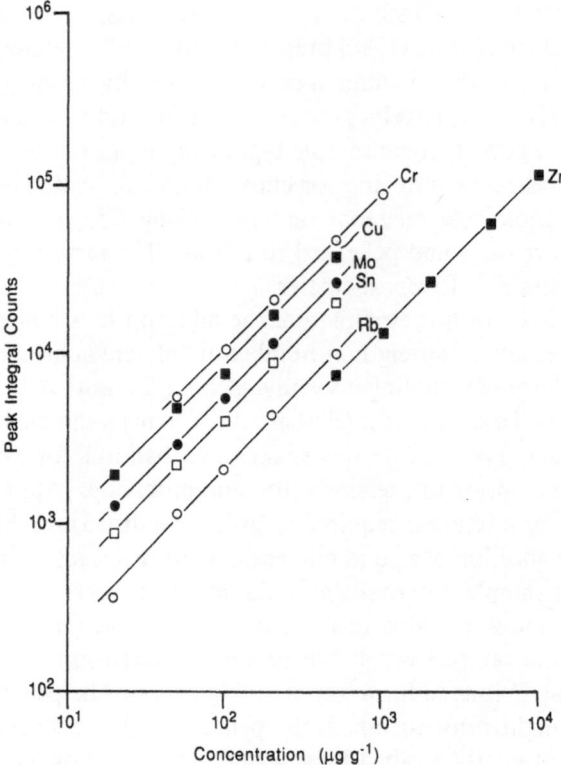

Figure 10.7 Calibration plots for BGS in house gel standards IP01–IP10. For each data point the laser is fired 10 times at 1 s intervals, 0.5 J per shot. Laser is operated in normal mode. Data aquisition time is 1 min per analysis.

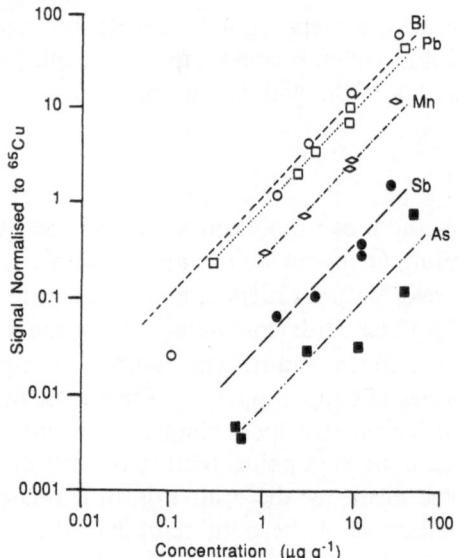

Figure 10.8 Analytical curves obtained for a series of copper SRMs with internal standardisation using attenuated ^{65}Cu signal. Laser operated in normal mode (after Arrowsmith, 1987a).

Table 10.8 Relative sensitivity factor (normalised to Ni for each sample type) obtained by laser ablation and solution nebulisation ICP–MS[a]

Isotope	Laser			Solution
	Silicate	Steel	Ni-alloy	
B	0.32	0.17	0.13	0.19
Mg	0.50	18	–[b]	0.82
Si	0.23	0.19	–	–
P	0.06	0.04	–	0.05
Ca	0.81	–	–	1.08
Sc	0.71	–	–	1.01
Ti	1.11	2.02	15.4	1.04
V	1.03	1.26	–	1.09
Cr	1.52	1.89	1.71	1.11
Mn	1.44	3.50	–	1.16
Fe	1.43	1.44	–	1.04
Co	1.27	1.38	–	1.06
Ni	1.00	1.00	1.00	1.00
Cu	1.54	1.42	–	0.92
Zn	1.47	15.1	–	0.78
Ga	2.52	–	–	1.25
As	1.14	0.27	–	0.24
Se	3.83	–	–	0.25
Rb	2.56	–	–	1.07
Sr	1.22	–	–	1.19
Y	1.08	–	–	1.22
Zr	1.03	4.85	1.97	1.13
Nb	0.95	2.09	1.28	1.12
Mo	4.32	1.82	–	1.11
Sn	6.79	5.09	–	1.08
Sb	2.63	4.63	–	0.72
Te	–	7.03	3.29	0.32
Cs	2.61	–	–	0.85
Ba	1.68	–	–	0.78
La	1.34	11.9	–	0.99
Ce	2.17	12.3	–	1.02
Pr	2.15	–	–	1.02
Nd	1.87	–	–	1.07
Sm	2.17	–	–	1.02
Eu	2.22	–	–	1.02
Gd	1.86	–	–	1.12
Tb	2.00	–	–	1.01
Dy	2.32	–	–	0.98
Ho	2.20	–	–	0.94
Er	2.32	–	–	0.96
Tm	1.79	–	–	0.93
Yb	2.14	–	–	0.95
Lu	1.89	–	–	0.86
Hf	2.41	2.16	3.26	1.23
Ta	2.21	5.56	1.57	1.07
W	–	3.15	1.93	1.17
Pb	4.31	51.3	18.7	0.58
Bi	3.52	21.8	16.4	0.49
Th	1.64	–	–	0.71
U	3.25	–	–	1.01

[a] From Mochizuki *et al.* (1990).
[b] –, not determined.

chemical matrix as the unknown samples, quantitative data may be obtained. The importance of matched matrices is demonstrated by the data shown in Table 10.8. The relative sensitivity factors (RSF) for a number of elements in a silicate, steel and Ni alloy by laser ablation are compared with those obtained by solution analysis. The sensitivity of each element is ratioed to that of ^{60}Ni in the same matrix. Using solution analysis, the sensitivity and mass bias of all elements is relatively uniform, the exception being those elements with a high ionisation potential. By contrast, the RSF in the three matrices measured using laser ablation, vary by an order of magnitude. For example, Mg is over 30 times more sensitive (relative to ^{60}Ni) in the steel compared with a silicate matrix. The importance of matrix matching is clear.

Both the precision and accuracy of the analytical data may be improved by use of an internal standard. An internal standard may be used to correct for differences in the amount of material ablated during successive analyses or for a variation in power input from the laser. For an element to be useful as an internal standard, its concentration must be known in each sample and it should respond in the same way as the element which it is being used to correct. It should also be present at a reasonable concentration to ensure that counting statistics are not a limiting factor. Jarvis and Williams (unpublished data) found that Ba could be successfully used as an internal standard for the determination of the REE in geological samples (Figure 10.9). Without internal correction there is considerable bias in the data set shown and low recoveries are observed. Using ^{137}Ba as an internal standard successfully corrected this bias giving a high correlation between measured and reference values (Figure 10.9).

A possible alternative method of calibration used with laser ablation ICP–AES, involves the introduction of solution standards rather than solids (Thompson *et al.*, 1989). The relative sensitivity factors between solution and solid are recorded and used for calibration. This method may prove useful for quantitative measurements by ICP–MS.

10.3.7 *Interferences*

Introduction of samples into the ICP–MS by laser ablation generates a dry vapour, which is low in the components which form polyatomic ions and refractory oxide species, i.e. O^+ and H^+. Typical levels of oxide and doubly charged ion species for Ba, Ce, Th and U are shown in Table 10.9. The data were determined for a BGS in-house gel standard, PC1, using a ruby laser operated in normal mode with 5×0.5 J shots. Although the recorded level of 2^+ ions is only a little lower than with solution introduction, the oxide and hydroxide levels are almost 2 orders of magnitude lower. In addition, the signal due to the formation of $^{40}Ar^{16}O$ was indistinguishable from background. The reduction of these interferences is therefore highly significant and should allow for the low level determination of elements such

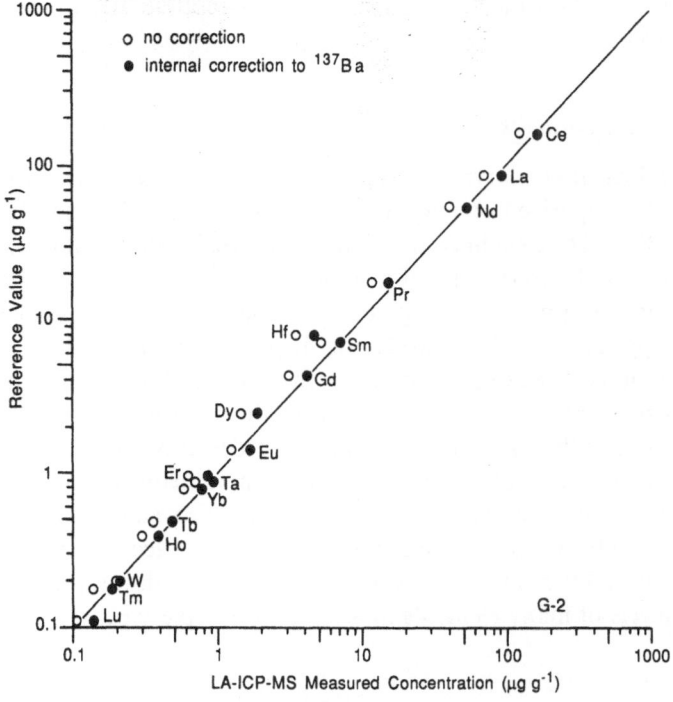

Figure 10.9 Quantitative determination of REE, Hf, W and Ta in USGS granite G-2. A comparison of data with and without internal standardisation.

Table 10.9 Percentage levels of oxide and doubly charged ions in laser ablation ICP−MS[a]

	%
Barium	
BaO/Ba	0.0225
BaOH/Ba	0.015
Ba²⁺	0.018
Cerium	
CeO/Ce	0.63
Thorium	
ThO/Th	0.74
Uranium	
UO/U	0.65

[a]Data for normal mode operation, 5 shots of 0.1 J duration on BGS gel standard PC1. $^{40}Ar^{16}O$ indistinguishable from background.

as Fe, S and Ti which are particularly susceptible to polyatomic ion interferences.

10.3.8 Detection limits

The overall sensitivity is typically poorer using laser ablation than for solution analysis. A comparison of detection limits for the REE, Th and U are given in Table 10.10. The data in column 1 are calculated from the analysis of 1000 scans from 135 to 243 m/z (with a skipped mass region 180–227 m/z) and a total integration time per isotope of 1.4 s. The absolute values for the laser work (in pg) are calculated assuming 12 μg of sample is ablated by five successive shots. The range of figures given for the detection limits in solution has been re-calculated as concentration in the solid assuming either a 200 or 1000 times dilution factor. The absolute detection limit in picograms, assumes there is between 2 and 21 mg of sample nebulised in the solution method. In absolute terms, detection limits are improved using laser ablation but if the practical aspects of analysis are taken into account, this is not the case. The detection limits are, however, low enough to allow the determination of trace levels of many elements in a range of sample types.

Table 10.10 Detection limits for REE, Th and U in silicated SRM, JB-3[a]

Isotope	Abundance (%)	Laser		Solution	
		DL relative[b]	DL absolute[c]	DL relative[d]	Dl absolute[e]
^{139}La	99.9	–	–	0.03	320
^{140}Ce	88.48	0.75	9.0	0.03	300
^{141}Pr	100	0.09	1.1	0.03	320
^{146}Nd	17.26	0.86	10	0.19	1900
^{152}Sm	26.63	0.33	4.0	0.29	2900
^{153}Eu	52.23	0.07	0.8	0.10	960
^{158}Gd	24.87	0.21	2.5	0.3	1300
^{159}Tb	100	0.02	0.2	0.05	170
^{163}Dy	24.97	0.57	6.8	0.2	680
^{165}Ho	100	0.09	1.1	0.05	140
^{166}Er	33.41	0.16	1.9	0.2	466
^{169}Tm	100	0.02	0.2	0.03	96
^{172}Yb	21.82	0.24	2.9	0.2	466
^{175}Lu	97.4	0.06	0.7	0.05	117
^{232}Th	100	0.11	1.3	ND	ND
^{238}U	99.28	0.09	1.1	ND	ND

[a] After Mochizuki et al. (1990). Laser data assumes 12 μg of sample ablated by five successive shots. Solution data assumes a dilution factor of 1000.
[b] Detection limit in μg g^{-1}, La internal standard.
[c] Absolute detection limit in pg.
[d] Detection limit in solid in μg ml^{-1}.
[e] Absolute detection limit in pg assuming 21 mg of sample nebulised during analysis.

10.3.9 *Practical considerations*

When using laser ablation for the analysis of solid samples a number of practical considerations arise in choosing operating parameters and interpreting the data collected, which are not necessarily present in solution analysis.

The nature and quantity of material transported to the ICP varies greatly between different sample matrices and with different modes of laser operation. The extent of plasma loading by the ablated sample varies over a wider range and is less readily controlled than with solutions. For the greatest stability of the plasma equilibrium, the more uniform delivery rate of sample obtained with a high repetition rate of small pulses is preferably and the 10–20 Hz obtainable with Nd:YAG lasers is advantageous. Operation in the Q switched mode produces a high proportion of small particulates from the sample which are more efficiently transported to the plasma than the larger particles produced in the normal mode. A 100 mJ Q switched pulse produces about 1–10 ng of material from a sample and operation at 10 Hz results in a sample loading at the plasma equivalent to a total dissolved solids (TDS) concentration in solution analysis of less than 0.05% ($<500\,\mu g\,ml^{-1}$). At such a level, matrix effects due to disturbance of the plasma equilibrium and drift due to solids deposition on aperture and skimmer are unlikely. Normal mode (free running) operation at the same rate and energy level, however, produces between 10 and 100 times as much material per pulse, depending on the nature of the sample. This greatly increases the probability of solid condensation and deposition in the expansion stage. Although more of the material is produced in the form of larger particles, and more is therefore lost by gravitational deposition in the cell and transfer tube, the plasma loading may still be great enough to disturb the plasma and alter its ionisation equilibrium. The problem is exacerbated if a lower repetition rate and higher pulse energy are used to maintain the rate of ablation. This may be seen in Figure 10.10 which shows the peak integrals obtained on the ablation of a nickel sample in the limiting cases of single ruby laser shots of increasing energy in both normal and Q switched modes. The response is shown for the two most abundant isotopes $^{58}Ni^{+}$ and $^{60}Ni^{+}$ (67.88 and 26.23% respectively). At the lowest energy in the normal mode (circles) the ratio of integrals for the two isotopes is approximately correct, but as the pulse energy is increased, the response increases more slowly and above 0.4 J it saturates. For both isotopes in the Q switched mode (squares), the response with increasing energy is as close to linear as can be expected up to an energy of about 0.5 J, since the measurement of laser energy at low levels using the installed monitor is relatively imprecise. Above this, however, the response drops for both isotopes. Since that for $^{58}Ni^{+}$ is linear up to integrals of just over 2.5×10^{5} counts, the drop in $^{60}Ni^{+}$ response above 10^{5} counts is most likely due to a shift in plasma equilibrium, rather than counting losses. In the normal mode, at an energy of 35 mJ, the ratio between the isotopes is approximately

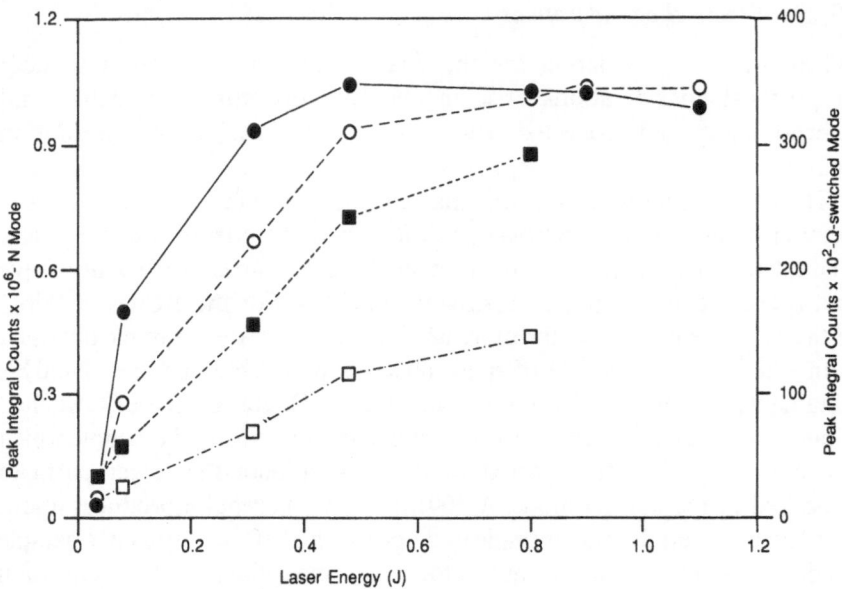

Figure 10.10 Peak integrals for Q-switched and normal mode of operation versus laser energy for two Ni isotopes. ●, N mode ^{58}Ni; ○, N mode ^{60}Ni; ■, Q-switched ^{58}Ni; □, Q-switched ^{60}Ni.

correct but even by 85 mJ there is some compression, the count integral on ^{58}Ni$^+$ approaching 3×10^5, at which counting losses were also apparent in the Q switched mode. The shift of plasma equilibrium which will also be occurring would not affect the ratio since both isotopes would behave in the same way, but it would of course affect the linearity of the response to increasing energies.

For this particular system therefore peak integrals of more than about 2.5×10^5 counts, even on a major isotope, from a single shot, imply some plasma disturbance in either mode. The extent of the disturbance may be expected to be very matrix dependent, and be affected by the ionisation energy distribution between the major components of the matrix and by the energy required from the plasma to volatilise and dissociate the introduced material. However, provided that the matrix is correctly matched between successive samples, the response to minor and trace elements should still be linear with concentration in the sample, even though the plasma will not be entirely dominated by Ar but will be affected by the volatilised matrix. Any differences in the matrix between samples, or sample and standards, may well yield different values of elemental sensitivity and hence the often experienced problems of calibration in laser ablation analysis. The transport system used for this experiment had a relatively high efficiency and therefore delivered more material per pulse than some other designs. Thus the example shown in Figure 10.10 presents an extreme case. In most commercial systems transport efficiency is somewhat lower and the use of smaller pulses, of 100 mJ or less at repetition rates of $10 \, \mathrm{s}^{-1}$ or more, greatly reduces plasma disturbance. However, for some sample types, particularly those of larger grain sizes such as rocks and pelleted materials, the larger ablation volumes of normal mode operation do offer some advantages. In such cases care must be taken that the detector response does not incur counting losses for the wanted isotopes. This is only likely to occur on elements of high concentration, which are often useful as internal standards, and for which good counting statistics are required to ensure adequate precision. It is an advantage to use a multi-isotopic element for this, such as Fe, so that the isotope ratios may be used as a check on counting losses before making the choice of an internal standard. Instrument data handling software usually includes a rapid check of isotope ratios, which enables this to be performed on screen during setting up.

Although disturbance of the plasma equilibrium by the introduced material is a true matrix effect, and will cause a change of sensitivity which depends on the matrix elements, a second matrix effect, peculiar to laser ablation, arises from the dependence of the ablation process on the interaction between the laser beam and the sample surface. This contributes greatly to the problems of calibration and is usually more serious in normal mode operation. In the Q switched mode much of the ablation results from the interaction between the plasma produced in the gas above the sample surface, by the very high energy density of this mode, and the sample itself. The penetration of the surface is small and the ablation pit shallow. In the normal mode this plasma is absent and energy transfer from the laser beam and its propagation through the sample, to produce ablation from a much larger volume, depends greatly on the optical and thermal characteristics of the material surface and interior. In multi-phase materials, trapped gas and inclusions in grain boundaries also play a part in the ablation process. Thus for normal mode operation, matrix matching of sample and standards becomes especially important, although in many cases it is not really practicable.

It is similarly impracticable, in most cases, for either mode of operation, to produce true blanks for naturally occurring samples, since ideally they should have the same physical characteristics as the sample without the analyte elements being determined. However, a major source of blank interferences arises from residual particles from previous ablations, deposited in the cell and transfer line, which are disturbed by the acoustic pulse in the carrier gas generated by the laser shot and carried into the ICP during the blank integration. This is much more pronounced in Q switched operation. The practice of taking a blank without pulsing the laser, a 'gas' blank, is not satisfactory as it only provides the blank integrals from the plasma gas, which are not usually wanted analytes. A better estimate of the blank levels from detritus disturbance may be obtained by using a normal laser cycle on the most inert material available, or at least one known to be free of the elements of interest. A simple PTFE surface has been found convenient in some cases. At the least this can demonstrate the need to clean the cell and transfer line at frequent intervals to reduce memory effects. These may also be minimised by ensuring that the lines are as smooth and as free of discontinuities as possible as these can provide locations for detritus collection.

10.3.10 *Applications*

Materials suitable for laser ablation are wide ranging and those analysed so far include rocks, minerals, plants, ceramics, alloys, steels, semiconductors, nuclear materials and foodstuffs. The spectra accumulated for the BGS in-house reference material PN1 are shown in Figure 10.11. The blank spectrum, shown for comparison, contains only the major peaks of argon plus O, N and C derived from the sample matrix. Using only two laser shots (normal mode) of 0.5 J, some of the major elements are clearly visible. With the scale expanded from about $60\,m/z$ ($\times$ 16), some trace elements such as Zr become clear. If a greater number of laser shots are used (10) and the scale expanded ($\times$ 16) a range of trace elements becomes visible including Nb, REE, Ta, Pb, Th and U. A further spectrum, accumulated using 10 successive 1 J laser shots is shown for the mass range 198–242 m/z for the same sample (Figure 10.12). The three largest Pb isotopes are clearly seen and the channel count for ^{208}Pb is 11 044 counts ($50\,\mu g\,g^{-1}$ Pb). With the scale expanded,

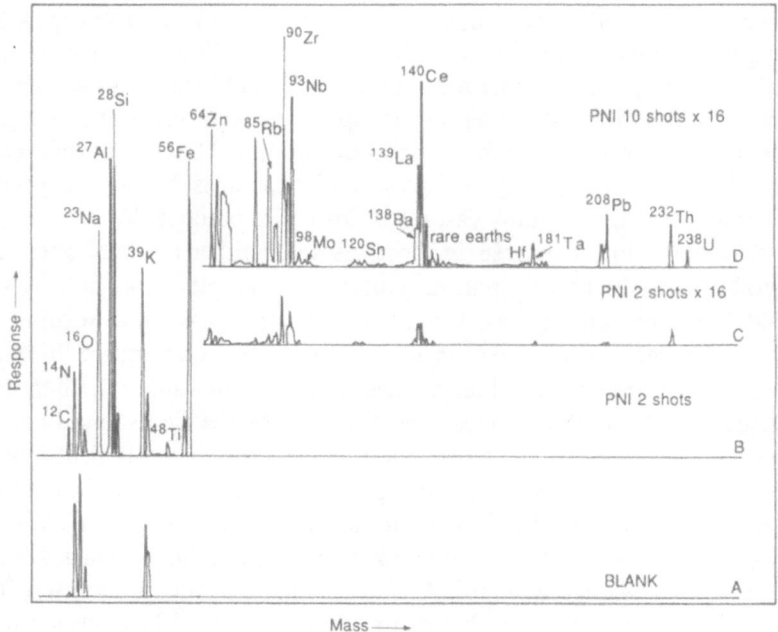

Figure 10.11 Full mass range spectra for BGS in-house SRM PN1 using 0.5 J laser shots. (A) blank spectra, (B) 2 shots, ^{28}Si peak height 209 220 counts s^{-1}, (C) as B but vertical scale expanded $\times$ 16 above $60\,m/z$, (D) 10 shots vertical scale as C (from Gray, 1985b).

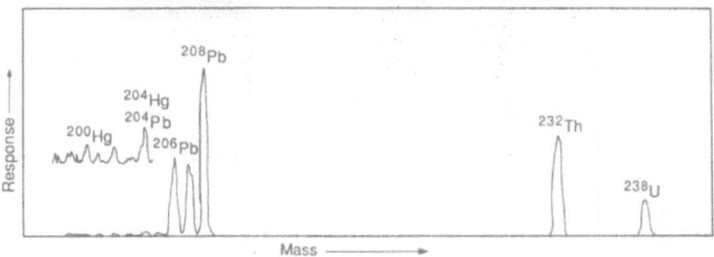

Figure 10.12   Spectrum over the mass range 198–242 m/z for BGS SRM PN1. Accumulated signal from 10 × 1 J shots. Peak channel count ^{208}Pb 11044 (from Gray, 1985b).

even the minor ^{204}Pb isotope is visible along with the small Hg isotopes. The potential for the direct measurement of Pb isotope ratios is therefore possible. It is this ability to successively build up a visual picture of the sample which is one of the real benefits of laser ablation ICP–MS.

The determination of heavy metals in plant samples has been investigated in our laboratory. The spectrum for NIST 1571 orchard leaves (Figure 10.13) illustrates the sensitivity of the technique. The peak for U at only 29 ng g^{-1} is clearly seen while the many isotopes of Hg at 155 ng g^{-1} are also visible adjacent to the Pb isotopes which are shown off scale.

Quantitative determination in some matrices has been made with some degree of success. The data shown in Figure 10.14 are for two NIST SRM steels and two BCR Ni-alloys which cover a range of analyte concentrations from < 10 to 10^6 μg g^{-1}. All four samples were calibrated against another reference steel NIST-662. A comparison of measured with certified values shows excellent agreement despite differences in the sample matrix. Some

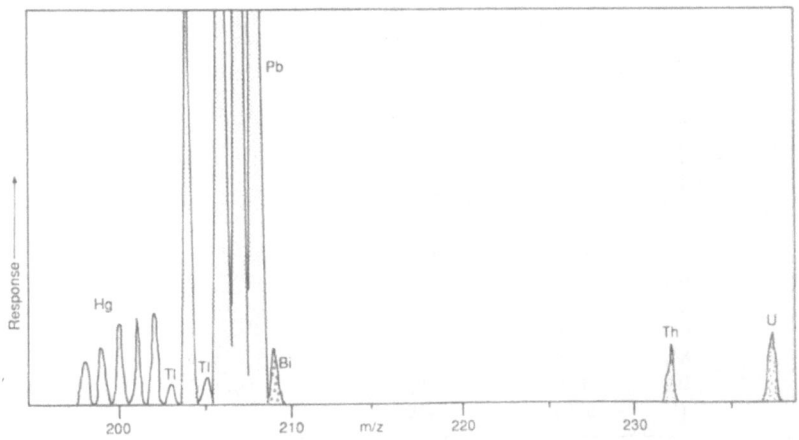

Figure 10.13   Mass spectrum for NIST 1571 Orchard leaves showing Hg at 155 ng g^{-1}, U at 29 ng g^{-1} and peaks for Tl, Bi and Th.

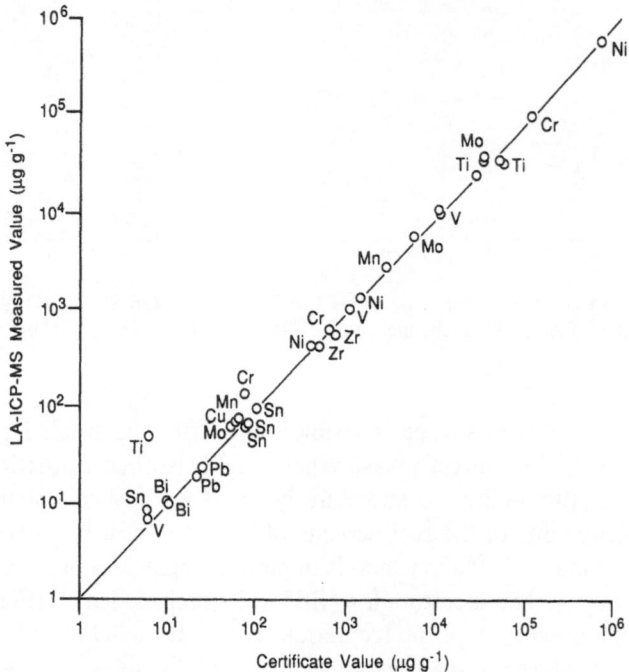

Figure 10.14 Laser ablation data for two SRM steels NIST 664, 665 and two SRM Ni-alloys BCR 345, 346 calibrated against steel NIST 662.

Table 10.11 Analysis of Chinese sediment GSD-1 by laser ablation ICP–MS[a]

Element	Laser ablation	Standard deviation	Reference
Sb	0.23	0.17	0.22
W	0.3	0.40	1.04
Eu	1.64	0.49	1.8
As	2.22	1.39	1.96
Yb	2.28	0.60	2.36
U	5.22	1.62	4.4
Dy	3.59	1.48	4.4
Cs	3.9	1.24	5.1
Sm	5.61	1.34	7.2
Sc	12.3	8.69	15.6
Co	18.9	5.66	20.4
Th	34.0	8.85	28
Nd	35.6	10.5	39
La	30.8	7.74	43
Ce	93.5	26.0	81
Rb	114	35.0	116
V	117	35.5	121
Cr	166	47.1	194
Mn	786	191	920
Ba	1090	252	950
Tl	5210	1190	5870

[a] From Durrant (pers. commun.). Concentration in $\mu g\,g^{-1}$. Calibrated against IAEA Soil 5. Laser operated in normal mode 4 runs of 5 shots at 0.1 J. Pellets pressed without binder.

data for the Chinese sediment GSD-1 are shown in Table 10.11. The instrument response for the sample was calibrated against IAEA soil 5 for a range of trace elements. The data were acquired using a ruby laser operating in normal mode, 5×0.1 J shots and four runs per analysis. The sample was prepared as a free standing pellet without the use of a binder. The agreement of the reference and measured data is reasonably good although the precision is a little poor, typically 20–30% RSD.

In summary, laser ablation is a useful method of sample introduction for a wide range of sample types. For qualitative and semi-quantitative analysis it is invaluable. However, the problems of calibration are, for many sample types, extensive and matrix matching is essential in most cases. If calibration problems can be overcome, fully quantitative data measurement should be possible.

10.4 Direct sample insertion

Direct insertion of solid samples into the ICP has received some attention (e.g. Boomer *et al.*, 1986; Brenner *et al.*, 1987; Sing and Salin, 1989) although to date only a single ICP–MS application has been documented (Blain *et al.*, 1989). Metal powders, salts and geological samples (Brenner *et al.*, 1987) melt or sinter into globules when they are directly inserted into the ICP. The direct insertion of 30 mg of pure copper powder, for example, will produce a steady signal that may last as long as 1 h, as the metal on the surface of the globule is slowly boiled off. If the sample is mixed with graphite powder, a shorter but more intense signal can be produced (Blain *et al.*, 1989). For some forms of direct sample insertion, sample powders are placed in a graphite cup which is inserted into the ICP. A modification of this design, which has proved successful, is to mix the sample powder with graphite which is subsequently pressed to form a small (typically 250 mg) briquette. This is then inserted into the plasma using either a wire loop or platform support. Unlike other probes, such as graphite cups, the whole body of the pellet is composed of sample mixture, so a relatively large sample can be introduced into the plasma at once. The detection limit for Pb in a mixture of metal oxides using this method of introduction, is around $0.02 \, \mathrm{ng \, g}^{-1}$.

10.5 Powdered solids

The feasibility of delivering powdered solid samples into an ICP–MS instrument has been demonstrated by Pfannerstill *et al.* (1990). Samples are ground to a fine powder (1–10 μm) and mixed with a diluent such as Li_2CO_3. A specially designed disperser generates an aerosol of the sample, which is contained in an inflated latex balloon. Controlled deflation of the balloon allows uniform aerosol delivery to the plasma over a period of time ($\sim$ 30 s). The technique has the potential for qualitative analysis of solid powders providing they are non-hygroscopic and non-conductive. Conductive solids such as graphite or metallic powders for example, may be analysed if the particles are small enough (Pfannerstill *et al.*, 1990).

10.6 Arc nebulisation

The technique is suitable only for substances which are electrically conducting, e.g. metals, or to which a conductor can be added since the sample acts as a cathode in an intermittent arc. The eroded sample condenses into particulates which are then transported into the ICP. The technique has been applied with some success to the analysis of steels (Jiang and Houk, 1986). The reduced oxide levels which occur with this method may prove beneficial in some applications.

11 Isotope ratio measurement

11.1 Introduction

The potential use of ICP–MS for the determination of isotope ratios was demonstrated early in its development history (Date and Gray, 1983a). Indeed, the original remit for the technique was that it should be capable of both elemental detection and isotope ratio measurements. ICP–MS is already being used for isotope ratio determinations in a wide range of fields including earth sciences, medical applications, environmental studies and in the nuclear industry.

Species which have the same atomic number (i.e. the same number of protons) and therefore similar chemical properties, but different atomic weights, are called isotopes. Of the thousands of nuclides known, less than 300 are categorised as 'naturally occurring'. The patterns of these naturally occurring isotopes are summarised by Russ (1989). Although a few elements have only one isotope, the majority have two or more. There are potentially, therefore, many elements for which isotope ratio determinations can be made. Studies of isotopic variation in 'natural samples' can be divided into two distinct categories. Those isotopes which originate from the creation of the galaxy, are said to be 'stable' such that their isotopic composition has remained essentially constant throughout geological time. The members of the second group result from the radioactive decay of an unstable parent, and are termed 'radiogenic'. In addition to naturally occurring species, man's activity may also produce isotopes which fall into both of these catagories.

During the generation of the galaxy, extensive mixing of the elements occurred to produce an essentially homogeneous isotopic composition in terrestrial, lunar and meteoric materials. However, a number of subsequent processes have led to a modification of some isotope ratios. The most important of these processes are mass fractionation (often by thermal and biological activity) and radioactive decay of either natural or anthropogenic origin. The effects of mass fractionation are generally seen in the light elements, i.e. those lighter than calcium, although the extent of these is still relatively small, typically in the parts per thousand range. Most variation in isotope ratios produced by radioactive decay is small although Pb is an exception to this. Extremely precise measurements are therefore usually required to detect differences in isotopic ratio between samples.

The effects of man's activity on isotopic abundances can be significant in the field of both stable and radioactive isotopes. Many can be readily synthesised and are widely used as tracers in medical research particularly in nutritional studies. The waste products from such studies create the risk of local environmental 'contamination' and therefore modification of natural isotopic ratios. In addition, man's influence on the generation of radio-isotopes may be highly significant, with the production of short lived species from nuclear explosions or longer lived ones from spent fuel re-processing, for example.

The precise measurement of isotope ratios can be used indirectly for the determination of elemental concentration via the process of 'isotope dilution' (ID). Many of the instrumental and analytical limitations discussed below apply equally to the measurement of natural isotope ratios and to isotope dilution analysis. However, providing that ICP–MS operating conditions are carefully controlled, ID should provide the most accurate method for measuring elemental concentration data when compared with external calibration for elemental analysis. The methods used and specific requirements for isotope dilution have been discussed in detail in chapter 6.

11.1.1 Traditional methods of isotope ratio determination

Mass spectrometric methods have been developed using a variety of ion sources. Among these are thermal ionisation (TIMS), the spark source (SSMS), secondary ionisation sources (SIMS), electron impact and field desorption. Some sample types and preferred compounds for analysis, particularly in the life sciences, are given in Table 11.1. An excellent discussion of the limitations and advantages of some of these techniques is given by Russ (1989). The ICP source offers a number of advantages over many of the above techniques. Most importantly, sample throughput is rapid. Determinations by TIMS, for example, typically involve counting times of 1 h per sample. By contrast, measurements by ICP–MS are typically 1–5 min and usually involve less sample pre-treatment. A single instrument can be used for such determination of a very wide range of elements. In addition, ICP–MS

Table 11.1 Ionisation methods for isotope ratio determination[a]

Ionisation method	Sample	Preferred compound analysed
Thermal ionisation	Inorganic solids	Metals with low ionisation energy (+ ve ions), metals with high electron affinity (− ve ions)
Spark source	Inorganic solids	All elements
Electron impact	Low molecular gases	Noble gases, H_2, CO_2, N_2, SO_2
	Volatile compounds after gas chromatographic separation	Metals after chelation
ICP	Inorganic solutions	All elements
Field desorption	Inorganic solids	Alkali and alkaline earths

[a] From Heumann (1985).

can be used for the determination of, for example, elements with high ionisation potentials, e.g. Re, Os, where determination is precluded by other mass spectrometric techniques. However, the advantages of ICP–MS are offset by relatively limited precision.

11.2 Instrument performance

ICP–MS has been used for the determination of isotope ratios from lithium to uranium and the elements which have received the most attention are discussed individually below. There are, however, a number of important factors which are pertinent to the determination of any isotope ratio by ICP–MS. These include (a) sensitivity and counting statistics, (b) dead time, (c) resolution and abundance sensitivity and (d) mass bias.

11.2.1 *Sensitivity and counting statistics*

In pulse counting systems, the precision of any measurement is greatly affected by the counting (Poisson) statistics. The uncertainty of the measurement is given by the square root of the number of counts. Precision of the measurement is therefore defined by the combined relative standard deviation which is given by

$$a/b = (\sqrt{(1/a + 1/b)}) \times 100$$

where a and b are the integrals for the two isotopes of interest. This equation holds true providing that the background signal is a negligible component of the total integral (see Russ, 1989). The precision on the ratio will be limited by the abundance of the smallest isotope. Thus, when figures of merit are given by instrument manufacturers, silver isotope ratio precision figures are usually quoted since the isotopes are of equal abundance. In practice, for 'real' samples, authors rarely report precision values which are equal to counting statistics. Factors such as noise from the sample introduction system and the plasma itself, lead to precisions which are poorer than would be predicted from counting statistics alone. It is interesting to note that even if sufficient counts are available, it is not currently possible to achieve better than about 0.1% RSD in routine analysis, for any isotope ratio, even when the ratio is unity.

11.2.2 *Dead time*

At high count rates, the channel electron multiplier and pulse counting systems used in ICP–MS instruments, began to count fewer events than actually occur. The interval during which the detector and its associated counting electronics are unable to resolve successive pulses is termed the

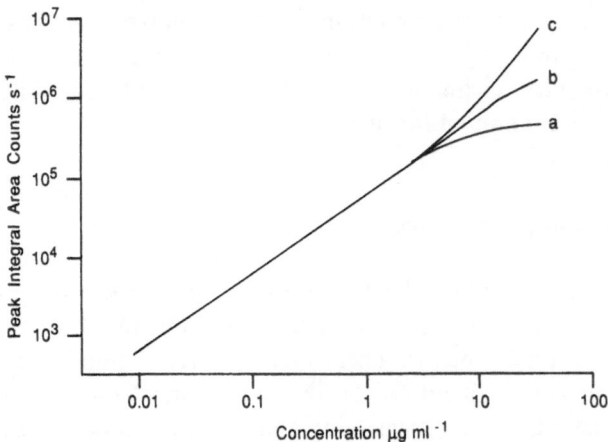

Figure 11.1 Signal response against Ni concentration with and without dead time correction. (a) no dead time, (b) 110 ns dead time and (c) 150 ns dead time (from Williams, 1989).

'dead time'. A consequence of dead time is that instrument signal response becomes non-linear above a certain count rate, typically about 1×10^6 counts s^{-1}. Some correction for dead time (T) is made by the instrument software during data processing. If the true rate is much less than $1/T$ then $m = n(1 - nT)$ where m is the observed rate. The graph shown in Figure 11.1 illustrates the effect of over and under correction for dead time. The problem is particularly difficult when dealing with two isotopes of unequal abundance. Maintaining sufficient counts on the smaller isotope may result in serious dead time correction to the larger. This can be particularly serious in the scanning mode where integration times on the two isotopes are equal. Even when peak hopping is used to increase the integral on the small isotope, care must be taken to avoid dead time error on the larger. Dead time correction on most early systems was about 110 ns, however modern instruments tend to require only about 20 ns to correct for response curvature.

11.2.3   Resolution and abundance sensitivity

One important characteristic of a mass spectrometer is its resolution, the ability to distinguish peaks of adjacent mass. In the quadrupole, as the resolution is increased transmission of the ions, and therefore sensitivity, is reduced. Using a scanning mode of operation, the accuracy of an isotope ratio measurement may be limited by overlap from an adjacent peak if the resolution is insufficient. To avoid this problem, the quadrupole is normally adjusted to operate with baseline resolution between peaks (Figure 11.2), which should be set to distinguish between two adjacent peaks of equal intensity and in preference, close in mass to the isotopes of interest. The resolution may be defined in terms of the peak width, measured at a specified fraction of the maximum height, usually 5 or 10%. For a quadrupole analyser

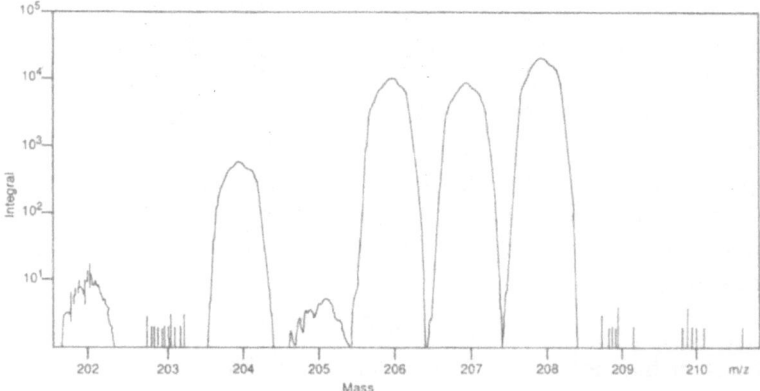

Figure 11.2 Spectrum for 100 ng ml^{-1} Pb. Minor amounts of Tl and Hg are also seen showing baseline resolution between peaks (after Russ, 1989).

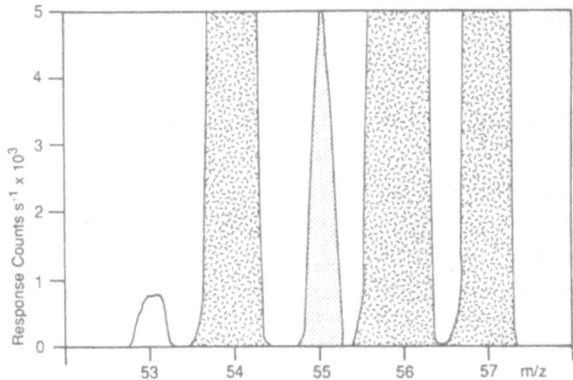

Figure 11.3 Spectrum obtained for a solution containing 1000 μg m^{-1} Fe and 10 μg ml^{-1} Mn. Sensitivity of ^{55}Mn is 5×10^5 counts s^{-1} per μg ml^{-1} (from Williams, 1989).

the resolution is normally expressed as the ratio of the separation between peaks (unit mass) divided by the peak width ΔM. Using a 5% peak width definition, the resolution in Figure 11.2 is about 2.3 M.

The peak shapes defined in the scanning mode of operation illustrate another important property of a mass spectrometer, abundance sensitivity which refers to the overlap between adjacent peaks caused by tailing above and below the peak. Abundance sensitivity is illustrated in Figure 11.3 for a solution containing 10 ng ml^{-1} Mn and 1000 μg ml^{-1} Fe. Taking into account the difference in concentration of the two elements (10^5) and the abundance of ^{56}Fe, the top of the Mn peak represents an abundance sensitivity of 1.09×10^{-5}. Most of the channels between 54 and 56 m/z are below the 10^{-6} level, and the area response for Mn is thus free from interference from Fe.

11.2.4 *Mass bias*

In mass spectrometric analysis, the observed or measured isotope ratio may deviate from the 'true' value as a function of the difference in mass between the two isotopes. This effect is termed mass bias and may be a positive or negative feature. True mass fractionation, such as that recorded in TIMS for example, results in a preferential loss of the light isotope. In addition, the term bias implies a systematic and predictable difference. The true ratio of isotopes A and B, $(A/B)_t$, can be related to the measured ratio, $(A/B)_m$, by the equation

$$(A/B)_m = (A/B)_t(1 + an)$$

where a is the bias per mass unit and n is the mass difference between isotopes A and B and the amount of bias is determined by comparision with a standard of known isotopic composition. Further discussion of origin and causes of mass bias is given by Russ and Bazan (1987). A correction for mass bias can be made to measured data by using a standard of known isotopic composition. Ideally the standard used should be close in composition to that of the unknown samples and contain a similar elemental concentration.

11.3 Applications and methods of isotopic anslysis

The range of isotope ratio measurements which can be made by ICP–MS is large. However, there are a number of points which are relevant to particular elements in certain sample types and this information is highlighted below for those elements which have received the most attention to date.

11.3.1 *Lithium*

Lithium has two isotopes ^{6}Li (7.52%) and ^{7}Li (92.48%) which are free from both isobaric overlap and polyatomic ion interferences. Mass bias is, however, likely to be significant since the relative mass difference between the two isotopes is very large. A correction must be made to the measured data using either a natural lithium standard or a series of standards prepared by spiking with ^{6}Li. Although lithium isotope ratios have not been widely measured, they have been determined by TIMS (e.g. Chan and Edmund, 1988).

Lithium plays an important role in the medical field as a treatment for manic depressive illnesses. The doses required for effective therapy are relatively high and may be close to the levels at which the element becomes toxic. Monitoring the uptake of lithium using stable isotopes is of great potential importance and the limitations and optimum operating conditions for ICP–MS determination of Li isotope ratios have been examined by Sun *et al.* (1987). All of their experimental data were obtained using the peak hopping mode of measurement on a Sciex Elan 250 and their operating

Table 11.2 ICP–MS data aquisition conditions
for lithium isotope ratio determination using the
peak hopping mode[a]

Instrument	Sciex Elan
Resolution	M
Measurements per peak	3
Scanning mode	1
Measurement mode	M
Measurement time	1.0005
Repeats per integration	10
Dwell time	50 ms
Cycle time	0.400 s

[a] From Sun *et al.* (1987).

conditions are given in Table 11.2. Sample types investigated in this study include red blood cells, serum and urine. In each case, lithium was extracted from the sample matrix using an anion-cation double exchange column. Enriched lithium in the form $^6Li_2CO_3$ ($^6Li = 94.88\%\,m/m$) was used as an isotopic spike while accuracy of the isotope ratio measurements was made by comparison with a natural lithium standard. It is interesting to note in this context, that many lithium compounds available are depleted in ^6Li since this isotope is used in the manufacture of nuclear weapons.

One of the physical characteristics of lithium in solution form is its affinity for glass and plastic surfaces, and therefore memory effects are an important consideration during an analytical run. The amount of time required to remove lithium from the system is shown in Figure 11.4 for Li concentrations up to $0.5\,\mu g\,ml^{-1}$. Using deionised water for washing, the Li signal decayed by 3 orders in about 75 s. Similarly it took the same length of time for the measured $^6Li:^7Li$ ratio to stabilise after introduction of the Li sample.

The effect of sodium concentration in the matrix is also an important factor for consideration particularly in biological samples such as serum or urine. The effects of sodium at concentrations from 2 to $100\,\mu g\,ml^{-1}$ (a Na:Li ratio of 1000) on both the ion count rate and the measured $^6Li:^7Li$ ratio are shown in Table 11.3. Although there was some evidence for suppression of the ion count rate for ^6Li at $100\,\mu g\,ml^{-1}$, there appeared to be no modification of the measured $^6Li:^7Li$ ratio. In addition, the precision on both the ion count rate and the measured ratios was not degraded due to the presence of Na.

11.3.2 *Boron*

Boron has two naturally occurring isotopes ^{10}B (20%) and ^{11}B (80%) both of which are free from isobaric overlap and polyatomic ion interferences. The sensitivity of boron is reasonably good with detection limits of a few $ng\,ml^{-1}$. The determination of boron isotopes has traditionally been carried out using

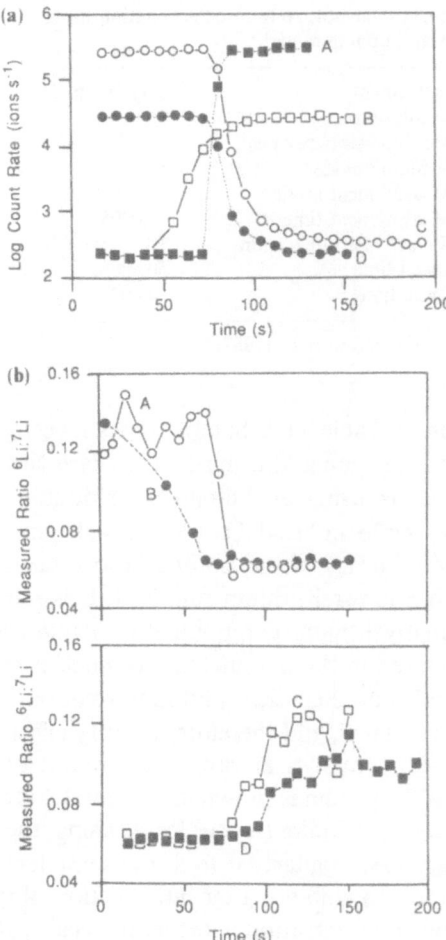

Figure 11.4 (a) Data showing memory effects of Li. Abscissa indicates the time from introduction of the particular solution. Ordinate indicates the log of the count rate (ions s^{-1}) at $7m/z$. (A) deionised water to $0.5 \mu g\, m^{-1}$ Li, (B) deionised water to $0.05 \mu g\, m^{-1}$ Li, (C) $0.5 \mu g\, m^{-1}$ Li to deionised water, (D) $0.05 \mu g\, m^{-1}$ Li to deionised water (from Sun *et al.*, 1987). (b) Data showing the overall memory effect on measured ratio 6:7. (A) deionised water to $0.5 \mu g\, m^{-1}$ Li, (B) deionised water to $0.05 \mu g\, m^{-1}$ Li, (C) $0.5 \mu g\, m^{-1}$ Li to deionised water, (D) $0.05 \mu g\, m^{-1}$ Li to deionised water (from Sun *et al.*, 1987).

TIMS where precisions of 0.02–0.2% have been reported (e.g. Spivack and Edmund, 1986). However, extensive sample preparation and pretreatment are required since the isotopic ratios are measured on the dicaesium metaborate molecule at 308 and 309 m/z, and analysis times are long. ICP–MS offers a number of advantages over TIMS. Sample preparation is straightforward and relatively rapid with analysis times of between 1 and 5 min per sample. In addition, the precision achievable is only a little poorer than that

Table 11.3 Effect of Na concentration on ion intensity and isotope ratio for Li ($0.10 \mu g \, ml^{-1}$). Data shown are the mean + SD of 100 sequential measurements[a]

Time (min)	Na ($\mu g \, ml^{-1}$)	^6Li integral	Measurement ratio $^6Li:^7Li$
0–30	0	11680 ± 303	0.0681 ± 0.0009
30–60	2	11550 ± 187	0.0680 ± 0.0007
60–90	2	11210 ± 206	0.0682 ± 0.0007
90–120	2	11350 ± 153	0.0681 ± 0.0007
120–150	100	10120 ± 149	0.0681 ± 0.0007
150–180	100	9943 ± 172	0.0680 ± 0.0007
180–240	100	9745 ± 125	0.0683 ± 0.0007
240–300	10	10750 ± 185	0.0686 ± 0.0007
300–360	10	10520 ± 190	0.0686 ± 0.0006
360–420	10	10340 ± 195	0.0686 ± 0.0007
420–480	0	10590 ± 182	0.0684 ± 0.0007

[a] From Sun et al. (1987).

obtained by TIMS, typically 0.1–0.2% RSD. However, B is difficult to flush from the surface of glassware and washout times may be relatively long. Contamination between samples may therefore be a serious problem.

The main potential applications of B isotope ratio determinations lie firstly in the field of earth sciences where fractionation of B isotopes may be very large, and secondly in the nuclear industry where ^{10}B (as boric acid) is used in reactors as a neutron absorber. To date, the determination of boron isotopes by ICP–MS has received relatively little attention. Russ and Bazan (1987) determined boron isotopes in synthetic solutions and found that the isotope ratio was not impaired in the presence of a Pb or Li matrix. However, the measured ratio differed from the reference value of NIST 951 by 15%. Gregoire (1987b) investigated a number of parameters relevant to the determination of boron isotopes in geological samples, including instrumental mass discrimination, matrix induced mass discrimination, sample preparation and separation procedures. The instrumental operating conditions used in this study are given in Table 11.4. The resolution of the mass spectrometer is a vital consideration for boron isotopes analysis since ^{11}B is located adjacent to the very large ^{12}C isotope. Carbon is present not only in many sample types but also from CO_2 entrained in the plasma itself and as dissolved CO_2 in solution. The resolution must therefore be set to prevent tailing of ^{12}C into 11 m/z. Under the scan conditions used by Gregoire (1987b), a longer time was spent counting the smaller B isotope such that the counting statistics for the two B isotopes were normalised.

Matrix effects may be severe for a light element such as boron. The effect of added Na, Cs and Pb on the ^{11}B count rate is shown in Table 11.5. The recovery is calculated from the analyte ion count rate obtained for ^{11}B in the presence of a concomitant element divided by the analyte ion count rate obtained for ^{11}B alone. Although the suppression effects are relatively minor

Table 11.4 ICP–MS operating conditions and signal measurement parameters for the determination of B isotope ratios[a]

Instrument	Sciex Elan
Forward power	1100 W
Coolant gas flow rate	$13.0 \, l \, min^{-1}$
Nebuliser gas flow rate	$1.1 \, l \, min^{-1}$
Auxiliary gas flow rate	$2.2 \, l \, min^{-1}$
Sample uptake rate	$1.0 \, l \, min^{-1}$
Sampling distance	18 mm

Data aquisition parameters

Resolution mode	High
Measurements per peak	3
Scanning mode	Isotope
Measurement mode	Multi-channel
Measurement time	5 s
Repeats per integration	5
Dwell time	300 ms
Cycle time	5 s

[a] After Gregoire (1987b).

Table 11.5 Effect of Na, Cs and Pb on the recovery of ^{11}B ion count rate[a]

Molar ratio $M/^{11}B$[b]	Recovery in the presence of concomitant element at molar concentration		
	Na	Cs	Pb
50			0.88
100		1.00	0.63
200			0.59
300		1.00	0.41
500	1.00		0.26
600		0.67	
1000	1.00	0.43	0.12
1500		0.24	
2000	1.00	0.18	0.05
5000	1.00		
10000	0.97		
15000	0.93		
20000	0.88		

[a] From Gregoire (1987b).
[b] Boron concentration $100 \, ng \, ml^{-1}$.

at Na concentrations of $< 400 \, \mu g \, ml^{-1}$ (15 000 molar ratio), matrix elements with a higher mass, e.g. Cs, cause acute loss of signal at a molar ratio of about 1000. Similar observations have been made in the author's laboratory where $2000 \, \mu g \, ml^{-1}$ Ca causes a B signal suppression of about 10%.

Instrumental mass bias is likely to be significant for boron isotopic

Table 11.6 Matrix induced mass discrimination effects on B isotope ratios[a]

Na$(\mu g\,ml^{-1})$	Measured $^{11}B:^{10}B$ ratio
1000	4.044
1500	4.040
2000	4.042
3000	4.124
4000	4.205
5000	4.286
6000	4.367

[a] From Gregoire (1987b).

measurements due to the large relative mass difference (10%) between the two isotopes. The tuning of the ion lenses can critically affect the magnitude and direction of the bias. To correct for bias, Gregoire (1987b) suggested using a mixture of two reference solutions of differing isotopic composition (e.g. NIST 951, $^{11}B:^{10}B = 4.0436$ and NIST 952, $^{11}B:^{10}B = 0.2678$) which can be mixed to provide isotopic ratios spanning the range of the two end members. The instrumental mass discrimination was typically 4–5% (Gregoire, 1987b).

In addition the introduction of a matrix also influences the mass bias (Gregoire, 1987b). The effect of added Na on the $^{11}B:^{10}B$ ratio is shown in Table 11.6. Up to $2000\,\mu g\,ml^{-1}$ Na there is no effect on the bias. However at $> 3000\,\mu g\,ml^{-1}$ an apparent loss of the light isotope is observed, with an 8% total bias between measured and actual ratio at $6000\,\mu g\,ml^{-1}$ Na. Two approaches can be taken to alleviate the problem of matrix induced discrimination. The first is by dilution of the sample providing that boron is sufficiently high in concentration, and the second by separation of boron from the matrix. Gregoire (1987b) evaluated a number of different ion exchange resins (Amberlite XE-243, Dowex 50W-X8 and Chelex 100) for the separation of boron from minerals, rocks and fossil materials. The reproducibility of the techniques was tested using NRCC reference seawater, NASS-1. Although this sample contains $5\,\mu g\,ml^{-1}$ B, it was necessary to separate B from the NaCl matrix to reduce the levels of Na and minimise matrix effects. A correction for mass bias was made using NIST 951 as the calibration standard. Measured $^{11}B:^{10}B$ are reported in Table 11.7 for a range of sample types and the $\delta^{11}B$ (i.e. variation in parts per thousand relative to a standard) values are reported relative to NASS-1. The data illustrate the considerable variation in $^{11}B:^{10}B$ ratios in nature which range from 3.906 to 4.169. Typical precision values for three replicates were 0.7% RSD.

11.3.3 Iron

Iron has four naturally occurring isotopes, ^{54}Fe (5.8%), ^{56}Fe (91.7%), ^{57}Fe (2.14%) and ^{58}Fe (0.31%). Two of these, however, have an isobaric overlap

Table 11.7 Boron isotope ratios in geological materials determined by ICP–MS[a]

Sample	$^{11}B/^{10}B$ ratio	$\delta^{11}B$[b]
Oceanic sediment	4.065 ± 0.036	-31.7
	4.027 ± 0.022	-40.7
	4.072 ± 0.029	-30.0
	4.128 ± 0.042	-16.7
	4.074 ± 0.014	-29.5
Tin-bearing silicate	4.014 ± 0.025	-40.9
	3.906 ± 0.023	-66.7
	3.942 ± 0.005	-58.1
	3.946 ± 0.022	-57.1
	4.026 ± 0.014	-38.0
Serpentinite	4.011 ± 0.007	-39.7
	3.997 ± 0.021	-43.1
	3.967 ± 0.069	-50.3
	4.039 ± 0.041	-33.0
	3.991 ± 0.045	-44.5
Basalt	3.949 ± 0.037	-57.5
	3.927 ± 0.028	-62.8
Altered hyaloclastite	3.969 ± 0.025	-52.7
	4.024 ± 0.035	-39.6
Mn/Fe oxide mineral	3.982 ± 0.037	-49.7
Sulfide	3.982 ± 0.017	-49.7
	3.922 ± 0.065	-64.0
Conodont fossil	4.169 ± 0.035	-1.9
Alder wood	4.061 ± 0.023	-27.8

[a] From Gregoire (1987b).
[b] Relative to NASS-1 reference seawater.

with ^{54}Cr (2.36%) and ^{58}Ni (67.8%) and therefore separation of these elements from iron is sometimes necessary. In addition to isobaric overlap, a number of polyatomic ion interferences occur on each of the iron isotopes ^{54}Fe ($^{40}Ar\,^{14}N$), $^{56}Fe(^{40}Ar\,^{16}O)$, $^{57}Fe(^{40}Ar\,^{16}O^1\,H)$ and $^{58}Fe(^{40}Ar\,^{18}O)$. The largest of these is the interference from $^{40}Ar^{16}O$ on ^{56}Fe which occurs at an equivalent Fe concentration of $10-100\,\text{ng}\,\text{ml}^{-1}$. Due to the magnitude of the interfering peak, ^{56}Fe has not yet been determined in samples which are introduced into the ICP–MS by solution nebulisation.

A simple separation procedure involving selective precipitation of $Fe(OH)_3$ was used by Ting and Janghorbani (1986) for human faecal material. Using a separation procedure such as this, it is possible to adjust the concentration of iron in the final solution, to between 1 and $5\,\mu\text{g}\,\text{ml}^{-1}$. At this concentration, the polyatomic ion interferences on ^{57}Fe are insignificant.

Iron bioavailability in infants and absorption of iron during pregnancy (Ting and Janghorbani, 1986; Janghorbani et al., 1986; Whittaker et al., 1989) are two areas where precise and accurate determination of iron isotope ratios have been employed. Previously, radioactive iron tracers (^{55}Fe and ^{59}Fe) were used for this type of study but their use poses serious ethical problems and the use of stable isotope tracers is now preferred. When determining iron isotope ratios in faecal material, Ting and Janghorbani (1986) used ^{58}Fe (spike) for *in vivo* labelling, ^{57}Fe (spike) as an *in vitro* tracer and ^{54}Fe as a reference isotope.

One of the advantages of using ICP–MS for IR determinations lies in the rapid sample throughput. If separation procedures are employed, the actual time taken to process a single sample may be considerable. An alternative approach has been used with considerable success by Whittaker *et al.* (1989). Samples of blood serum, taken from patients who have been given a double spiked dose of ^{54}Fe and ^{57}Fe, are injected directly onto the graphite rod of an ETV system. The temperature of the rod is gradually increased firstly to 'dry' the sample (thus removing oxygen) and then to volatilise the sample into the ICP. The sample size used is very small, typically about 5 μl. Samples were analysed using the scanning mode of operation and the ICP–MS operating conditions are given in Table 11.8. A narrow range scan is set which includes only the isotopes of interest using a 1 min analysis time, and this is rapidly scanned. The spectrum for a solution blank (introduced by solution nebulisation), containing 10 ng ml^{-1} of Mn (55 m/z) for reference, is shown in Figure 11.5a. The peaks at 56 m/z (^{40}Ar^{16}O) and 54 m/z (^{40}Ar^{14}N) are equivalent to almost 10 ng ml^{-1}. The background counts measured at 57 m/z are low at only 5. For comparison, the spectrum obtained from the graphite rod blank is also shown (Figure 11.5b). Using ETV for sample introduction (Figure 11.5b), the 56 and 54 peaks are reduced by over an order of magnitude.

Table 11.8   ICP–MS operating conditions for Fe isotope ratio determination by ETV[a]

Forward power	1300 W
Coolant gas flow rate	14 l min^{-1}
Auxiliary gas flow rate	0.5 l min^{-1}
Carrier gas flow rate:	
ETV unit	0.5 l min^{-1}
By-pass	0.2 l min^{-1}
Sampling cone	1.00 mm diameter Ni
Skimmer cone	0.7 mm diameter Ni
Data aquisition parameters	
Mode of aquisition	Scanning
Dwell time	100 μs
Sweeps	100
Channels	1024
Mass range	53–58 m/z

[a] After Whittaker *et al.* (1989).

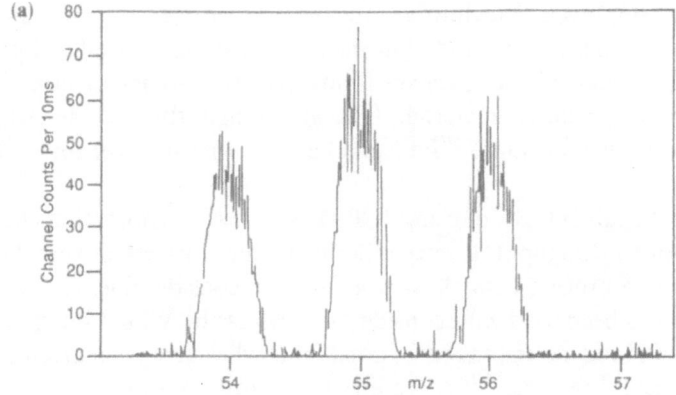

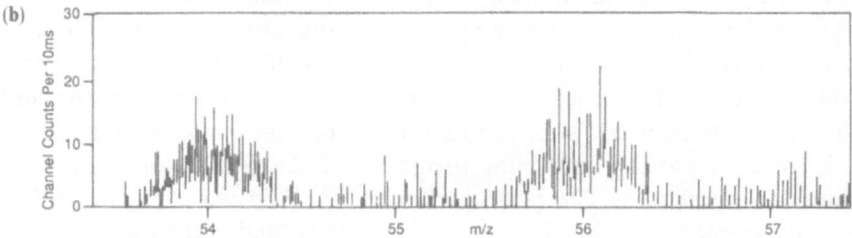

Figure 11.5 (a) Spectrum of solution blank containing $10 \, ng \, ml^{-1}$ of ^{55}Mn. Integrated counts: $54 \, m/z$ 3900; $55 \, m/z$ 5300; $56 \, m/z$ 4300; $57 \, m/z$ 5 (from Whittaker *et al.*, 1989). (b) Spectrum of the graphite rod blank. Integrated counts: $54 \, m/z$ 315; $56 \, m/z$ 378; $57 \, m/z$ 7 (from Whittaker *et al.*, 1989).

Optimisation of a transient signal in an essentially dry plasma, such as that provided by the ETV method of sample introduction, is an important consideration. For this work, the authors optimised the system firstly by monitoring the signal of ^{12}C (partly present as an impurity in the argon but also entrained CO_2 in the plasma) then the fine tuning of the ion optics using ^{114}Cd which was slowly volatilised from the graphite rod. The optimum rod vaporisation temperature was assessed by continually increasing the rod temperature until maximum Fe signal was obtained. Typical integrals obtained in this study for a $5 \, \mu l$ injection of enriched serum are 24 100 (^{54}Fe) and 321 000 (^{56}Fe) with a precision of 0.9% RSD on the measured ^{54}Fe:^{56}Fe ratio. This compares well with the precision predicted by counting statistics alone of 0.67% RSD. The use of ETV is therefore a viable alternative to the use of separation procedures for the determination of iron isotope ratios, whilst having the advantage of almost completely eliminating the polyatomic ion interferences at 56 and $57 \, m/z$.

11.3.4 *Copper*

Copper has two stable isotopes, ^{63}Cu (69.09%) and ^{65}Cu (30.91%) and is an important element for consideration in nutritional studies particularly in conjunction with Zn and Fe metabolism (Ting and Janghorbani, 1987). In biological applications, ^{65}Cu has been used as an *in vitro* tracer while ^{63}Cu is used as a reference isotope (Ting and Janghorbani, 1987). There are few major background peaks or isobaric overlaps which generate any significant interferences on either of the copper isotopes. However, in matrices high in phosphorus, there is the possibility of formation of PO_2 and in a sodium matrix of ^{23}Na^{40}Ar. To overcome problems from a sodium matrix, and to increase the concentration of Cu in the final solution, separation of copper from the matrix was performed for the analysis of faecal material in a study by Ting and Janghorbani (1987) using selective precipitation with ammonium pyrrolidinedithiocarbamate (APDC). These authors compared the precision obtained for ^{65}Cu:^{63}Cu using two modes of data acquisition termed the 'fast' and 'slow' scan mode. The fast scan mode uses a single point of measurement per peak with a dwell time of 50 ms repeated 600 times. The slow scan method uses three points per peak, with a 3 s dwell time per point, repeated 10 times. The results show little difference in either the measured ratio or precision for either mode of data acquisition. The authors recommend a forward power setting of 1.25 kW, coolant gas flow rate 12 l min^{-1}, auxiliary gas flow rate 1.9 l min^{-1}, with a nebuliser pressure of 42 lb in^{-2}. Under these instrumental operating conditions, 0.11 μg ml^{-1} of Cu gave a signal of $\sim 35\,000$ and background of 400 counts s^{-1} on the ^{63}Cu isotope.

11.3.5 *Zinc*

Zinc has five naturally occurring isotopes, the most abundant of which has an isobaric overlap with ^{64}Ni (Table 11.9). Zinc is one of the essential elements required in the human diet and consequently monitoring of elemental uptake has received some attention (e.g. Serfass *et al.*, 1989). The use of ICP–MS for the determination of zinc isotope ratios was first demonstrated in 1983 by Date and Gray for the analysis of blood plasma and faeces. For this study, subjects were administered zinc enriched in the ^{67}Zn isotope. The accuracy of the measured ratios was assessed using a natural zinc laboratory reagent.

Table 11.9 Abundance of naturally occurring zinc isotopes

Mass *m/z*	Natural abundance (%)	Isobaric overlap
64	48.9	Ni (0.95%)
66	27.8	–
67	4.1	–
68	18.6	–
70	0.62	Ge (20.7%)

An extensive optimisation study was later carried out by Serfass *et al.* (1986) to examine the effects of varying a number of instrumental operating parameters for zinc isotope ratio determination. The parameters studied included the effect of (a) dwell time on each peak, (b) total analysis time, (c) zinc concentration, (d) nebuliser gas flow rate, (e) forward power, and importantly for many 'human' derived samples, (f) the effects of sodium concentration. The optimum operating conditions derived from this study were used to analyse faeces samples which were prepared by ashing and dissolution in HCl. The zinc was then complexed and extracted using diethylammonium diethyldithiocarbamate (DDDC) and trichloromethane. The authors used a double spike method, i.e. two added isotopes (^{67}Zn and ^{70}Zn) to improve accuracy. The experimental results showing the effect of dwell time per mass with total measurement time, are shown in Table 11.10. The results clearly illustrate that better precision is achieved using short dwell times (50–100 ms) and is further improved if the total measurement times are extended. Although, in general, those operating conditions which give maximum signal also generate the lowest % RSD, this is not always so. In many cases, precision is limited by small instrumental fluctuations, rather than by counting statistics. The authors conclude that the precision of isotope ratio measurements is controlled by two factors, (a) the total number of ions reaching the detector (usually determined by the total analysis time) and (b) the variation in the rate at which they arrive at the detector, i.e. dwell time, scan rate, fluctuations in the ICP generator.

The occurrence of matrix elements such as sodium can be particularly

Table 11.10 Effect of dwell time and total measurement time on zinc isotope ratio precision[a]

Measure-ment time (s)	Dwell time					
	50 ms		100 ms		200 ms	
	^{67}Zn:^{68}Zn	^{70}Zn:^{68}Zn	^{67}Zn:^{68}Zn	^{70}Zn:^{68}Zn	^{67}Zn:^{68}Zn	^{70}Zn:^{68}Zn
0.2	0.2147 (1.1)[a]	0.0335 (1.2)	0.2144 (2.1)	0.0343 (2.3)	0.2148 (2.9)	0.0336 (3.5)
0.4	0.2152 (1.5)	0.0337 (1.4)	0.2114 (1.3)	0.0340 (1.6)	0.2157 (2.0)	0.0336 (2.9)
0.8	0.2147 (0.8)	0.0333 (1.1)	0.2123 (0.7)	0.0343 (1.3)	0.2153 (1.4)	0.0334 (1.8)
1.0	0.2151 (0.5)	0.0335 (1.3)	0.2130 (0.6)	0.0341 (1.0)	0.2163 (1.7)	0.0336 (1.3)
2.0	0.2147 (0.3)	0.0333 (0.7)	0.2132 (0.4)	0.0339 (0.3)	0.2161 (1.6)	0.0334 (1.3)
5.0	– –	– –	0.2140 (0.4)	0.0338 (0.4)	0.2154 (0.7)	0.0333 (0.6)

[a] After Serfass *et al.* (1986).
[b] % RSD for 10 measurements in parentheses.

Table 11.11 Effect of Na matrix on zinc isotope ratios[a]

Na concentration (M)	^{68}Zn[b] (counts s^{-1})	$^{67}Zn:^{68}Zn$	$^{70}Zn:^{68}Zn$
0	565 000	0.2183(0.9)[c]	0.0319(1.4)
1×10^{-4}	530 000	0.2184(1.4)	0.0319(1.0)
5×10^{-4}	446 000	0.2195(1.2)	0.0315(0.8)
1×10^{-3}	372 200	0.2198(0.5)	0.0316(1.3)
5×10^{-3}	145 000	0.2232(0.7)	0.0307(0.7)
1×10^{-2}	76 450	0.2264(0.6)	0.0310(1.4)

[a] From Serfass et al. (1986).
[b] Zinc concentration 5 μg ml^{-1}.
[c] % RSD in parentheses.

important in biological matrices. The effect of Na matrices at concentrations from 0.0001 M to 0.01 M (2.3–230 μg ml^{-1}) on 5 μg ml^{-1} Zn were examined by Serfass et al. (1986). The results shown in Table 11.11 show suppression of the ion count rate in the presence of only a few ppm of Na. There is an apparent loss of the heavy isotope even at 10 μg ml^{-1} Na. However if the data are plotted to take into account the errors on the data, the picture is not so clear (Figure 11.6). The error bars overlap up to $\sim 25 \mu$g ml^{-1} Na and more data points are needed to establish whether the accuracy of the isotope ratios are being systematically modified at high concentrations of sodium. One thing is clear from these data, however, that precision is not degraded by the presence of Na even at concentrations of about 200 μg ml^{-1}.

Matrix matching of standards and samples was investigated by Ting and Janghorbani (1987). The authors concluded that for Zn, matrix matching was not required, although in this case Zn was extracted from faecal material

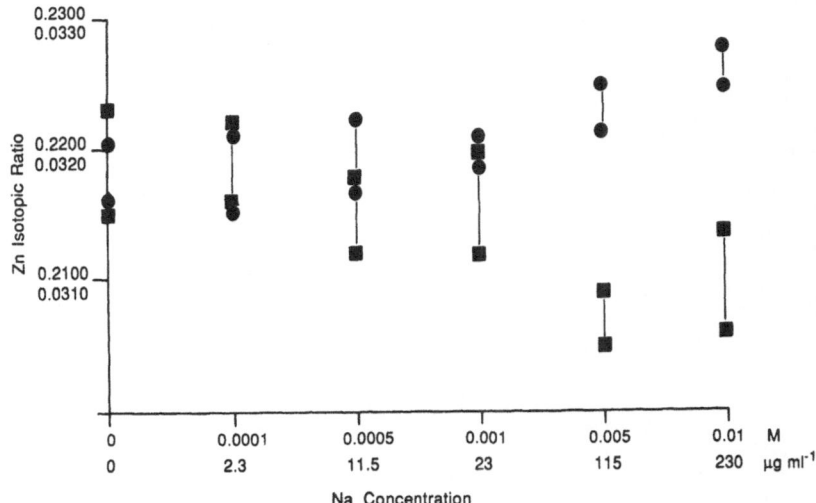

Figure 11.6 Effect of increasing Na concentration on measured Zn isotope ratios. Upper ratio on x-axis is $^{67}Zn:^{68}Zn$ (●), lower ratio is $^{70}Zn:^{68}Zn$ (■) (data from Serfass et al., 1986).

using APDC and therefore most matrix elements were removed from the sample prior to analysis.

A number of general recommendations can therefore be made with respect to the measurement of Zn isotope ratios. For medical applications it is useful to measure three of the available Zn isotopes. Ting and Janghorbani (1987) used ^{70}Zn (spiked) as an *in vitro* tracer, ^{67}Zn to perform isotope dilution analysis, i.e. as a second spike and ^{68}Zn as a reference isotope. Unless zinc concentrations are exceptionally high, it is necessary to separate the element from the matrix using either the APDC method of Ting and Janghorbani (1987) or DDDC and tetrachloromethane which can be back extracted into 1.2 M HCl (Serfass *et al.*, 1986). An assessment of mass bias and accuracy may be made by analysis of a natural abundance zinc standard such as that sold as a laboratory reagent. The instrumental operating parameters for a Sciex Elan 250 ICP–MS are discussed by Serfass *et al.* (1986) for the analysis of faecal matter using the peak hopping mode of measurement. These are typical of those used by other workers.

11.3.6 *Rhenium and osmium*

Rhenium has two isotopes ^{185}Re (37.4%) and ^{187}Re (62.6%) and displays similar chemical behaviour to that of W and Mo. By contrast the behaviour of Os is more closely allied to the other platinum group metals (PGM). Osmium has seven isotopes five of which have an isobaric overlap with either W, Re or Pt. The abundances of the isotopes in the mass range 184–194 *m/z* are shown in Figure 11.7. Due to differences in their physical properties, Re

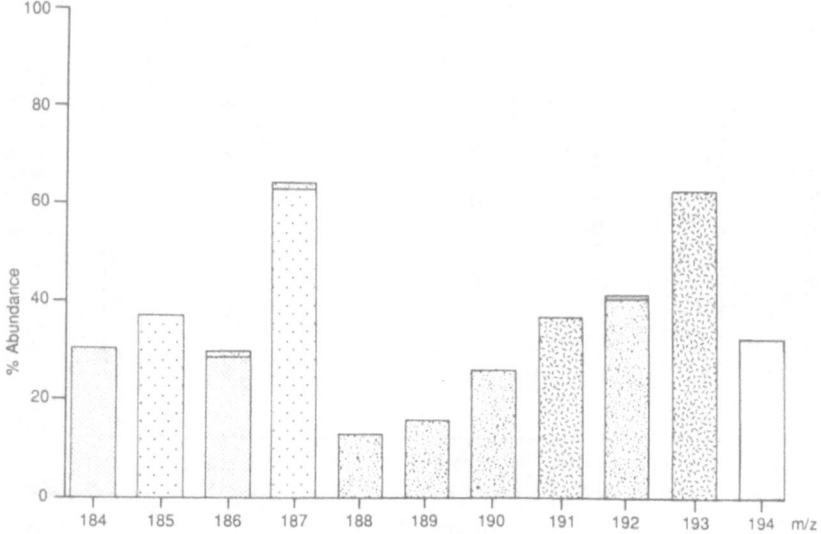

Figure 11.7 Natural isotopic abundances from 184 to 194 *m/z* showing ^{185,187}Re, ^{184,186}W, $^{186-190,192}$Os, ^{192,194}Pt and ^{191,193}Ir (from Richardson *et al.*, 1990).

and Os are fractionated during natural processes and therefore a range of Re/Os ratios are found in nature. These elements are of particular interest in geological studies and have been used to determine the age and genesis of ore deposits (e.g. Hart and Kinloch, 1989), the origin of meteorites and as indicators of catastrophic events in the geological record (Lichte *et al.*, 1986). Their determination is precluded by TIMS because of the low degree of ionisation of both Re and Os and the volatility of Os. This is not, however, a limiting factor in ICP–MS, where detection limits are better than 0.02 ng ml^{-1}.

11.3.6.1 *Rhenium* The first ICP–MS Os and Re isotopic measurements were made by Lindner *et al.* (1986, 1989) for the determination of the ^{187}Re half-life. Two main methods are used for the preparation of geological samples for isotopic determination. The first is via a dissolution stage using mixed acids followed by an anion exchange procedure to separate Re from Os and thereby eliminate the problem of isobaric overlap of ^{187}Re with ^{187}Os. The second method is to use a fire assay procedure to separate Re, Os and other PGM from the sample. Rhenium is preferentially partitioned into the flux phase which can then be dissolved and subsequently passed through an anion exchange separation. In either cases, Re is concentrated in a small final volume (~ 5 ml) containing low levels of TDS and can be determined in solution. The ICP–MS operating conditions used by Richardson *et al.* (1989) are given in Table 11.12. The scan range used, 182–220 m/z, was relatively wide and included a number of background positions (181, 183, 203 m/z), Re and Os isotopes (^{185}Re, ^{187}Re, ^{190}Os, ^{192}Os) and peak positions for Pt and W. For samples containing 30 ng ml^{-1} Re, the signal to background ratio was about 1000. Typical values of precision reported for ^{185}Re:^{187}Re are from 0.1–1% RSD (2σ) (Richardson *et al.*, 1990).

Table 11.12 ICP–MS operating and measurement parameters for Os and Re isotope ratio determinations[a]

Instrument	Sciex Elan
Forward power	1200 W
Coolant gas flow rate	12.0 l min^{-1}
Carrier gas flow rate	1.8 l min^{-1}
Auxiliary gas flow rate	2.0 l min^{-1}
Data aquisition parameters	
Resolution mode	High
Measurements per peak	5
Scanning mode	Isotope
Measurement mode	Multi-channel
Measurement time	0.06 s
Dwell time	60 m
Cycle time	10 s
Number of scans	300 (Re), 100–150 (Os)

[a] After Richardson *et al.* (1989).

11.3.6.2   *Osmium*   Osmium is one of the least abundant elements in nature and the radiogenic isotope makes up only 1–2% of the total Os. Counting statistics are therefore the major limiting factors for precise Os isotope ratio determination. Although Os has been introduced into the ICP in solution form using pneumatic nebulisation (Masuda *et al.*, 1986) and a glass frit nebuliser (Lichte *et al.*, 1986), the most favoured method is to distil Os to form the highly volatile tetroxide species and introduce this gas directly into the ICP (Bazan, 1987; Russ *et al.*, 1987; Dickin *et al.*, 1988). The formation of the tetroxide gas has a number of advantages. Firstly the sensitivity of Os is increased over conventional introduction techniques by about two orders of magnitude and secondly there is almost total separation of ^{187}Os from ^{187}Re during the generation process such that interference corrections are minimal. A mass spectrum illustrating the enhanced sensitivity for a 10 ng sample of Os using vapour generation is shown in Figure 11.8. The practical limitations of the tetroxide generator system are given in detail in Russ *et al.* (1987), Bazan (1987), Dickin *et al.* (1988) and Richardson *et al.* (1989). Further practical considerations are given in chapter 4.

The scan conditions for Os are essentially the same as those for Re (Table 11.12) with the exception that the number of scans is reduced to 100–150 (Richardson *et al.*, 1989). Five Os isotopes and four other full mass positions are monitored during the determinations of Os (Table 11.13). To improve analytical precision on ^{187}Os, this isotope is counted for almost as long as ^{188}Os, ^{189}Os, ^{190}Os and ^{192}Os combined. Although using an Os generator results in the less volatile species (Pt, W, Ta and Re) remaining in the sample chamber, masses 181, 182, 185 and 194 are routinely monitored. Background signals are measured at 181 and 183 m/z, while ^{194}Pt is used to assess the extent of the Pt interference at ^{190}Os and ^{192}Os, and ^{185}Re to

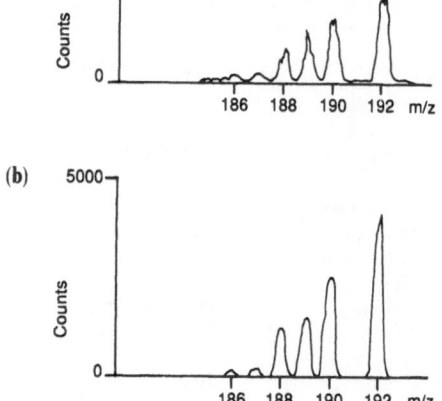

Figure 11.8   Mass spectra of the osmium region obtained for 10 ng of Os by (a) solution nebulisation, and (b) OsO$_4$ vapour generation (from Russ *et al.*, 1987).

Table 11.13   Mass positions measured for Os isotopic analyses[a]

Isotope	Actual measurement time (%)[b]	Estimated measurements per cycle	Relative amount of time taken (%)
^{181}Ta	100	2	6
^{183}W	100	2	6
^{185}Re	100	2	7
^{187}Re–^{187}Os	43	10 ⎫	
^{188}Os	30	7 ⎪	
^{189}Os	9	2 ⎬	75
^{190}Os	9	2 ⎪	
^{192}Os	9	2 ⎭	
^{194}Pt	100	2	6
Total	–	31	–

[a] From Richardson *et al.* (1989).
[b] Portion of the time allotted to a particular element for measuring a given isotope.

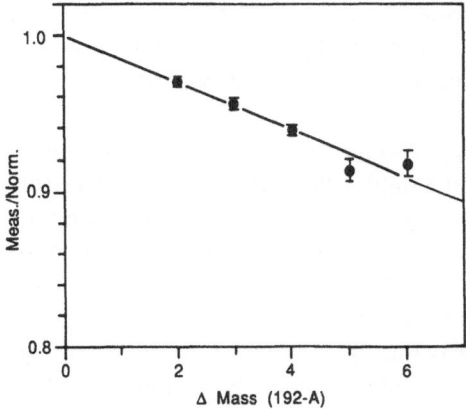

Figure 11.9   Deviation of measured isotope ratios from the accepted values as a function of the difference in mass between the isotope plotted and the reference isotope (192 m/z) (from Russ *et al.*, 1987).

correct for the interference at ^{187}Os. Under these operating conditions, the total analysis time for 60–100 scans is between 10 and 20 min per sample.

Mass bias was reported for Os isotope ratio determinatons by both Russ *et al.* (1987) and Hirata *et al.* (1988a). The observed ratios, calculated relative to the signal at 192 m/z, deviated from the accepted values linearly as a function of the mass difference between the isotopes (Figure 11.9). The deviation was about 1.5% per mass unit, suggesting loss of the lighter isotopes, and varied slightly from sample to sample. The data shown in Figure 11.9 are for a 5 ng sample. Mass bias on more modern instruments may be reduced below 1.5%.

Long term precision was assessed by Dickin *et al.* (1988) over a period of 3 months. One sigma RSD values range typically from 0.6 to 2.2% for a $10 \, \mathrm{ng \, ml^{-1}}$ pure Os solution with an average value of 0.3% for $^{190}\mathrm{Os}:^{188}\mathrm{Os}$ over that period. Accuracy was assessed by comparison with data obtained by SIMS giving values of 0.1–0.2%.

11.3.7 *Lead*

Lead has four naturally occurring isotopes ($^{204}\mathrm{Pb}\, 1.4\%$, $^{206}\mathrm{Pb}\, 24.1\%$, $^{207}\mathrm{Pb}\, 22.1\%$ and $^{208}\mathrm{Pb}\, 52.3\%$), three of which are radiogenic decay products of either uranium or thorium. The fourth, $^{204}\mathrm{Pb}$, has a very long half-life such that it may be considered 'stable' on a geological time scale. For this reason, in geological applications, Pb isotopes are frequently ratioed to $^{204}\mathrm{Pb}$ for comparative purposes. Unfortunately, the $^{204}\mathrm{Pb}$ isotope is the least abundant and frequently shows the poorest precision of measurement by ICP–MS (NB: a Pb standard solution may not necessarily have the 'common' Pb isotope signature listed in the abundance tables). Overall precision is inferior to TIMS and Pb isotope ratio determination by ICP–MS is not precise enough to be used for geological dating. However, the identification of Pb sources in geological and environmental studies is, in many cases, feasible.

A number of applications concerning the determination of Pb isotope ratios in geological samples have been published (Strong and Longerich, 1985; Date and Cheung, 1987; Smith *et al.*, 1984; Gray, 1988) principally on their determination in galena (PbS). Since lead is the matrix element in such samples, the total lead concentration available is not a limiting factor. Both the peak hopping and scanning modes of measurement have been used with some success. Some of the above studies were carried out on early instruments and the large amounts of mass bias, for example ($\sim 4\%$ per m/z) described by Strong and Longerich (1985) were probably a feature of the early instrument used for that work. Later work (e.g. Date and Cheung, 1987; Gray, 1988) reports mass bias of typically less than 1% and this figure is typical of more modern instruments. An interesting method of correcting for mass bias on lead isotope ratios was suggested by Longerich *et al.* (1987b) using $^{203}\mathrm{Tl}:^{205}\mathrm{Tl}$. However, the method had limited success and should be adopted with caution.

The second main area of application is in the field of environmental studies, particularly the involvement of lead additives in petrol and their effect on human health (Delves and Campbell, 1988; Campbell and Delves, 1989; Hinners *et al.*, 1987; Viczian *et al.*, 1990a). Typical analysis times of 15 min per sample were reported by Delves and Campbell (1988) with precisions of 0.9% RSD for $^{208:207}\mathrm{Pb}$. Hinners *et al.* (1987) reported a non-linear response at about 180 000 counts (presumably because no dead time correction was applied) which decreased the precision significantly of the $^{208:204}\mathrm{Pb}$ ratio. Dean *et al.* (1987b) carried out an extensive comparison of the relative merits of acquiring data under the scanning and peak jumping modes of measurement. They concluded that scanning mode gave better precision (under the particular conditions used) but many of their observations were inconclusive. Lead isotope ratio measurements of paint, blood, soil and dust by Viczian *et al.* (1990a) showed some correlation between blood ratio and Pb source.

Despite the wide range of often contradictory observations made in the literature, a number of conclusions can be drawn which are pertinent to the precise and accurate determination of Pb isotope ratios. The precision of measurement at low concentrations is limited by counting statistics. In the case of galena samples, the concentration of Pb in the final solution can be adjusted to give high count rates (typically $1-5 \mu g \, ml^{-1}$; e.g. Date and Cheung, 1987) without the TDS exceeding $1000-2000 \mu g \, ml^{-1}$ in the solution. In solutions containing low concentrations of total Pb, either poorer precision must be accepted (and probably poorer accuracy) or the Pb must be separated from the solution matrix and pre-concentrated, in order that the TDS do not limit the Pb concentration available for measurement. Adequate peak resolution is particularly important for lead isotope determination since three of the isotopes occur adjacent to each other $(206, 207, 208 \, m/z)$. In general 'over resolved' peaks are preferable to 'under' resolved peaks, since if the latter occurs, tailing of the ^{208}Pb peak for example could significantly affect the measurement of ^{207}Pb.

A typical set of operating parameters is given in Table 11.14, for the scanning mode of operation, for an ICP–MS system using an MCA for data transfer. As with most isotopic measurements, the scan rate is rapid to improve the precision of the measurements. The quadrupole rest mass is set to $203 \, m/z$ so that a minimum amount of time is spent changing from the rest mass to the start position of the scan. Thallium is present at low

Table 11.14 ICP–MS operating conditions for Pb isotope ratio determination using the scanning mode

Instrument	VG Elemental PlasmaQuad
Forward power	1400 W
Coolant gas flow rate	$14 l \, min^{-1}$
Auxiliary gas flow rate	$0.5 l \, min^{-1}$
Nebuliser gas flow rate:	$0.75 l \, min^{-1}$
Sampling cone	1.00 mm diameter Ni
Skimmer cone	0.7 mm diameter Ni
Data aquisition parameters	
Mode of aquisition	Scanning
Dwell time	$80 \mu s$
Sweeps	1600
Channels	512
Scan range	$200-209 \, m/z$
Rest mass	$203 \, m/z$

Table 11.15 Lead isotope ratio measurements for NIST 981[a]

	$^{204}Pb:^{206}Pb$	$^{207}Pb:^{206}Pb$	$^{208}Pb:^{206}Pb$
Certificate value	0.05904	0.9146	2.168
Mean measured value	0.05965	0.9179	2.159
Bias (%)	+ 1.03	+ 0.36	− 0.40
Experimental % RSD	1.29	0.35	0.32
Count rate % RSD	1.10	0.38	0.31

[a] From Gray (1988).

concentrations in many sample types, however if Tl is expected to occur at significant concentrations ($> 100\,ng\,ml^{-1}$) an alternative rest position should be chosen. The scan range is set sufficiently wide to include at least two Hg isotopes. A blank correction for ^{204}Hg on ^{204}Pb is frequently required (either due to Hg in the sample or from the dissolution acids) and ^{202}Hg can be used to calculate the relative contribution from Hg at 204 m/z. If the sample contains W as a major component, it may be necessary to use ^{201}Hg for correction of 204, since $^{186}W^{16}O$ may cause a significant interference at 202 m/z. Using the scan conditions in Table 11.14, five replicate analyses per sample can be made in just over 5 min. Data acquired under these operating conditions are shown in Table 11.15 for NIST 981. The effects of mass bias should be monitored (and corrected if necessary) using a standard of known isotopic composition, preferably one which is close in isotopic composition to that of the unknown samples. The most commonly used reference standard is NIST 981 although NIST 982 and 983 are sometimes used. The standard should be analysed frequently during an analytical run to monitor changes in mass bias with time. A typical run procedure would include multiple determinations of reagent blanks, reference standards and unknown samples. Washout times between samples depend partly upon the concentration of Pb present in the sample, but for concentrations less than $\sim 1\,\mu g\,ml^{-1}$, 60–120 s wash-out with the acid used to prepare the standards and samples, is usually adequate.

For more complex organic based matrices, such as the determination of Pb in blood, the protocol outlined by Delves and Campbell may be followed. Whole blood samples were diluted in a mixture of ammonia, EDTA, $NH_4H_2PO_4$ and Triton X-100. The final concentration of Pb as measured was between 20 and $700\,ng\,ml^{-1}$. Wash-out between samples was achieved using firstly ammonia solution (2 min) followed by deionised water (2 min). The measurement conditions are given in Table 11.16. Platinum was added to all standards and samples at a concentration of $100\,ng\,ml^{-1}$ to allow monitoring of $^{194}Pt:^{195}Pt$ as a check of instrument stability. A summary of some results from Delves and Campbell are given in Table 11.17. The ICP–MS data, when compared with either an in-house quality control standard or NIST 981, show a positive bias. The magnitude of the bias is not dependent

Table 11.16 Data aquisition parameters for the determination of Pb in blood[a]

	Determination of total Pb	Determination of Pb isotope ratios
Mass range	204.5–215	189–215
Sweeps	240	1500
Channels	512	512
Dwell time (μs)	500	250
Total run time (s)	61	192
Internal standard	^{209}Bi	–[b]

[a] From Delves and Campbell (1988).
[b] Pt was added at 100 ng ml^{-1} to monitor ^{194}Pt:^{195}Pt as a check of instrument stability.

Table 11.17 Measurements of Pb isotope ratios in a series of quality control blood specimens[a]

						% Error of ICP–MS data	
Sample	Total lead	Pb ratio	IDMS (A)	ICP–MS (B)[b]	ICP–MS (C)[b]	$\frac{B-A}{A} \times 100$	$\frac{C-A}{A} \times 100$
F	70	208:206	2.130	2.133	2.133	0.14	0.16
		206:207	1.125	1.107	1.121	−1.60	−0.36
		206:204	17.40	17.78	16.12	2.18	−7.4
A	142	208:206	2.144	2.171	2.140	1.26	−0.19
		206:207	1.113	1.096	1.112	−1.53	−0.09
		206:204	17.26	17.12	17.30	−0.81	0.23
Z	648	208:206	2.095	2.105	2.094	0.48	−0.05
		206:207	1.159	1.151	1.170	−0.69	0.95
		206:204	18.07	17.89	18.72	−1.00	3.60

[a] From Delves and Campbell (1988).
[b] ICP–MS data obtained using Z as calibrant (B) and NIST 981 as calibrant (C).

on the mass difference of the ratioed isotopes as noted by Russ and Bazan (1987). However, the average bias values for the quality control specimen are smaller than the expected difference in the sample set chosen for analysis, indicating that this bias should not be a limiting factor for the generation of accurate data.

11.3.8 *Uranium*

Uranium has three naturally occurring isotopes ^{234}U (0.06%), ^{235}U (0.72%) and ^{238}U (99.28%). However, it should be noted that most laboratory reagent uranium is depleted in ^{235}U since it is this isotope which is used for weapons

Table 11.18 Uranium isotopic ratio data for NIST U005 (uncertainties are 1σ where $n = 12$)[a]

Date	Conc. ($\mu g\,ml^{-1}$)	^{238}U:^{235}U	^{234}U:^{235}U	^{236}U:^{235}U
18/12	0.5	202.35 ± 0.97	0.00578 ± 0.00068	0.00871 ± 0.00061
20/12	0.5	203.55 ± 0.99	0.00399 ± 0.00069	0.01044 ± 0.00098
20/12	4.6	204.14 ± 0.94	0.00441 ± 0.00023	0.01018 ± 0.00038
20/12	4.6	201.04 ± 0.56	0.00446 ± 0.00016	0.00998 ± 0.00019
20/12	0.5	202.62 ± 0.95	0.00361 ± 0.00091	0.00930 ± 0.00115
21/12	0.5	202.56 ± 1.03	0.00380 ± 0.00071	0.00972 ± 0.00062
21/12	1.0	203.20 ± 0.86	0.00478 ± 0.00028	0.01005 ± 0.00032
21/12	4.6	202.57 ± 0.97	0.00458 ± 0.00019	0.00962 ± 0.00014

[a] From Russ and Bazan (1987).

manufacture (Russ and Bazan, 1987). In addition to the naturally occurring isotopes, ^{236}U is often of interest and typically occurs at levels about 10 times lower than ^{235}U. Interferences on the uranium isotopes are negligible, expected for the formation of $^{235}U^1H$ which may need correction. The principal limitations for the precise measurement of U isotope ratios are (a) the large difference in abundance between the natural isotopes, and (b) the low concentrations which often occur, e.g. in urine analysis for personnel monitoring in nuclear establishments.

At low levels of determination, blank correction becomes a critical factor. Russ and Bazan (1987) used an average of 233, 240 and 241 m/z as a blank correction for each of the U isotopes. In this way, memory effects and wash-out times do not adversely affect blank measurements. These authors developed a rigorous data reduction scheme which corrects for background, mass bias and dead time. The results obtained for NIST U005 are shown in Table 11.18 for data accumulated over 4 days. The total uranium concentration ranges from 0.5 to 4.6 $\mu g\,ml^{-1}$. A correction was made for mass bias (2–5%) using the measured values for NIST U500 run under the same operating conditions. At a total U concentration of 0.5 $\mu g\,ml^{-1}$, the precision of ^{236}U:^{235}U is about 20% RSD, a figure which is comparable with counting statistics. Unfortunately the precise scan conditions are not given.

An alternative to sample introduction by pneumatic nebulisation, is the use of ETV. The feasibility of determining femtogramme levels of uranium and plutonium using ETV was investigated by Hall et al. (1990b). At these low concentrations ($< 1\,pg\,ml^{-1}$), U was found to be unstable in solution and had to be used within 1 h of preparation. Fifty microlitre samples were injected onto the graphite rod, dried and vaporised into the ICP. Replicate determinations of ^{235}U at the 50 fg level gave an RSD of $\sim 9\%$. The detection limit for ^{235}U calculated in solution was close to 15 fg ml^{-1} with an absolute detection limit of 0.8 fg (2.0×10^6 atoms).

11.3.9 *Other isotopic ratios determined by ICP–MS*

A number of other elements have been determined for their isotopic composition by ICP–MS including Mg, K, Cr, Se and Br. The interest in these elements lies principally in their potential as stable isotope tracers in biological studies.

Magnesium has three stable isotopes, ^{24}Mg (78.99%), ^{25}Mg (10.00%) and ^{26}Mg (11.01%). A method is described by Schuette *et al.* (1988) for the accurate determination of ^{25}Mg:^{24}Mg and ^{26}Mg:^{24}Mg in a range of biological samples including bone, brain, kidney, liver, muscle, plasma, red cells and urine. The precision for each matrix was in the range 0.1–1% RSD. A complete analytical scheme for the accurate determination of the three Mg isotopes, which is based on precipitation with ammonium phosphate, is described. The accuracy of the method was tested using reference materials NIST 1577a bovine liver and IAEA H-5 animal bone and was about 7% RSD for both $^{24:25}$Mg and $^{25:26}$Mg.

The interference problems associated with potassium isotope ratio determination were investigated by Jiang *et al.* (1988). Although potassium has three isotopes ^{39}K (93.26%), ^{40}K (0.012%) and ^{41}K (6.73%), isotope ratio measurements are hampered by the intense background peaks from Ar and ArH. Jiang *et al.* (1988) established a set of operating conditions which almost eliminated these interferences. The sampling cone was positioned relatively far (35 mm) from the load coil while the forward power was reduced to only 0.75 kW. In addition, the nebuliser gas flow rate was increased to 1.6 l min^{-1}. These modifications caused the background spectrum to become dominated by NO and suppressed Ar and ArH ions leaving 39 and 41 m/z free for K isotope ratio measurement. The precision for ^{39}K:^{41}K was found to be 0.3–0.9% RSD for K concentrations in the range 1–50 μg ml^{-1}. The mass bias was significant at $\sim$9% for ^{39}K:^{41}K and varied with the concentration of Na present. The authors provide a note of caution stating that the load coil geometry on the experimental instrument used for this work was not necessarily the same as that used on all commercial instruments. It is unlikely that this technique can be directly applied to some instruments without significant modification.

Chromium isotopic measurements were made by Dever *et al.* (1989) to compare the performance of radioactive ^{51}Cr with that of stable ^{50}Cr for the determination of red blood cell survival. Chromium has four stable isotopes, ^{50}Cr (4.31%), ^{52}Cr (83.76%), ^{53}Cr (9.5%) and ^{54}Cr (2.38%) two of which have an isobaric overlap with ^{50}Ti, ^{50}V and ^{54}Fe. Experimental work was carried out by Dever *et al.* (1989) using CrO$_3$ enriched in ^{50}Cr. The precision of the measurements for ^{50}Cr:^{52}Cr was about 0.7% while the mass bias relative to NIST 979 was about 7%.

Selenium has six isotopes ^{74}Se (0.9%), ^{76}Se (9.0%), ^{77}Se (7.5%), ^{78}Se (23.5%), ^{80}Se (50%) and ^{82}Se (9.0%). Unfortunately, all but one isotope (^{77}Se) have an isobaric overlap from either Ar, Ge or Kr while ^{77}Se suffers from a polyatomic ion interference from ^{40}Ar^{37}Cl and ^{38}Ar^{35}Cl. A method is described by Ting et al. (1989) which allows the determination of Se isotope ratios using hydride generation with isotope dilution. Using this method, they compared the isotopic signal resulting from a selenite solution containing 5 ng ml^{-1} Se with the total sample blank contribution at 74, 77 and 82 m/z. The blank was found to be less than 5% of the respective isotopic signal. The precision of the measured ^{82}Se:^{77}Se and ^{74}Se:^{77}Se ratios was typically 1% RSD. Details of the methods employed are described in full in Ting et al. (1989).

Naturally occurring bromine has two isotopes, ^{79}Br (50.7%) and ^{81}Br (49.3%) which are of almost equal abundance. Although both isotopes are free from isobaric overlap, there is significant background at 81 m/z from ^{40}Ar^{40}Ar^{1}H and correction must therefore be made for this interference. The sensitivity of bromine is rather low, due to its high first ionisation energy (11.85 eV) with a detection limit of 5–10 ng ml^{-1}. The level of bromine in many sample types, particularly biological fluids of interest in human metabolic studies, is typically too low for direct determination, and pre-concentration is usually required. Two chemical separation schemes are given by Janghorbani et al. (1988) based on either a cation exchange or distillation procedure. The techniques were applied to the analysis of urine samples using isotope dilution where a precision and accuracy of about 2% RSD was achieved.

Chlorine has two naturally occurring isotopes, ^{35}Cl (75.8%) and ^{37}Cl (24.2%). The determination of chlorine isotope ratios presents a number of problems both in relation to sensitivity (detection limit 12.7 ng ml^{-1}) and to the occurrence of polyatomic ion interference from ^{36}Ar^{1}H on ^{37}Cl. This interference may be alleviated if samples are prepared in D_2O rather than H_2O so that ^{36}ArD is formed at 38 m/z thus reducing the background species at ^{36}Ar^{1}H (Smith and Houk, 1991).

Appendix 1

Originators of reference materials cited in the text

ANRT: Association Nationale de la Recherche Technique, France

FK-N	K-feldspar
UB-N	serpentine

BAS: Bureau of Analysed Samples, England

345	Ni based alloy
346	Ni based alloy

CANMET: Canada Centre for Energy and Mineral Technology, Canada

PTC-1	sulphide concentrate
SU-1a	nickel-copper-cobalt ore

CCRMP: Canadian Certified Reference Materials Project, Canada

SY-2	syenite

GSJ: Geological Survey of Japan, Japan

JB-3	basalt

IAEA: International Atomic Energy Agency, Austria

A-11	milk powder freeze dried
H-4	animal muscle
H-5	animal bone
SOIL-5	soil

IGGE: Institute of Geophysical and Geochemical Prospecting, People's Republic of China

GSD-1	stream sediment
GSD-4	stream sediment

MINTEK: Council for Mineral Technology, South Africa

NIM-G	granite
SARM-7	platinum ore

NIST: National Institute of Standards and Technology (formerly National Bureau of Standards, NBS), USA

1b	argilaceous limestone
1c	argilaceous limestone
70a	potash feldspar
120b	phosphate rock (Florida)
365	electrolytic iron
662	steel
671	nickel oxide 1
688	basalt rock
951	boric acid (isotopic)
979	chromium (isotopic)
981	common lead (isotopic)
982	equal atom lead (isotopic)
983	radiogenic lead (isotopic)
1566	oyster tissue (replaced by 1566a)
1567	wheat flour (replaced by 1567a)
1568	rice flour (replaced by 1568a)
1570	spinach
1571	orchard leaves
1572	citrus leaves
1573	tomato leaves (replaced by 1573a)
1575	pine needles
1577	bovine liver (replaced by 1577a)
1634a	trace elements in fuel oil (replaced by 1634b)
1643	trace elements in water
1643a	trace elements in water (replaced by 1643b)
1648	urban particulate matter

NRCC: National Research Council of Canada, Canada

BCSS-1	coastal marine sediment
CASS-1	coastal seawater
CASS-2	coastal seawater
DOLT-1	dogfish liver
DORM-1	dogfish muscle
MESS-1	estuarine sediment
NASS-1	open ocean seawater
NASS-2	open ocean seawater
PACS-1	marine harbour sediment
SLEW-1	estuarine water
SLRS-1	river water
SLRS-2	river water
TORT-1	lobster hepatopancreas

USGS: United States Geological Survey, USA

AGV-1	andesite

BHVO-1	basalt
MAG-1	marine mud
NOD-A-1	Mn nodule
NOD-P-1	Mn nodule
PCC-1	peridotite
RGM-1	rhyolite
SCo-1	Cody shale

Appendix 2

Naturally-occurring isotopes—useful data

Atomic no.	Element	Symbol	Mass no.	Relative abundance	Atomic mass	Ionization energy (1)	(2)
1	Hydrogen	H	1	99.9855	1.00797	1312	
	Deuterium	D	2	0.0145			
2	Helium	He	3	0.00014	4.0026	2372.3	5250.4
			4	99.99986			
3	Lithium	Li	6	7.50	6.939	513.3	7298.0
			7	92.50			
4	Beryllium	Be	9	100	9.0122	899.4	1757.1
5	Boron	B	10	19.78	10.811	800.6	2427
			11	80.22			
6	Carbon	C	12	98.888	12.0111	1086.2	2352
			13	1.112			
7	Nitrogen	N	14	99.633	14.0067	1402.3	2856.1
			15	0.367			
8	Oxygen	O	16	99.759	15.9994	1313.9	3388.2
			17	0.037			
			18	0.204			
9	Fluorine	F	19	100	18.9984	1681	3374
10	Neon	Ne	20	90.92	20.183	2080.6	3952.2
			21	0.257			
			22	8.82			
11	Sodium	Na	23	100	22.9898	495.8	4562.4
12	Magnesium	Mg	24	78.70	24.312	737.7	1450.7
			25	10.13			
			26	11.17			
13	Aluminium	Al	27	100	26.9815	577.4	1816.6
14	Silicon	Si	28	92.21	28.086	786.5	1577.1
			29	4.70			
			30	3.09			
15	Phosphorus	P	31	100	30.9738	1011.7	1903.2
16	Sulphur	S	32	95.018	32.064	999.6	2251
			33	0.760			
			34	4.215			
			36	0.014			
17	Chlorine	Cl	35	75.53	35.453	1251.1	2297
			37	24.47			

Atomic no.	Element	Symbol	Mass no.	Relative abundance	Atomic mass	Ionization energy (1)	(2)
18	Argon	Ar	36	0.337	39.948	1520.4	2665.2
			38	0.063			
			40	99.600			
19	Potassium	K	39	93.10	39.102	418.8	3051.4
			40	0.0118			
			41	6.88			
20	Calcium	Ca	40	96.97	40.08	589.7	1145
			42	0.64			
			43	0.145			
			44	2.06			
			46	0.003			
			48	0.185			
21	Scandium	Sc	45	100	44.956	631	1235
22	Titanium	Ti	46	7.93	47.90	658	1310
			47	7.28			
			48	73.94			
			49	5.51			
			50	5.34			
23	Vanadium	V	50	0.24	50.942	650	1414
			51	99.76			
24	Chromium	Cr	50	4.31	51.996	652.7	1592
			52	83.76			
			53	9.55			
			54	2.38			
25	Manganese	Mn	55	100	54.9381	717.4	1509.0
26	Iron	Fe	54	5.82	55.847	759.3	1561
			56	91.66			
			57	2.19			
			58	0.33			
27	Cobalt	Co	59	100	58.9332	760.0	1646
28	Nickel	Ni	58	67.88	58.71	736.7	1753
			60	26.23			
			61	1.19			
			62	3.66			
			64	1.08			
29	Copper	Cu	63	69.09	63.54	745.4	1958
			65	30.91			
30	Zinc	Zn	64	48.89	65.37	906.4	1733.3
			66	27.81			
			67	4.11			
			68	18.57			
			70	0.62			
31	Gallium	Ga	69	60.4	69.72	578.8	1979
			71	39.6			
32	Germanium	Ge	70	20.52	72.59	762.1	1537
			72	27.43			
			73	7.76			
			74	36.54			
			76	7.76			

Atomic no.	Element	Symbol	Mass no.	Relative abundance	Atomic mass	Ionization energy (1)	energy (2)
33	Arsenic	As	75	100	74.9216	947.0	1798
34	Selenium	Se	74	0.87	78.96	940.9	2044
			76	9.02			
			77	7.58			
			78	23.52			
			80	49.82			
			82	9.19			
35	Bromine	Br	79	50.537	79.909	1139.9	2104
			81	49.463			
36	Krypton	Kr	78	0.35	83.80	1350.7	2350
			80	2.27			
			82	11.56			
			83	11.55			
			84	56.90			
			86	17.37			
37	Rubidium	Rb	85	72.15	85.47	403.0	2632
			87	27.85			
38	Strontium	Sr	84	0.56	87.62	549.5	1064.2
			86	9.86			
			87	7.02			
			88	82.56			
39	Yttrium	Y	89	100	88.905	616	1181
40	Zirconium	Zr	90	51.46	91.22	660	1267
			91	11.23			
			92	17.11			
			94	17.40			
			96	2.80			
41	Niobium	Nb	93	100	92.906	664	1382
42	Molybdenum	Mo	92	15.84	95.94	685.0	1558
			94	9.04			
			95	15.72			
			96	16.53			
			97	9.46			
			98	23.78			
			100	9.63			
44	Ruthenium	Ru	96	5.51	101.07	711	1617
			98	1.87			
			99	12.72			
			100	12.62			
			101	17.07			
			102	31.63			
			104	18.58			
45	Rhodium	Rh	103	100	102.905	720	1744
46	Palladium	Pd	102	0.96	106.4	805	1875
			104	10.97			
			105	22.23			
			106	27.33			
			108	26.71			
			110	11.81			

Atomic no.	Element	Symbol	Mass no.	Relative abundance	Atomic mass	Ionization energy (1)	(2)
47	Silver	Ag	107	51.817	107.87	731	2073
			109	48.183			
48	Cadmium	Cd	106	1.22	112.40	867.6	1631
			108	0.88			
			110	12.39			
			111	12.75			
			112	24.07			
			113	12.26			
			114	28.86			
			116	7.58			
49	Indium	In	113	4.28	114.82	558.3	1820.6
			115	95.72			
50	Tin	Sn	112	0.96	118.69	708.6	1411.8
			114	0.66			
			115	0.35			
			116	14.30			
			117	7.61			
			118	24.03			
			119	8.58			
			120	32.85			
			122	4.72			
			124	5.94			
51	Antimony	Sb	121	57.25	121.75	833.7	1794
			123	42.75			
52	Tellurium	Te	120	0.089	127.60	869.2	1795
			122	2.46			
			123	0.87			
			124	4.61			
			125	6.99			
			126	18.71			
			128	31.79			
			130	34.48			
53	Iodine	I	127	100	126.9044	1008.4	1845.9
54	Xenon	Xe	124	0.096	131.30	1170.4	2046
			126	0.090			
			128	1.919			
			129	26.44			
			130	4.08			
			131	21.18			
			132	26.89			
			134	10.44			
			136	8.87			
55	Caesium	Cs	133	100	132.905	375.7	2420
56	Barium	Ba	130	0.101	137.34	502.8	965.1
			132	0.097			
			134	2.42			
			135	6.59			
			136	7.81			
			137	11.32			
			138	71.66			

Atomic no.	Element	Symbol	Mass no.	Relative abundance	Atomic mass	Ionization energy (1)	(2)
57	Lanthanum	La	138	0.089	138.91	538.1	1067
			139	99.911			
58	Cerium	Ce	136	0.193	140.12	527.4	1047
			138	0.250			
			140	88.48			
			142	11.07			
59	Praseodymium	Pr	141	100	140.907	523.1	1018
60	Neodymium	Nd	142	27.11	144.24	529.6	1035
			143	12.17			
			144	23.85			
			145	8.30			
			146	17.22			
			148	5.73			
			150	5.62			
62	Samarium	Sm	144	3.09	150.35	543.3	1068
			147	14.97			
			148	11.24			
			149	13.83			
			150	7.44			
			152	26.72			
			154	22.71			
63	Europium	Eu	151	47.82	151.96	546.7	1085
			153	52.18			
64	Gadolinium	Gd	152	0.20	157.25	592.5	1167
			154	2.15			
			155	14.73			
			156	20.47			
			157	15.68			
			158	24.87			
			160	21.90			
65	Terbium	Tb	159	100	158.925	564.6	1112
66	Dysprosium	Dy	156	0.052	162.50	571.9	1126
			158	0.090			
			160	2.29			
			161	18.88			
			162	25.53			
			163	24.97			
			164	28.18			
67	Holmium	Ho	165	100	164.930	580.7	1139
68	Erbium	Er	162	0.136	167.26	588.7	1151
			164	1.56			
			166	33.41			
			167	22.94			
			168	27.07			
			170	14.88			
69	Thulium	Tm	169	100	168.934	596.7	1163

Atomic no.	Element	Symbol	Mass no.	Relative abundance	Atomic mass	Ionization energy (1)	(2)
70	Ytterbium	Yb	168	0.135	173.04	603.4	1176
			170	3.03			
			171	14.31			
			172	21.82			
			173	16.13			
			174	31.84			
			176	12.73			
71	Lutetium	Lu	175	97.41	174.97	523.5	1340
			176	2.59			
72	Hafnium	Hf	174	0.18	178.49	642	1440
			176	5.20			
			177	18.50			
			178	27.14			
			179	13.75			
			180	35.24			
73	Tantalum	Ta	180	0.012	180.948	761	(1500)
			181	99.988			
74	Tungsten	W	180	0.14	183.85	770	(1700)
			182	26.41			
			183	14.40			
			184	30.64			
			186	28.41			
75	Rhenium	Re	185	37.07	186.2	760	1260
			187	62.93			
76	Osmium	Os	184	0.02	190.2	840	(1600)
			186	1.59			
			187	1.64			
			188	13.3			
			189	16.1			
			190	26.4			
			192	41.0			
77	Iridium	Ir	191	37.3	192.2	880	(1680)
			193	62.7			
78	Platinum	Pt	190	0.013	195.09	870	1791
			192	0.78			
			194	32.9			
			195	33.8			
			196	25.3			
			198	7.21			
79	Gold	Au	197	100	196.967	890.1	1980
80	Mercury	Hg	196	0.146	200.59	1007.0	1809.7
			198	10.02			
			199	16.84			
			200	23.13			
			201	13.22			
			202	29.80			
			204	6.85			
81	Thallium	Tl	203	29.5	204.37	589.3	1971.0
			205	70.5			

Atomic no.	Element	Symbol	Mass no.	Relative abundance	Atomic mass	Ionization energy (1)	(2)
82	Lead	Pb	204	1.48	207.19	715.5	1450.4
			206	23.6			
			207	22.6			
			208	52.3			
83	Bismuth	Bi	209	100	208.98	703.2	1610
90	Thorium	Th	232	100	232.04	587	1110
92	Uranium	U	234	0.0057	238.03	584	1420
			235	0.72			
			238	99.27			

( ) suggested value
Ionization energy in $kJ\,mol^{-1}$

Appendix 3

Glossary

Absorption   To take into.

Abundance sensitivity   The amount of overlap between peaks which is defined as the ratio of the signal at $M-1$ or $M+1$ to that at M.

Accuracy   A measure of how close the analysed data lie to the true value.

Actinides   Elements of atomic number 89–102, that is actinium to lawrencium, all of which are radioactive.

ADC   Analogue to digital conversion.

Adsorption   To take onto.

Adduct ion   See polyatomic ion.

Aerosol   System of particles dispersed in a gas.

Analyte   The element, isotope or species to be determined.

Anneal   To toughen a substance, usually glass or metal, by heating and controlled cooling.

Anthropogenic   Originated by man.

Aqua regia   1 part $16\,M$ HNO_3 with 3 parts $12\,M$ HCl by volume.

Atomic number or proton number   The number of protons in the nucleus or the number of electrons revolving around the nucleus. All isotopes of an element have the same atomic number.

Atomisation   The formation of free ground state atoms.

Azeotrope   A constant boiling point mixture.

Background   The unwanted part of a signal which usually contains no analytically useful information.

Bond energy   Energy required to break a chemical bond between two atoms in a molecule which is dependent on the type of atoms and on the nature of the molecule.

Boundary layer   A cool layer of gas which forms across the sampling cone face.

Calibration line   The relationship between the output signal from the instrument and the concentration of the element to be measured. Under ideal conditions, measured data would form an exact linear function of concentration.

Central channel   Cooler, central part of the plasma along which the sample aerosol passes and from which ions are extracted.

Chromatography   Chemical separation of substances by the distribution of the component between a stationary phase and a mobile phase.

Concomitant elements   Matrix elements accompanying an analyte.

Constant width integration   Peak integration over a fixed, pre-defined mass width.

Correlation coefficient   A measure of the goodness of fit of a line to a set of measured data.

Counting (Poisson) statistics   Used to describe the number of events in a given time providing those events occur randomly with time.

Crystal controlled generator   One in which the operating frequency is determined by an oscillating quartz crystal.

DAC   Digital to analogue conversion.

Dead time   The inability of a detector and its associated electronics to resolve successive pulses which occur at less than a given interval.

Deionised water   Water which has been purified by its passage over ion exchange resins. A purity of typically better than 18 Mohm can be obtained.

Demountable torch   One that may be readily dismantled into its component parts for cleaning or replacement of damaged parts.

Desolvation   Removal of solvent usually by heating.

Diffusion (or vapour) pump   Non-mechanical vacuum pump which relies on the condensation of hot oil vapour to extract gases and thereby create a vacuum. Diffusion pumps can only operate at pressures of $< 1 \times 10^{-1}$ mbar and are therefore backed by mechanical pumps.

Digestion bomb   Sealed vessel, composed of a rigid material, which is generally able to withstand substantial internal pressure.

Dimer   Compound formed by the combination of two molecules of a monomer.

Dissociation   The process whereby a molecule is split into simpler fragments which may be smaller molecules, atoms, free radicals or ions.

Doubly charged ion   An atom or group of atoms which have lost or gained two electrons, and which occur in the mass spectrum at $M/2$ where M is the mass of the parent ion.

Element   A substance that cannot be further subdivided by chemical methods.

External calibration   Determination of elemental concentration by comparison with a set of solutions (or solids) containing known amounts of the elements of interest. The number of standards used for external calibration is usually 2 or more.

Flow injection   Introduction of a discrete volume of sample into a continuously flowing carrier stream.

Free electron path   The distance an electron can travel before collision with an atom occurs.

Free running generator   One in which the frequency of the system is controlled by an oscillating circuit and by the load coil parameters.

Freon   Series of halogenated methyl gases, e.g. Freon 23, trifluoromethane.

Fuming nitric acid   Nitric acid of 20–23 M.

Hertz (Hz)   SI unit of frequency defined as the frequency of a phenomenon that has a period of 1 s.

Hydride forming elements   The elements As, Bi, Ge, Pb, Sb, Se, Sn and Te form hydrides which are gaseous at ambient temperatures and which can be generated easily from aqueous solution.

Hydrolysis   Term used to signify reactions due to the presence of the hydrogen or hydroxyl ion of water.

Initial radiation zone   A hollow, bullet shaped region of the ICP of intense atomic emission.

Integral torch   One of fixed configuration.

Inter (between) sample precision   A measure of the reproducibility of data from several separate preparations of the same sample.

Internal standardisation   Method of data correction used to compensate for non-spectroscopic interferences.

Intra (within) sample precision   A measure of the reproducibility of replicate determinations from a single sample preparation.

Ion   An electrically charged atom, molecule or group of atoms or molecules.

Ion detector   Electronic device used to detect ions. In the channel electron multiplier each ion hitting the detector releases electrons. The number of electrons are multiplied during their passage through the detector.

Ion exchange resins   Synthetic, cross-linked polymers carrying acidic or basic side groups which have high exchange capacity.

Ion lens   Device used to form as many ions as possible, from the cloud at the rear of the skimmer, into an axial beam of circular cross-section at the entrance to the quadrupole mass analyser.

Ionisation   The process of forming ions.

Ionisation potential (IP)   The first IP is the minimum energy necessary to remove the least strongly bound electron from a given atom or molecule to infinity. The second IP is the minimum energy required to remove the second least strongly bound electron from a neutral atom.

Isobaric overlap   Two isotopes having similar mass which cannot be resolved by the mass spectrometer.

Isotope   One of two or more forms of an element differing from each other in atomic weight but not atomic number and in nuclear but not chemical properties.

Isotope dilution   A calibration method in which the measurement of the change of signal intensity ratio for two selected isotopes of an element, after the addition of a known quantity of spike enriched in one of those isotopes, is made.

Isotope ratio   Ratio of two isotopes of the same or different elements.

Kinetic energy   The energy possessed by virtue of motion.

Lanthanides   Rare earth elements of atomic number 51–71, which have chemical properties similar to lanthanum.

Laser   Light amplification by stimulated emission of radiation.

Limit of quantitation   A concentration equal to 10 times the standard deviation of the background signal. The ideal error in measurements at this level is 10% relative standard deviation (1 sigma).

Linear dynamic range   The range over which the relationship between concentration and response is linear.

Linear regression   Statistical procedure used to calculate a best fit line for a series of standards.

Load coil   Cooled coiled tube, often made from copper, along which a high frequency current flows. The load coil is located around the open end of the plasma torch.

Mass bias   The apparent deviation of the observed or measured isotope ratio from the true ratio as a function of the difference in mass between the two isotopes.

Mass calibration   Procedure to determine the digital to analogue conversion setting of the quadrupole system to set it at any given mass.

Mass discrimination   The process, in the mass spectrometer, which results in a mass bias being observed.

Mass fractionation   The separation of two or more isotopes by some physical process. In TIMS, for example, this normally results in a loss of the lighter isotope.

Matrix   Those elements which comprise the bulk of a sample.

Matrix matched standards   Those which contain the same elemental matrix as the unknown samples.

Matrix modification   Method of stabilising an analyte by addition of an element or compound usually at high concentration.

Mixed gas plasma   Plasma composed of two or more gases usually Ar, N, H or He.

Multi-channel analyser   A fast digital buffer store.

Nd:YAG   Neodymium-yttrium aluminium garnet.

Nebuliser   Device for generating a gas borne aerosol.

Non-spectral (spectroscopic) interferences Those which result due to the matrix of a sample. The interferences may manifest themselves in several ways, including analyte signal suppression or enhancement and signal loss with time.

On-line separation Separation of analytes from a matrix during sample introduction.

Peak hopping (jumping) mode The mode in which the mass spectrometer is set to sequentially transmit selected masses only. The time spent at each mass position may be different for each isotope.

Peak integration A method of adding the component parts of the signal at a given mass position.

Peristaltic pump Liquid sample delivery device using a set of rollers which successively constrict a plastic tube causing the liquid to be passed along (peristalsis). Sample delivery by this method is pulsed rather than continuous.

Phase separator Device used in hydride generation which allows separation of the gaseous from liquid phase.

Photon stop Metal disc, positioned on axis in the ion lenses, to prevent direct photons from the plasma reaching the detector and thereby contributing to the background signal.

Plasma A highly ionised gas containing approximately equal numbers of electrons and positive ions.

Plasma potential The electrical potential (in volts) assumed by the ICP above ground.

Platinum group metals The elements Ru, Rh, Pd, Os, Ir and Pt.

Polyatomic or adduct ion Results from ion molecule reactions between major species in the plasma, e.g. ArO.

Potential energy The energy possessed by virtue of position.

Precision A measure of analytical repeatability.

Procedural blank One which has been passed through the same preparation procedure as the sample itself and which contains all of the reagents used in that procedure.

Q-switched Mode of laser operation whereby the lasing procedure is delayed and a pulse is emitted of high power and short duration.

Radio frequency Electromagnetic radiation in the frequency band 3 kHz to 300 GHz.

Radioactive decay The transformation of a radioactive nuclide, the parent, into its daughter product by disintegration, resulting in the exponential decrease in the activity of the parent.

Radiogenic Resulting from radioactive decay.

Rare earth elements Often used as synonymous with lanthanide although strictly also includes yttrium and scandium.

Refractory materials   Those which are not damaged by heating to at least 1500°C in a clean oxidising atmosphere, e.g. lime, magnesia, alumina, dolomite and most of the rarer refractory oxides particularly zirconia.

Refractory oxide ion   One which is formed from a refractory element and oxygen, and occurs in the mass spectrum $M + 16$, $M + 32$ or $M + 48$ mass units above the parent ion.

Resolution   The ability of the mass analyser to separate ions of similar mass.

Rotary (or vane) pump   Mechanical vacuum pump which compresses gases from intake to atmospheric pressure in a single operation.

Saha equation   Describes the degree of ionisation for an element based on the ionisation energy, plasma temperature and electron density.

Sapphire   Form of corundum (Al_2O_3) used in the manufacture of HF resistant torch injectors and in some types of nebuliser.

Scanning mode   The mode in which the mass spectrometer is set to sequentially transmit all masses over a given range. The time spent at each mass position is the same.

Sensitivity   Number of ions measured per unit of concentration.

Signal   The physical response containing analytical information related to the concentration of the elemental or isotopic component of a sample.

Signal enhancement   Increase in the signal obtained for the given concentration of an element in the presence of a matrix, compared with that measured in the absence of that matrix.

Signal suppression   Reduction in the signal obtained for the given concentration of an element in the presence of a matrix, compared with that measured in the absence of that matrix.

Single mass monitor   The mass spectrometer is set to transmit only one isotope of the element of interest.

Slurry   A uniform suspension of small particles.

Solvent extraction   Process whereby an element or compound is preferentially separated from its matrix into a solvent which is usually organic in nature.

Space charge   Occurs in a region where the net charge density is significantly different from zero.

Spectral (spectroscopic) interferences   Those which result from the nearly direct overlap of two mass peaks, whether of elemental (e.g. ^{48}Ti and ^{48}Ca) or molecular (e.g. ^{40}Ar^{16}O at 56 m/z) origin.

Spectrum   Any distribution of energies, momenta, velocities, etc. in a system of particles.

Spray chamber   Device for sorting aerosol droplets. Usually designed to remove droplets of $> 4\,\mu$m.

Standard   Substance containing a known amount of an element or isotope.

Standard additions calibration   A method in which a sample is divided into several aliquots. To each of these are added a different (increasing) quantity of a reagent containing the elements of interest.

Standard reference materials   Substances which contain a range of elements for which agreed concentrations have been derived.

Tesla coil   Device used to produce 'free' or 'seed' electrons necessary to initiate the plasma.

Torch   A series of concentric, usually glass, tubes to contain the plasma.

Total dissolved solids   Concentration of solid material dissolved in a solvent. For example 0.2 g of solid in 100 ml of solvent gives a solution containing 0.2% or 2000 μg ml^{-1} total dissolved solids.

Transient signal   One of short duration (typically 2–10 s) such as that produced by ETV.

Transport efficiency (in nebulisers and spraychambers)   Is defined as the analyte mass reaching the plasma compared with the analyte mass aspirated.

Turbomolecular pumps   The only purely mechanical vacuum pump which, in conjunction with a backing pump, can achieve pressures of $< 1 \times 10^{-10}$ mbar. These pumps contain a series of rotors turning at very high speeds which are used to create the vacuum.

USEPA   United States Environmental Protection Agency.

Valley integration   Peak integration to a preset minimum background signal.

Vapour   A substance in gaseous form.

Vitreous graphite   Non-porous, glassy, fused form of graphite used in the manufacture of digestion crucibles.

Wash-out or clean-out time   Time taken for the signal to diminish to a pre-determined level after removal of sample from the introduction system.

Wet ashing   Breakdown of organic materials by addition of a strong oxidising agent such as concentrated HNO_3.

Zone of silence   Region of freely expanding gas behind the sampling cone.

References

Abu-Samra, A., Morris, J.S. and Koiryohann, S.R. (1975) Wet ashing of some biological samples in a microwave oven. *Anal. Chem.* **47**, 1475–1477.

Allain, P., Mauras, Y., Douge, C., Jaunault, L., Delaporte, T. and Beaugrand, C. (1990) Determination of iodine and bromine in plasma and urine by inductively coupled plasma mass spectrometry. *Analyst* **115**, 813–815.

Allenby, P. (1987) A user's view of inductively coupled plasma mass spectrometry in routine analysis. *Anal. Proc.* **24**, 12–13.

Al Swaidan, H.M. (1988) Simultaneous determination of trace metals in Saudi Arabian crude oil products by inductively coupled plasma mass spectrometry (ICP–MS). *Anal. Lett.* **21** 1487–1497.

Anderson, H., Kaiser, H. and Meddings, B. (1981) High precision (< 0.5% RSD) in routine analysis by ICP using a high pressure (200 psig) cross flow nebuliser. In *Developments in Atomic Spectrochemical Analysis, Proc. Int. Winter Conf.*, ed. Barnes, R.M., Heyden, Philadelphia, pp. 251–277.

Arrowsmith, P. (1987a) Laser ablation of solids for elemental analysis by inductively coupled plasma mass spectrometry. *Anal. Chem.* **59**, 1437–1444.

Arrowsmith, P. (1987b) Analysis of solids by laser ablation plasma source mass spectrometry. *ICP Inform. Newslett.* **12**, 579–608.

Arrowsmith, P and Hughes, S.K. (1988) Entrainment and transport of laser ablated plumes for subsequent elemental analysis. *Appl. Spectrosc.* **42**, 1231–1239.

Asif, M. and Parry, S.J. (1989) Elimination of reagent blank problems in the fire-assay preconcentration of the platinum group elements and gold with a nickel sulphide bead of less than one gram mass. *Analyst* **144**, 1057–1059.

Babington, R.S. (1973) It's Superspray. *Popular Sci.* **May**, 102–104.

Bacon, W.G. Hawthorn, G.W. and Poling, G.W. (1989) Gold analyses—myths, frauds and truths. *Can. Inst. Mining Bull.* **82**, 29–36.

Baginski, B.R. and Meinhard, J.E. (1984) Some effects of high-solids matrices on the sample delivery system and the meinhard concentric nebulizer during ICP emission analyses. *Appl. Spectrosc.* **38**, 568–572.

Bailey, A.G. (1984) Electrostatic spraying of liquids. *Phys. Bull.* **35**, 146–148.

Bakowska, E., Falkner, K., Barnes, R.M. and Edmond, J.M. (1989) Sample handling of gold at low concentration in limited volume solutions preconcentrated from seawater for inductively coupled plasma mass spectrometry. *Appl. Spectrosc.* **43**, 1283–1286.

Baumann, H. and Pavel, J. (1989) Determination of trace impurities in electronic grade quartz: comparison of inductively coupled plasma mass spectrometry with other analytical techniques. *Mikrochim. Acta* **111**, 413–422.

Baxter, M.J., Burrell, J.A., Crews, H.M., Massey, R.C. and McWeeny, D.J. (1989) A procedure for the determination of lead in green vegetables at concentrations down to 1 μg/kg. *Food Additives Contam.* **6**, 341–349.

Bazan, J.M. (1987) Enhancement of osmium detection in inductively coupled plasma atomic emission spectrometry. *Anal. Chem.* **59**, 1066–1069.

Beamish, F.E. (1966) *The Analytical Chemistry of the Noble Metals*, Pergamon, Oxford.

Beauchemin, D. (1989) The ICP–MS approach to environmental studies. *Mikrochim. Acta* **111**, 273–281.

Beauchemin, D. and Berman, S.S. (1989) Determination of trace metals in reference water standards by inductively coupled plasma mass spectometry with on-line preconcentration. *Anal. Chem.* **61**, 1857–1862.

Beauchemin, D. and Craig, J.M. (1990) Investigations on mixed gas plasmas produced using a sheathing device in ICP–MS. In *Plasma Source Mass Spectrometry*, eds. Jarvis, K.E., Gray, A.L., Jarvis, I. and Williams, J.G., Special Publication 85 RSC, pp. 25–42.

Beauchemin, D., McLaren, J.W. and Berman, S.S. (1987a) Study of the effects of con-comitant elements in inductively coupled plasma mass spectrometry. *Spectrochim. Acta* **42B**, 467–490

Beauchemin, D., McLaren, J.W., Mykytiuk, A.P. and Berman, S.S. (1987b) Determination of trace metals in a river water reference material by inductively coupled plasma mass spectrometry. *Anal. Chem.* **59**, 778–783.

Beauchemin, D., Bednas, M.E., Berman, S.S., McLaren, J.W., Siu, K.W.M and Sturgeon, R.E. (1988a) Identification and quantification of arsenic species in a dogfish muscle reference material for trace elements. *Anal. Chem.* **60**, 2209–2212.

Beauchemin, D., McLaren, J.W. and Berman, S.S. (1988b) Use of external calibration for the determination of trace metals in biological materials by inductively coupled plasma mass spectrometry. *J. Anal. Atom. Spectrom.* **3**, 775–780.

Beauchemin, D., McLaren, J.W., Mykytiuk, A.P. and Berman, S.S. (1988c) Determination of trace metals in an open ocean water reference material by inductively coupled plasma mass spectrometry. *J. Anal. Atom. Spectrom.* **3**, 305–308.

Beauchemin, D., McLaren, J.W., Willie, S.N. and Berman, S.S. (1988d) Determination of trace metals in marine biological reference materials by inductively coupled plasma mass spectrometry. *Anal. Chem.* **60**, 687–691.

Beauchemin, D., Siu, K.W.M. and Berman, S.S. (1988e) Determination of organomercury in biological reference materials by inductively coupled plasma mass spectrometry using flow injection analysis. *Anal. Chem.* **60**, 2587–2590.

Beauchemin, D., Siu, K.W.M., McLaren, J.W. and Berman, S.S. (1989) Determination of arsenic species by coupled high-performance liquid chromatography-inductively coupled plasma mass spectrometry. *J. Anal. Atom. Spectrom.* **4**, 285–289.

Beck, G.L. and Farmer, O.T. (1988) Applications of inductively coupled plasma mass spectrometry to the production control of aerospace and nuclear materials. *J. Anal. Atom. Spectrom.* **3**, 771–773.

Beijerink, H.C.W., Van Gerwen, R.J.F., Kerstel, E.R.T., Martens, J.F.M., Van Vliembergen, E.J.W., Smits, M.R.Th. and Kaashoek, G.M. (1985) *Chem. Phys.* **96**, 153–173.

Berglund, R.N. and Liu, B.Y.H. (1973) Generation of monodisperse aerosol standards. *Environ. Sci. Technol.* **7**, 147–153.

Bernas, B. (1968) A new method for decomposition and comprehensive analysis of silicates by atomic absorption spectrometry. *Anal. Chem.* **42**, 1682–1686.

Bettinelli, M., Baroni, U. and Pastorelli, N. (1989) Microwave oven sample dissolution for the analysis of environmental and biological materials. *Anal. Chim. Acta* **225**, 159–174.

Blain, L., Salin, E.D. and Boomer, D.W. (1989) Probe design for the direct insertion of solid samples in the inductively coupled plasma for analysis by atomic emission and mass spectrometry. *J. Anal. Atom. Spectrom.* **4**, 721–725.

Blair, P.D. (1986) The application of inductively coupled plasma mass spectrometry in the nuclear industry. *Trends Anal. Chem.* **5**, 220–223.

Bock, R. (1979) *A Handbook of Decomposition Methods in Analytical Chemistry*, Blackie, Glasgow, London, 444 pp.

Bolton, A., Hwang, J. and Vander Voet, A. (1982) The determination of the rare-earth elements in geological materials by ICP emission spectrometry. *ICP Inform. Newslett.* **7**, 498–500.

Boomer, D.W. and Powell, M.J. (1986) The analysis of acid precipitation samples by inductively coupled plasma mass spctrometry. *Can. J. Spectrosc.* **31**, 104–109.

Boomer, D.W. and Powell, M.J. (1987) Determination of uranium in environmental samples using inductively coupled plasma mass spectrometry. *Anal. Chem.* **59**, 2810–2813.

Boomer, D.W., Powell, M., Sing, R.L.A. and Salin, E.D. (1986) Application of a wire loop direct sample insertion device for inductively coupled plasma mass spectrometry. *Anal. Chem.* **58**, 975–976.

Boomer, D.W., Powell, M.J. and Hipfner, J. (1990) Characterization and optimization of HPIC for on-line preconcentration of trace metals with detection by ICP-mass spectrometry. *Talanta* **37**, 127–134.

Boorn, A., Fulford, J.E. and Wegscheider, W. (1985) Determination of trace elements in organic material by inductively coupled plasma mass spectrometry. *Mikrochim. Acta* **11**, 171–178.

Boumans, P.W.J.M. (1966) *Theory of Spectrochemical Excitation*, Hilger, London.

Boumans, P.W.J.M., ed. (1987) *Inductively Coupled Plasma Emission Spectroscopy* Parts 1 and 2, *Chemical Analysis* Vol. 90, Wiley-Interscience, New York.

Bowman, W.S. (Compiler) (1990) Certified reference materials, Canadian Centre for Mineral and Energy Technology Report, **CCRMP 90-1E**, 65 pp.

Bozic, J., Maskery, D., Maggs, S., Susil, H. and Smith, F.E. (1989) Rapid procedure for the dissolution of a wide variety of ore and smelter samples prior to analysis by inductively coupled plasma atomic emission spectrometry. *Analyst* **114**, 1401–1403.

Bradshaw, N., Hall, E.F.H. and Sanderson, N.E. (1989) Inductively coupled plasma as an ion source for high-resolution mass spectrometry. *J. Anal. Atom. Spectrom.* **4** 801–803.

Branch, S., Bancroft, K.C.C., Ebdon, L. and O'Neill, P. (1989a) The determination of arsenic species by coupled high-performance liquid chromatography-atomic spectrometry. *Anal. Proc.* **26**, 73–75.

Branch, S., Ebdon, L., Hill, S. and O'Neill, P. (1989b) Liquid chromatography—inductively coupled plasma mass spectrometry for monitoring tributyltin in waters. *Anal. Proc.* **26**, 401–403.

Branch, S., Corns, W.T., Ebdon, L., Hill, S. and O'Neill, P. (1991) The determination of arsenic by hydride generation-inductively coupled plasma mass spectrometry using a tubular membrane gas liquid separator. *J. Anal. Atom. Spectrom.* **6**, 155–158.

Brenner, I.B., Watson, A.E., Russell, G.M. and Goncalves, M. (1980) A new approach to the determination of the major and minor constituents in silicate and phosphate rocks. *Chem. Geol.* **28**, 321–330.

Brenner, I.B., Lorber, A. and Goldbart, Z. (1987) Trace element analysis of geological materials by direct solids insertion of a graphite cup into an inductively coupled plasma. *Spectrochim. Acta* **42B**, 219–225.

Brotherton, T.J., Shen, W.L. and Caruso, J. (1989) Use of a Hildebrand grid nebuliser for analysis of high matrix solutions containing easily ionisable elements with inductively coupled plasma mass spectrometry. *J. Anal. Atom. Spectrom.* **4**, 39–44.

Brown, P.G., Davidson, T.M. and Caruso, J.A. (1988) Application of helium microwave-induced plasma mass spectrometry to the detection of high ionisation potential gas phase species. *J. Anal. Atom. Spectrom.* **3**, 763–769.

Burman, J.O. (1981) ICP–OES applications in steel industry steel and slag analysis. In *Developments in Atomic Plasma Spectrochemical Analysis, Proc. Int. Winter Conf.*, ed. Barnes, R.M., Heyden, Philadelphia, pp. 564–574.

Busch, K.W. and Busch, M.A. (1990) *Multielement Detection Systems for Spectrochemical Analysis.* Wiley, New York, pp. 392–403.

Bushee, D.S. (1988) Speciation of mercury using liquid chromatography with detection by inductively coupled plasma mass spectrometry. *Analyst* **113**, 1167–1170.

Bushee, D.S., Moody, J.R. and May, J.C. (1989) Determination of thimerosal in biological products by liquid chromatography with inductively coupled plasma mass spectrometric detection. *J. Anal. Atom. Spectrom.* **4**, 773–775.

Campargue, R. (1966) High intensity supersonic molecular beam apparatus. In *IV Symposium on Rarified Gas Dynamics*, ed. De Leeuw, J.H., Academic Press, New York, pp. 279–298.

Campbell, M.J. and Delves, H.T. (1989) Accurate and precise determination of lead isotope ratios in clinical and environmental samples using inductively coupled plasma source mass spectrometry. *J. Anal. Atom. Spectrom.* **4**, 235–236.

Carius, L. (1860) Ueber die elementaranalyse organischer verbindungen. *Ann. Chem. Pharm.* **116**, 1–30.

Carr, J.W. and Horlick, G. (1982) Laser vaporisation of solid metal samples into an inductively coupled plasma. *Spectrochim. Acta* **37B**, 1–15.

Chambers, A., Fitch, R.K. and Halliday, B.S. (1989) *Basic Vacuum Technology.* Adam Hilger, Bristol.

Chan, L.-H. and Edmund, J.M. (1988) Variation of lithium isotope composition in the marine environment: a preliminary report. *Geochim. Cosmochim. Acta* **52**, 1711–1717.

Chapman, B. (1980) *Glow Discharge Processes.* Wiley, New York, Chaps. 3 and 5.

Choot, E.H. and Horlick, G. (1986) Vertical spacial emission profiles in Ar–N$_2$ mixed gas inductively coupled plasmas. *Spectrochim. Acta* **41B**, 889–906.

Chung, Y.S. and Barnes, R.M. (1988) Determination of gold, platinum, palladium and silver in geological samples by inductively coupled plasma atomic emission spectrometry after poly(dilithiocarbamate) resin pretreatment. *J. Anal. Atom. Spectrom.* **3**, 1079–1082.

Cotton, F.A. and Wilkinson, G. (1988) *Advanced Inorganic Chemistry*, 5th edition, Wiley, New York, 1455 pp.

Crain, J.S., Houk, R.S. and Smith, F.G. (1988) Matrix interferences in ICP–MS: some effects of skimmer orifice diameter and ion lens voltages. *Spectrochim. Acta* **43B**, 1355–1364.

Crain, J.S., Smith, F.G. and Houk, R.S. (1990) Mass spectrometric measurement of ionization temperature in an ICP. *Spectrochim. Acta* **45B**, 249–259.

Cremer, M. and Schlocker, J. (1976) Lithium borate decomposition of rocks, minerals and ores. *Am. Mineral.* **61**, 318–321.

Cresser, M. (1990) Sample preparation in environmental chemistry. *Anal. Proc.* **27**, 110–111.

Crews, H.M., Dean, J.R., Ebdon, L. and Massey, R.C. (1989) Application of high-performance liquid chromatography—inductively coupled plasma mass spectrometry to the investigation of cadmium speciation in pig kidney following cooking and *in vitro* gastro-intestinal digestion. *Analyst* **114**, 895–899.

Crock, J.G., Lichte, F.E. and Wildeman, T.R. (1984) The group separation of the rare earth elements and yttrium from geological materials by cation-exchange chromatography. *Chem. Geol.* **45**, 149–163.

Dahl, D.A. and Delmore, J.E. (1990) PC/PS2 SIMION Version 4.0, Idaho Nuclear Engineering Laboratory, EG & G Idaho, Inc. Idaho Falls, ID.

Darke, S.A., Pickford, C.J. and Tyson, J.F. (1989) Study of electrothermal vaporisation sample introduction for plasma spectrometry. *Anal. Proc.* **26**, 379–381.

Date, A.R. (1978) Preparation of trace element reference materials by a co-precipitation gel technique. *Analyst* **103**, 84–92.

Date, A.R. and Cheung, Y.Y. (1987) Studies in the determination of lead isotope ratios by inductively coupled plasma mass spectrometry. *Analyst* **112**, 1531–1540.

Date, A.R. and Gray, A.L. (1981) Plasma source mass spectrometry using an inductively coupled plasma and a high resolution quadrupole mass filter. *Analyst* **106**, 1255–1267.

Date, A.R. and Gray, A.L. (1983a) Development progress in plasma source mass spectrometry. *Analyst* **108**, 159–165.

Date, A.R. and Gray, A.L. (1983b) Isotope ratio measurements on solution samples using ICP–MS. *Int. J. Mass Spectrom. Ion Phys.* **48**, 357–360.

Date, A.R. and Gray, A.L. (1985) Determination of trace elements in geological samples by inductively coupled plasma source mass spectrometry. *Spectrochim. Acta* **40B**: 115–122.

Date, A.R. and Hutchison, D. (1987) Determination of the rare earth elements in geological samples by inductively coupled plasma mass spectrometry. *J. Anal. Atom. Spectrom.* **2**, 269–276.

Date, A.R. and Jarvis, K.E. (1989) Application of ICP–MS in the earth sciences. In *Applications of Inductively Coupled Plasma Mass Spectrometry*, ed. Date, A.R. and Gray, A.L., Blackie, Glasgow, pp. 43–70.

Date, A.R. and Stuart, M.E. (1987) Application of inductively coupled plasma mass spectrometry to the simultaneous determination of chlorine, bromine and iodine in National Bureau of standards reference material 1648 Urban Particulate. *J. Anal. Atom. Spectrom.* **3**, 659–665.

Date, A.R. and Stuart, M.E. (1988) Application of inductively coupled plasma mass spectrometry to the simultaneous determination of chlorine, bromine and iodine in the National Bureau of Standards standard reference material 1648 Urban Particulate. *J. Anal. Atom. Spectrom.* **3**, 659–665.

Date, A.R., Cheung, Y.Y. and Stuart, M.E. (1987a) The influence of polyatomic ion interferences in analysis by inductively coupled plasma mass spectrometry (ICP–MS). *Spectrochim. Acta* **42B**, 3–20.

Date, A.R., Davis, A.E. and Cheung, Y.Y. (1987b) The potential of fire assay and inductively coupled plasma mass spectrometry for the determination of platinum group elements in geological materials. *Analyst* **112**, 1217–1222.

Date, A.R., Cheung, Y.Y., Stuart, M.E. and Xiu-Hua, J. (1988) Application of inductively coupled plasma mass spectrometry to the analysis of iron ores. *J. Anal. Atom. Spectrom.* **3**, 653–658.

Dawson, P.H. ed. (1976) *Quadrupole Mass Spectrometry and its Applications*, Elsevier, Amsterdam.

Dawson, P.H. (1986) Quadrupole mass analyzers: performance, design and some recent applications. *Mass Spectrom. Rev.* **5**, 1–37.

Dean, J.R., Munro, S., Ebdon, L., Crews, H.M. and Massey, R.C. (1987a) Studies of metalloprotein species by directly coupled high-performance liquid chromatography inductively coupled plasma mass spectrometry. *J. Anal. Atom. Spectrom.* **2**, 607–610.

Dean, J.R., Ebdon, L. and Massey, R. (1987b) Selection of mode of measurement of lead isotope ratios by inductively coupled plasma mass spectrometry and its application to milk powder analysis. *J. Anal. Atom. Spectrom.* **2**, 369–374.

Dean, J.R., Ebdon, L., Crews, H.M. and Massey, R.C. (1988) Characteristics of flow injection inductively coupled plasma mass spectrometry for trace metal analysis. *J. Anal. Atom. Spectrom.* **3**, 349–354.

Dean, J.R., Crews, H.M. and Ebdon, L. (1989) Applications in food science. In *Applications of Inductively Coupled Plasma Mass Spectrometry*, eds. Date, A.R. and Gray, A.L., Blackie, Glasgow, pp. 141–168.

Dean, J.R., Parry, H.G.M., Massey, R.C. and Ebdon, L. (1990) Continuous and flow injection hydride generation coupled with inductively coupled plasma mass spectrometry. *ICP Inform. Newslett.* **15**, 569–572.

De Boer, J.L.M. and Maessen, F.J.M.J. (1983) A comparative examination of sample treatment procedures for ICAP–AES analysis of biological tissue. *Spectrochim. Acta* **38B**, 739–746.

Delves, H.T. and Campbell, M.J. (1988) Measurements of total lead concentrations and of lead isotope ratios in whole blood by use of inductively coupled plasma mass spectrometry. *J. Anal. Atom. Spectrom.* **3**, 343–348.

Denoyer, E., Ediger, R. and Hager, J. (1989) The determination of precious metals in geological samples by ICP-mass spectrometry. *Atom. Spectrosc.* **10**, 97–103.

Dever, M., Hausler, D.W. and Smith, J.E. (1989) Comparison between radioactive chromium-51 and stable isotope chromium-50 labels for the determination of red blood cell survival. *J. Anal. Atom. Spectrom.* **4**, 361–363.

Dickin, A.P., McNutt, R.H. and McAndrew, J.I. (1988) Osmium isotope analysis by inductively coupled plasma mass spectrometry. *J. Anal. Atom. Spectrom.* **3**, 337–342.

Dittrich, K. and Wennrich, R. (1984) Laser vaporization in atomic spectroscopy. *Prog. Anal. Atom. Spectrosc.* **7**, 139–198.

Doherty, W. (1989) An internal standardisation procedure for the determination of yttrium and the rare earth elements in geological materials by inductively coupled plasma mass spectrometry. *Spectrochim. Acta* **44B**, 263–280.

Doherty, M.P. and Hieftje, G.M. (1984) Jet-impact nebulisation for sample introduction in inductively coupled plasma spectrometry. *Appl. Spectrosc.* **38**, 405–412.

Doherty, W. and Vander Voet, A. (1985) The application of inductively coupled plasma mass spectrometry to the determination of rare earth elements in geological materials. *Can. J. Spectrosc.* **30**, 135–141.

Dolezal, J., Povondra, P. and Sulcek, Z. (1968) *Decomposition Techinques in Inorganic Analysis.* Elsevier, New York.

Douglas, D.J. (1983) ICP–MS: Technologies marry to produce better analyses. *Can. Res.* **16**, 55–60.

Douglas, D.J. (1985) Personal communication.

Douglas, D.J. (1991) Fundamental aspects of ICP–MS. In *ICPs in Analytical Atomic Spectrometry*, 2nd edn., eds. Montaser, A. and Golightly, D.W., VCH Publishers, New York, in press.

Douglas, D.J. and French, J.B. (1981) Elemental analysis with a microwave induced plasma quadrupole mass spectrometer system. *Anal. Chem.* **53**, 37–41.

Douglas, D.J. and French, J.B. (1986) An improved interface for ICP–MS. *Spectrochim. Acta* **41B**, 197–204.

Douglas, D.J. and French, J.B. (1988) Gas dynamics of the ICP–MS interface. *J. Anal. Atom. Spectrom.* **3**, 743–747.

Douglas, D.J. and Kerr, L.A. (1988) Study of solids deposition on inductively coupled plasma mass spectrometry samplers and skimmers. *J. Anal. Atom. Spectrom.* **3**, 749–752.

Douglas, D.J., Quan, E.S.K. and Smith, R.G. (1983a) Elemental analysis with an atmospheric pressure plasma (MIP, ICP)/quadrupole mass spectrometer system. *Spectrochim. Acta* **38B**, 39–48.

Douglas, D.J., Rosenblatt, G. and Quan, E.S.K. (1983b) Inductively coupled plasma-mass spectrometry—a new technique for trace element determination. *Trace Subst. Environ. Health* **17**, 385–390.

Dushman, S.M. and Lafferty, J.M. (1962) *Scientific Foundation of Vacuum Technique*, 2nd edn., Wiley, New York.

Ebdon, L. and Cave, M.R. (1982) A study of pneumatic nebulisation systems for inductively coupled plasma emission spectrometry. *Analyst* **107**, 172–178.

Ebdon, L. and Collier, A.R. (1988a) Particle size effects on kaolin slurry analysis by inductively coupled plasma-atomic emission spectrometry. *Spectrochim. Acta* **43B**, 355–369.

Ebdon, L. and Collier, A.R. (1988b) Direct atomic spectrometric analysis by slurry atomisation Part 5: Analysis of kaolin using inductively coupled plasma-atomic emission spectrometry. *J. Anal. Atom. Spectrom.* **3**, 557–561.

Ebdon, L. and Wilkinson, J.R. (1987a) Direct atomic spectrometric analysis by slurry atomisation Part 1: Optimisation of whole coal analysis by inductively coupled plasma-atomic emission spectrometry. *J. Anal. Atom. Spectrom.* **2**, 39–44.

Ebdon, L. and Wilkinson, J.R. (1987b) Direct atomic spectrometric analysis by slurry atomisation Part 3: Whole coal analysis by inductively coupled plasma-atomic emission spectrometry. *J. Anal. Atom. Spectrom.* **2**, 325–328.

Ebdon, L.C., Hill, S. and Ward, R.W. (1987) Directly coupled chromatography-atomic spectroscopy. Part 2. Directly coupled liquid chromatography-atomic spectroscopy. A review. *Analyst* **112**, 1–16.

Ebdon, L., Foulkes, M.E., Parry, H.G.M. and Tye, C.T. (1988) Direct atomic spectrometric analysis by slurry atomisation Part 7: Analysis of coals using inductively coupled plasma mass spectrometry. *J. Anal. Atom. Spectrom.* **3**, 753–761.

Ebdon, L., Foulkes, M.E. and Hill, S. (1989) Fundamental and comparative studies of aerosol sample introduction for solution and slurries in atomic spectroscopy. *Microchem. J.* **40**, 30–64.

Ekimoff, D., Van Norstrand, A.M. and Mowers, D.A. (1989) Semiquantitative survey capabilities of inductively coupled plasma mass spectrometry. *Appl. Spectrosc.* **43**, 1252–1257.

Emmett, S.E. (1988) Analysis of liquid milk by inductively coupled plasma mass spectrometry. *J. Anal. Atom. Spectrom.* **3**, 1145–1146.

Evans, E.H. and Ebdon, L. (1989) Simple approach to reducing polyatomic ion interferences on arsenic and selenium in inductively coupled plasma mass spectrometry. *J. Anal. Atom. Spectrom.* **4**, 299–300.

Evans, E.H. and Ebdon, L. (1990) Effect of organic solvents and molecular gases on polyatomic ion interferences in inductively coupled plasma mass spectrometry. *J. Anal. Atom. Spectrom.* **5**, 425–430.

Everett, K. and Graf, F.A. (1971) Handling perchloric acids and perchlorates. In *Handbook of Laboratory Safety*, ed. Steere, N.V., Chemical Rubber Company, Cleveland, pp. 265–276.

Falkner, K.K. and Edmond, J.M. (1990) Determination of gold at femtomolar levels in natural waters by flow injection inductively coupled plasma quadrupole mass spectrometry. *Anal. Chem.* **62**, 1477–1481.

Fassel, V.A. (1977) Current and potential applications of inductively coupled plasma (ICP)-atomic emission spectroscopy (AES) in the exploration mining and processing of materials. *Pure Appl. Chem.* **49**, 1533–1545.

Fassel, V.A. (1978) Quantitative elemental analyses by plasma emission spectrometry. *Science* **202**, 183–191.

Fassel, V.A. and Bear, B.R. (1986) Ultrasonic nebulization of liquid samples for analytical inductively coupled plasma-atomic spectroscopy: an update. *Spectrochim. Acta* **41B**, 1089–1113.

Fassett, J.D. and Paulsen, P.J. (1989) Isotope dilution mass spectrometry for accurate elemental analysis. *Anal. Chem.* **61**, 643A–649A.

Feldman, C. (1983) Behaviour of trace refractory minerals in the lithium metaborate-acid dissolution procedure. *Anal. Chem.* **55**, 2451–2453.

Friel, J.K., Skinner, C.S., Jackson, S.E and Longerich, H.P. (1990) Analysis of reference materials prepared by microwave dissolution using inductively coupled plasma mass spectrometry. *Analyst* **115**, 269–273.

Fulford, J.E. and D.J. Douglas (1986) Ion kinetic energies in inductively coupled plasma mass spectrometry (ICP–MS). *Appl. Spectrosc.* **40**, 971–974.

Fuller, C.W., Hutton, R.C. and Preston, B. (1981) Comparison of flame, electrothermal and inductively coupled plasma-atomic atomisation techniques for the direct analysis of slurries. *Analyst* **106**, 913–920.

Garbarino, J.R. and Taylor, H.E. (1987) Stable isotope dilution analysis of hydrologic samples by inductively coupled plasma mass spectrometry. *Anal. Chem.* **59**, 1568–1575.

Gillson, G.R., Douglas, D.J., Fulford, J.E., Halligan, K.W. and Tanner, S.D. (1988) Non-spectroscopic interelement interferences in inductively coupled plasma mass spectrometry. *Anal. Chem.* **60**, 1472–1474.

Gilman, L.B. and Engelhart, W.G. (1989) Recent advances in microwave sample preparation. *Spectroscopy* **4**(8), 14–21.

Gladney, E.S. and Roelandts, I. (1988) 1987 compilation of elemental concentration data for USGS BHVO-1, MAG-1, QLO-1, RGM-1, SCo-1, SDC-1, SGR-1 and STM-1. *Geostandards Newslett.* **12**, 253–362.

Gladney, E.S., O'Malley, B.T., Roelandts, I. and Gills, T.E. (1987) Standard reference materials: compilation of elemental concentration data for NBS clinical, biological, geological and environmental standard reference materials. *NBS Special Publication* **260–111**, US Department of Commerce, National Institute of Standards and Technology.

Gordon, J.S., van der Plas, P.S.C. and de Galan, L. (1988) Use of a water-cooled low-flow torch in inductively coupled plasma mass spectrometry. *Anal. Chem.* **60**, 375–377.

Gorsuch, T.T. (1970) *The Destruction of Organic Matter*, Pergamon, Oxford.

Govindaraju, K. (1989) 1989 compilation of working values and sample description for 272 geostandards. *Geostandards Newslett.* **13**, Special Issue, 113 pp.

Govindaraju, K. and Mevelle, G. (1987) Fully automated dissolution and separation methods for inductively coupled plasma atomic emission spectrometry rock analysis. Application to the determination of rare-earth elements. *J. Anal. Atom. Spectrom.* **2**, 615–621.

Gray, A.L. (1974) A plasma source for mass analysis. *Proc. Soc. Anal. Chem.* **11**, 182–183.

Gray, A.L. (1975) Mass spectrometric analysis of solutions using an atmospheric pressure ion source. *Analyst* **100**, 289–299.

Gray, A.L. (1978) Isotope ratio determination on solutions with a plasma ion source. *Dynamic Mass Spectrometry* **5**, ed., Todd, J.F.J., Heyden, London, pp. 106–113.

Gray, A.L. (1982) The use of an inductively coupled plasma as an ion source for atomic mass spectrometry. Unpublished PhD thesis, University of Surrey.

Gray, A.L. (1985a) The ICP as an ion source—origins, achievements and prospects. *Spectrochim. Acta* **40B**, 1525–1537.

Gray, A.L. (1985b) Solid sample introduction by laser ablation for inductively coupled plasma source mass spectrometry. *Analyst* **110**, 551–556.

Gray, A.L. (1986a) The evolution of the ICP as an ion source for mass spectrometry. *J. Anal. Atom. Spectrom.* **1**, 403–405.

Gray, A.L. (1986b) Mass spectrometry with an inductively coupled plasma as an ion source: the influence on ultra-trace analysis of background and matrix response. *Spectrochim. Acta* **41B**, 151–167.

Gray, A.L. (1986c) Influence of load coil geometry on oxide and double charged ion response in ICP–MS. *J. Anal. Atom. Spectrom.* **1**, 247–249.

Gray, A.L. (1988) Inductively coupled plasma source mass spectrometry. In *Inorganic Mass Spectrometry*, eds., van Grieken, R., Adams, F. and Gijbels, R., Chemical Analysis Series, Wiley, New York, pp. 257–300.

Gray, A.L. (1989a) The origins, realisation and performance of ICP–MS systems. In *Applications of Inductively Coupled Plasma Mass Spectrometry*, eds., Date, A.R. and Gray, A.L., Blackie, Glasgow, pp. 1–42.

Gray, A.L. (1989b) Visual observation of shock waves in an ICP–MS expansion stage. *J. Anal. Atom. Spectrom.* **4**, 371–373.

Gray, A.L. and Date, A.R. (1983) Inductively coupled plasma source mass spectrometry using continuum flow ion extraction. *Analyst* **108**, 1033–1050.

Gray, A.L. and Williams, J.G. (1987a) Oxide and doubly charged ion response of a commercial inductively coupled plasma mass spectrometry instrument. *J. Anal. Atom. Spectrom.* **2**, 81–82.

Gray, A.L. and Williams, J.G. (1987b) System optimisation and the effect on polyatomic, oxide and doubly charged ion response of a commercial inductively coupled plasma mass spectrometry instrument. *J. Anal. Atom. Spectrom.* **2**, 599–606.

Gray, A.L., Houk, R.S. and Williams, J.G. (1987) Langmuir probe potential measurements in the plasma and their correlation with mass spectral characteristics in inductively coupled plasma mass spectrometry. *J. Anal. Atom. Spectrom.* **2**, 13–20.

Greenfield, S. (1987) Common radiofrequency generators, torches, and sample introduction systems. In *Inductively Coupled Plasmas in Analytical Atomic Spectrometry*, eds., Montaser, A. and Golightly, D.W., VCH Publishers, New York.

Greenfield, S., Jones, I.Ll. and Berry, C.T. (1964) High pressure plasmas as spectroscopic emission sources. *Analyst* **89**, 713–720.

Gregoire, D.C. (1987a) Influence of instrument parameters on nonspectroscopic interferences in inductively coupled plasma mass spectrometry. *Appl. Spectrosc.* **41**, 897–903.

Gregoire, D.C. (1987b) Determination of boron isotopes in geological materials by inductively coupled plasma mass spectrometry. *Anal. Chem.* **59**, 2479–2484.

Gregoire, D.C. (1987c) The effect of easily ionizable concomitant elements on non-spectroscopic interferences in inductively coupled plasma-mass spectrometry. *Spectrochim. Acta* **42B**, 895–907.

Gregoire, D.C. (1988) Determination of platinum, palladium, ruthenium and iridium in geological materials by inductively coupled plasma mass spectrometry with sample intoduction by electrothermal vaporisation. *J. Anal. Atom. Spectrom.* **3**, 309–314.

Grillo, A.C. (1990) Microwave digestion using a closed-vessel system. *Spectroscopy* **5(1)**, 14–16.

Gromet, L.P., Dymet, R.F., Haskin, L.A. and Korotev, R.L. (1984) The 'North American shale composite': its compilation, major and trace element characteristics. *Geochim. Cosmochim. Acta* **8**, 2469–2482.

Grote, M. and Kettrup, A. (1987) *Anal. Chim. Acta* **201**, 95.

Gunn, A.M., Millard, D.L. and Kirkbright, G.F. (1978) Optical emission spectrometry with an inductively coupled radio-frequency argon plasma source and sample introduction with a graphite electrothermal vaporisation device. Part 1. Instrumentation assembly and performance characteristics. *Analyst* **103**, 1066–1073.

Gustavsson, A.G.T. (1979) Some aspects on nebulizer characteristics I. *ICP Inform. Newslett.* **5**, 312–328.

Gustavsson, A.G.T. (1987) Liquid sample introduction into plasmas. In *Inductively Coupled Plasmas in Analytical Atomic Spectrometry*, eds., Montaser, A. and Golightly, D.W., VCH Publishers, New York, pp. 399–430.

Hager, J.W. (1989) Relative elemental responses for laser ablation-inductively coupled plasma mass spectrometry. *Anal. Chem.* **61**, 1243–1248.

Hahn, M.H., Wolnik, K.A., Fricke, F.L. and Caruso, J.A. (1982) Hydride generation/condensation system with an inductively coupled argon plasma polychromator for determination of As, Bi, Ge, Sb, Se, and Sn in foods. *Anal. Chem.* **5**, 1048–1052.

Haines, J. and Robert, R.V.D. (1982) The determination by atomic absorption spectrophotometry using electrothermal atomisation of platinum, palladium, rhodium, ruthenium and iridium. Report MINTEK (South Africa), M34.

Halicz, L. and Brenner, I.B. (1987) Nebulisation of slurries and suspensions of geological materials for inductively coupled plasma-atomic emission spectrometry. *Spectrochim. Acta* **42B**, 207–217.

Hall, G.E.M. and Bonham-Carter, G.F. (1988) Review of methods to determine gold, platinum and palladium in production-orientated geochemical laboratories, with application of a statistical procedure to test for bias. *J. Geochem. Explor.* **30**, 255–286.

Hall, G.E.M. and Pelchat, J.C. (1990) Analysis of standard reference materials for Zr, Nb, Hf and Ta by ICP–MS after lithium metaborate fusion and cupferron separation. *Geostandards Newslett.* **1**, 197–206.

Hall, G.E.M., Park, C.J. and Pelchat, J.C. (1987) Determination of tungsten and molybdenum at low levels in gelogical materials by inductively coupled plasma mass spectrometry. *J. Anal. Atom. Spectrom.* **2**, 189–196.

Hall, G.E.M., Jefferson, C.W. and Michel, F.A. (1988a) Determination of W and Mo in natural spring waters by ICP–AES (inductively coupled plasma atomic emission spectrometry) and ICP–MS (inductively coupled plasma mass spectrometry): application to South Nahinni River area, N.W.T., Canada. *J. Geochem. Explor.* **30**, 63–84.

Hall, G.E.M., Pelchat, J.C., Boomer, D.W. and Powell, M. (1988b) Relative merits of two methods of sample introduction in inductively coupled plasma mass spectrometry: electrothermal vaporisation and direct sample insertion. *J. Anal. Atom. Spectrom.* **3**, 791–797.

Hall, G.E.M., Pelchat, J.C. and Loop, J. (1990a) Determination of zirconium, niobium, hafnium and tantalum at low levels in geological materials by inductively coupled plasma mass spectrometry. *J. Anal. Atom. Spectrom.* **5**, 339–349.

Hall, R.J.B., James, M.R., Wayman, T. and Hulmston, P. (1990b) The feasability of the use of electrothermal vaporization inductively coupled plasma mass spectrometry for the determination of femtogramme levels of plutonium and uranium. In *Plasma Source Mass Spectrometry*. eds. Jarvis, K.E., Gray, A.L., Jarvis, I. and Williams, J.G., Royal Society of Chemistry special publication no. 85, Cambridge, pp. 145–154.

Haraldsson, C., Westerlund, S. and Öhman, P. (1989) Determination of mercury in natural samples at the sub-nanogram level using inductively coupled plasma/mass spectrometry after reduction to elemental mercury. *Anal. Chim. Acta* **221**, 77–84.

Hart, S. (1990) *Econ. Geol.* **84**, 1651–1656.

Hausler, D. (1987) Trace element analysis of organic solutions using inductively coupled plasma mass spectrometry. *Spectrochim. Acta* **42B** 63–73.

Hayhurst, A.N. and Sugden, T.M. (1966) Mass spectrometry of flames. *Proc. R. Soc. London Ser. A* **293**, 36–50.

Hayhurst, A.N. and Telford N.R. (1971) The occurrence of chemical reactions in supersonic expansions of a gas into a vacuum and its relation to mass spectrometric sampling. *Proc. R. Soc. London Ser. A* **322**, 483–507.

Heithmar, E.M., Hinners, T.A., Rowan, J.T. and Riviello, J.M. (1990) Minimization of interferences in inductively coupled plasma mass spectrometry using on-line preconcentration. *Anal. Chem.* **62**, 857–864.

Heitkemper, D., Creed, J., Caruso, J. and Fricke, F.L. (1989) Speciation of arsenic in urine using high-performance liquid chromatography with inductively coupled plasma mass spectrometric detection. *J. Anal. Atom. Spectrom.* **4**, 279–283.

Hendel, Y., Ehrenthal, A. and Bernas, B. (1973) The rapid determination of cations in phosphate rocks by an acid pressure decomposition technique and atomic absorption spectroscopy. *Atom. Absorp. Newslett.* **12**, 130.

Henshaw, J.M., Heithmar, E.M. and Hinners, T.A. (1989) Inductively coupled plasma mass spectrometric determination of trace elements in surface waters subject to acidic deposition. *Anal. Chem.* **61**, 335–442.

Heumann, K.G. (1988) Isotope dilution mass spectrometry. In *Inorganic Mass Spectrometry*, eds. Adams, F., Gijbels, R. and Van Grieken, R., Chemical Analysis Series, 95, Wiley, New York, pp. 301–376.

van Heuzen, A. (1989) ICP–MS in the petroleum industry. In *Applications of Inductively Coupled Plasma Mass Spectrometry*, eds. Date, A.R. and Gray, A.L., Blackie, pp. 169–188.

van Heuzen, A. (1990) Inductively coupled plasma mass spectrometry: a spectrum. PhD thesis, Amsterdam, 178 pp.

Hickman, D.A., Rooke, J.M. and Thompson, M. (1986) Atomic spectrometry updade—minerals and refractories. *J. Anal. Atom. Spectrom.* **1**, 169R–200R.

Hinners, T.A., Heithmar, E.M., Spittler, T.M. and Henshaw, J.M. (1987) Inductively coupled plasma mass spectrometric determination of lead isotopes. *Anal. Chem.* **59**, 2658–2662.

Hirata, T. (1990) Development of a merging introduction technique for inductively coupled plasma mass spectrometry: some geochemical applications. *J. Anal. Atom. Spectrom.* **5**, 589–591.

Hirata, T. and Masuda, A. (1990) Determination of rhenium with enhanced sensitivity using inductively coupled plasma mass spectrometry. *J. Anal. Atom. Spectrom.* **5**, 627–630.

Hirata, T., Shimizu, H., Akagi, T. and Masuda, A. (1988a) Simultaneous determination of isotopic ratio and abundance of osmium by ICP–MS. Fundamental studies for the application of rhenium-osmium systematics. *ICP Inform. Newslett.* **13**, 731–735.

Hirata, T., Shimizu, H., Akagi, T., Sawatari, H. and Masuda, A. (1988b) Precise determination of rare earth elements in geological standards rocks by inductively coupled plasma source mass spectrometry. *Anal. Sci.* **4**, 637–643.

Hirata, T., Akagi, T., Shimizu, H. and Masuda, A. (1989) Determination of osmium and osmium isotope ratios by microelectrothermal vaporisation inductively coupled plasma mass spectrometry. *Anal. Chem.* **61**, 2263–2266.

Hoffman, E.L., Naldrett, A.J., van Loon, J.C., Hancock, R.G.V. and Manson, A. (1978) The determination of all the platinum group elements and gold in rocks and ore by neutron activation analysis after preconcentration by a nickel sulphide fire-assay technique on large samples. *Anal. Chim. Acta* **102**, 157–166.

Horlick, G., Tan, S.H., Vaughan, M.A. and Rose, C.A. (1985) The effect of plasma operating parameters of analyte signals in inductively coupled plasma mass spectrometry. *Spectrochim. Acta* **40B**: 1555–1572.

Horlick, G., Tan, S.H., Vaughan, M.A. and Shao, Y. (1987) Inductively coupled plasma-mass spectrometry. In *Inductively Coupled Plasmas in Analytical Atomic Spectrometry*, eds. Montaser A. and Golightly, D.W., VCH Publishers, New York, pp. 361–398.

Houk, R.S. (1986) Mass spectrometry of ICPs. *Anal. Chem.* **58**, 97A–105A.

Houk, R.S. (1990) Elemental analysis by atomic emission and mass spectrometry with ICPs. In *Handbook on the Physics and Chemistry of Rare Earths*, Vol. 13, eds. Gschneider, K.A. Jr., and Eyring, L., Elsevier, New York, pp. 385–421.

Houk, R.S. and Thompson, J.J. (1983) Trace metal analysis of microliter solution volumes by inductively coupled plasma mass spectrometry. *Biomed. Mass Spectrom.* **10**, 107–112.

Houk, R.S. and Thompson, J.J. (1988) Inductively coupled plasma mass spectrometry. In *Mass Spectrometry Reviews* 7, ed. Gross, H.L., Wiley, New York, pp. 425–461.

Houk, R.S., Fassel, V.A., Flesch, G.D., Svec, H.J., Gray, A.L. and Taylor, C.E. (1980) Inductively coupled argon plasma as an ion source for mass spectrometric determination of trace elements. *Anal. Chem.* **52**, 2283–2289.

Houk, R.S., Fassel, V.A. and Svec, H.J. (1981a) ICP–MS: Sample introduction, ionization, ion extraction and analytical results. In *Dynamic Mass Spectrometry* 6, eds. Price, D. and Todd, J.F.J., Heyden, London, pp. 234–251.

Houk, R.S., Svec, H.J. and Fassel, V.A. (1981b) Mass spectrometric evidence for suprathermal ionization in an ICP. *Appl. Spectrosc.* **35**, 380–384.

Houk, R.S., Schoer, J.K. and Crain, J.S. (1987) Plasma potential measurements for ICP–MS with a centre-tapped load coil. *J. Anal. Atom. Spectrom.* **2**, 283–286.

Huang, L.Q., Jiang, S.-J. and Houk, R.S. (1987) Scintillation-type ion detection for inductively coupled plasma mass spectrometry. *Anal. Chem.* **59**, 2316–2320.

Hulmston, P. (1983) A pneumatic recirculating nebuliser system for small sample volumes. *Analyst* **108**, 166–170.

Hulmston, P. and Hutton, R.C. (1990) Analytical capabilities of ETV–ICP–MS. *Spectrosc. Int.* **3**, 35–38.

Hutton, R.C. (1986) Application of inductively coupled plasma source mass spectrometry (ICP–MS) to the determination of trace metals in organics. *J. Anal. Atom. Spectrom.* **1**, 259–263.

Hutton, R.C. and Eaton, A.N. (1987) Role of aerosol water vapour loading in inductively coupled plasma mass spectrometry. *J. Anal. Atom. Spectrom.* **2**, 595–598.

Hutton, R.C. and Eaton, A.N. (1988) Analysis of solutions containing high levels of dissolved solids by inductively coupled plasma mass spectrometry. *J. Anal. Atom. Spectrom.* **3**, 547–550.

Hutton, R.C., Bridenne, M., Coffre, E., Marot, Y. and Simondet, F. (1990) Investigations into the direct analysis of semiconductor grade gases by inductively coupled plasma mass spectrometry. *J. Anal. Atom. Spectrom.* **5**, 463–466.

Ingamells, C.O. (1964) Rapid chemical analysis of silicate rocks. *Talanta* **11**, 665–666.

Ingamells, C.O. (1970) Lithium metaborate flux in silicate analysis. *Anal. Chim. Acta* **42**, 323–334.

Isoyama H., Uchida, T., Iida, C. and Nakagawa, G. (1990) Recycling nebulisation system with exchangeable spray chamber for inductively coupled plasma atomic emission spectrometry. *J. Anal. Atom. Spectrom.* **5**, 365–369.

Israel, Y., Lasztity, A. and Barnes, R.M. (1989) On-line dilution, steady-state concentrations for inductively coupled plasma atomic emission and mass spectrometry achieved by tandem injection and merging-stream flow injection. *Analyst* **114**, 1259–1265.

Ito, J. (1961) A new method of decomposition for refractory minerals and its application to the determination of ferrous iron and alkalis. *Bull. Chem. Soc. Jpn.* **35**, 225–339.

Jackson, S.E., Fryer, B.J., Gosse, W., Healey, D.C., Longerich, H.P. and Strong, D.F. (1990) Determination of the precious metals in geological materials by inductively coupled plasma-mass spectrometry (ICP–MS) with nickel sulphide fire-assay collection and tellurium co-precipitation. In *Microanalytical Methods in Mineralogy and Geochemistry*, eds. Potts, P.J., Dupuy, C. and Bowles, J.F.W., *Chem. Geol.* **83**, 119–132.

Jakubowski, N., Raeymaekers, B.J., Broekaert, J.A.C. and Stuewer, D. (1989) Study of plasma potential effects in a 40 MHz ICP–MS system. *Spectrochim. Acta* **44B**, 219–228.

Janghorbani, M. and Ting, B.T.G. (1989) Comparison of pneumatic nebulization and hydride generation inductively coupled plasma mass spectrometry for isotopic analysis of selenium. *Anal. Chem.* **61**, 701–708.

Janghorbani, M., Ting, B.T.G. and Fomon, S.J. (1986) Erythrocyte incorporation of ingested stable isotope of iron (^{58}Fe). *Am. J. Hematol.* **21**, 277–288.

Janghorbani, M., Davis, T.A. and Ting, B.T.G. (1988) Measurement of stable isotopes of bromine in biological fluids with inductively coupled plasma mass spectrometry. *Analyst* **113**, 405–411.

Jarvis, K.E. (1988) Inductively coupled plasma mass spectrometry, a new technique for the rapid or ultra-trace level determination of the rare earth elements in geological materials. *Chem. Geol.* **68**, 31–39.

Jarvis, K.E. (1989a) Chapter 3: Elemental analysis of the lanthanides. In *Lanthanide Probes in Life, Chemical and Earth Sciences. Theory and Practice*, eds. Bunzli, J.-C.G. and Choppin, G.R., Elsevier, Amsterdam, pp. 65–92.

Jarvis, K.E. (1989b) Determination of the rare earth elements in geological samples by inductively coupled plasma mass spectrometry. *J. Anal. Atom. Spectrom.* **4**, 563–570.

Jarvis, K.E. (1990) A critical evaluation of two sample preparation techniques for low-level determination of some geologically incompatible elements by inductively coupled plasma mass spectrometry. *Chem. Geol.* **83**, 89–103.

Jarvis, K.E. (1991) Role of slurry nebulisation for the analysis of geological samples by inductively coupled plasma spectrometry. In *Plasma Spectrometry in the Earth Sciences*, eds. Jarvis, I. and Jarvis, K.E., *Chem. Geol.* Special issue (in press).

Jarvis, I. and Jarvis, K.E. (1985) Rare-earth element geochemistry of standard sediments: a study using inductively coupled plasma spectrometry. *Chem. Geol.* **53**, 335–344.

Jarvis, K.E. and Williams, J.G. (1989) The analysis of geological samples by slurry nebulisation inductively coupled plasma-mass spectrometry (ICP–MS). *Chem. Geo.* **77**, 53–63.

Jarvis, K.E., Gray, A.L. and McCurdy, E. (1989) Avoidance of spectral interference on europium in inductively coupled plasma mass spectrometry by sensitive measurement of the doubly charged ion. *J. Anal. Atom. Spectrom.* **4**, 743–747.

Jefferey, P.G. and Hutchison, D. (1981) *Chemical Methods of Rock Analysis*, 3rd edn., Pergamon, Oxford, 379 pp.

Jenner, G.A., Longerich, H.P., Jackson, S.E. and Fryer, B.J. (1990) ICP–MS—a powerful tool for high-precision trace-element analysis in the Earth sciences: evidence from analysis of selected USGS reference samples. In *Microanalytical Methods in Mineralogy and Geochemistry*, eds. Potts, P.J., Dupuy, C. and Bowles, J.F.W., *Chem. Geol.* **83**, 133–148.

Jiang, S.J. and Houk, R.S. (1986) Arc nebulisation for elemental analysis of conducting solids by inductively coupled plasma mass spectrometry. *Anal. Chem.* **58**, 1739–1743.

Jiang, S.J. and Houk, R.S. (1988) Inductively coupled plasma mass spectrometric detection for phosphorus and sulfur compounds separated by liquid chromatography. *Spectrochim. Acta.* **43B**, 405–411.

Jiang, S.J., Palmieri, M.D., Fritz, J.S. and Houk R.S. (1987) Chromatographic retention of molybdenum, titanium and uranium complexes for removal of some interferences in inductively coupled plasma mass spectrometry. *Anal. Chim. Acta* **200**, 559–571.

Jiang, S.J., Houk, R.S. and Stevens, M.A. (1988) Alleviation of overlap interferences for the determination of potassium isotope ratios by inductively coupled plasma mass spectrometry. *Anal. Chem.* **60**, 1217–1221.

Johnson, W.M. and Maxwell, J.A. (1981) *Rock and Mineral Analysis*, 2nd edn., Wiley, New York, 489 pp.

Jolly, S.C. (1963) Metallic impurities in organic matter. In *Official, Standardised and Recommended Methods of Analysis*, Heffer, Cambridge, pp. 3–19.

Kammin, W.R. and Brandt, M.J. (1989a) ICP–OES evaluation of microwave digestion. *Spectroscopy* **4**(3), 49–55.

Kammin, W.R. and Brandt, M.J. (1989b) The simulation of EPA method 3050 using a high temperature and high-pressure microwave bomb. *Spectroscopy* **4**(6), 22–23.

Kantipuly, C.J., Longerich, H.P. and Strong, D.F. (1988) Application of inductively coupled argon plasma mass spectrometry (ICAP–MS) for the determination of uranium and thorium in tourmalines. *Chem. Geol.* **69**, 171–176.

Karanassios, V. and Horlick, G. (1989a) A computer controlled direct sample insertion device for inductively coupled plasma-mass spectrometry. *Spectrochim. Acta* **44B**, 1345–1360.

Karanassios, V. and Horlick, G. (1989b) Background spectral characteristics in direct sample insertion-inductively coupled plasma-mass spectrometry. *Spectrochim. Acta* **44B**, 1361–1385.

Karanassios, V. and Horlick, G. (1989c) Elimination of some spectral interferences and matrix effects in inductively coupled plasma-mass spectrometry using direct sample insertion techniques. *Spectrochim. Acta* **44B**, 1387–1396.

Kawaguchi, H., Tanaka, T., Nakamura, T., Morishita, M. and Mizuike, A. (1987) *Anal. Sci. Jpn.* **3**, 305.

Kawaguchi, H., Tanaka, T. and Mizuike, A. (1988a) Continuum background in ICP–MS. *Spectrochim. Acta* **43B**, 955–962.

Kawaguchi, H., Asada, K. and Mizuike, A. (1988b) Optical characteristics of an afterglow extracted from an ICP. *Mikrochim. Acta* III 143–153.

Kemp, A.J. and Brown, C.J. (1990) Microwave digestion of carbonate rock samples for chemical analysis. *Analyst* **115**, 1197–1199.

Kingston, H.M. and Jassie, L.B. (1986) Microwave energy for acid decomposition at elevated temperatures are pressures using biological and botanical samples. *Anal. Chem.* **58**, 2534–2541.

Kingston, H.M. and Jassie, L.B. eds. (1988a) *Introduction to Microwave Sample Preparation*, American Chemical Society, Washington, DC, 263 pp.

Kingston, H.M. and Jassie, L.B. (1988b) Monitoring and predicting parameters in microwave dissolution. In *Introduction to Microwave Sample Preparation*, eds. Kingston, H.M. and Jassie, L.B., American Chemical Society, Washington, DC, pp. 93–154.

Knapp, G. and Grillo, A. (1986) A high pressure asher for trace analysis. *Am. Lab.* **18**, 76–79.

Knewstubb, P.F. (1963) *Mass Spectrometry of Organic Ions*, Academic Press, New York, pp. 255–307.

Kniseley, R.H., Amenson, H., Butler C.C. and Fassel, V.A. (1974) An improved pneumatic nebuliser for use at low nebulising gas flows. *Appl. Spectrosc.* **28**, 285–286.

Koppenaal, D.W. and Quinton, L.F. (1988) Development and assessment of a helium inductively coupled plasma ionisation source for inductively coupled plasma mass spectrometry. *J. Anal. Atom. Spectrom.* **3**, 667–672.

Kurz, E.A. (1979) Channel electron multipliers. *Am. Lab.* **11**, 67–82.

LaFreniere, B.R., Houk, R.S. and Fassel, V.A. (1987) Direct detection of vacuum ultraviolet radiation through an optical sampling orifice: analytical figures of merit for the nonmetals, metalloids and selected metals by ICP atomic emission spectrometry. *Anal. Chem.* **59**, 2276–2282.

Lam, J.W.H and Horlick, G. (1990a) A comparison of argon and mixed gas plasmas for ICP–MS. *Spectrochim. Acta* **45B**, 1313–1326.

Lam, J.W.H. and Horlick, G. (1990b) Effects of sampler-skimmer separation in ICP–MS. *Spectrochim. Acta* **45B**, 1327–1338.

Lam, J.W. and McLaren, J.W. (1990a) Unpublished results obtained during the certification analysis of the NRCC freshwater reference material SLRS-2.

Lam, J.W. and McLaren, J.W. (1990b) The use of aerosol processing and N_2/Ar plasmas for oxide reduction in ICP–MS. *J. Anal. Atom. Spectrom.* **5**, 419–424.

Lamothe, P.J., Fries, T.L. and Consul, J.J. (1986) Evaluation of a microwave oven system for the dissolution of geologic samples. *Anal. Chem.* **58**, 1881–1886.

Langmyhr, F.J. and Paus, P.E. (1968) The analysis of inorganic siliceous materials by atomic absorption spectrophotometry and the hydrofluoric acid decomposition technique. Part 1: The analysis of silicate rocks. *Anal. Chim. Acta* **3**, 397–408.

Lasztity, A., Wang, X., Viczian, M., Israel, Y. and Barnes, R.M. (1989) Inductively coupled plasma spectrometry in the study of childhood soil ingestion. Part 2. Recovery. *J. Anal. Atom. Spectrom.* **4**, 737–747.

Layman, R.R. and Lichte, F.E. (1982) Glass frit nebuliser for atomic spectroscopy. *Anal. Chem.* **54**, 638–641.

Leybold-Heraeus Vacuum Products, Inc. (1990) *Product and Vacuum Technology Reference Book*, Export, PA.

Lichte, F.E., Wilson, S.M., Brooks, R.R., Reeves, R.D., Holzbecher, J. and Ryan, D.E. (1986) New method for the measurement of osmium isotopes applied to a New Zealand Cretaceous/ Tertiary boundary shale. *Nature* **332**, 816–817.

Lichte, F.E., Meier, A.L. and Crock, J.G. (1987) Determination of the rare earth elements in geological materials by inductively coupled plasma mass spectrometry. *Anal. Chem.* **59**, 1150–1157.

Lim, H.B. and Houk, R.S. (1990) Langmuir probe measurement of electron temperature in a supersonic jet extracted from an ICP. *Spectrochim. Acta* **45B** 453–461.

Lim, H.B., Houk, R.S., Edelson, M.C. and Carney, K.P. (1989a) Some fundamental characteristics of a reduced-pressure plasma extracted from an ICP. *J. Anal. Atom. Spectrom.* **4**, 365–370.

Lim, H.B., Houk, R.S. and Crain, J.S. (1989b) Langmuir probe measurements of potential inside a supersonic jet extracted from an ICP. *Spectrochim. Acta* **44B**, 989–998.

Lindblad, N.R. and Schneider, J.M. (1965) Production of uniform-sized liquid droplets. *J. Sci. Instrum.* **42**, 635–638.

Lindner, M., Leich, D.A., Borg, R.J., Russ, G.P., Bazan, J.M., Simmons, D.S. and Date, A.R. (1986) Direct laboratory determination of the [187]Re half-life. *Nature* **320**, 246–248.

Lindner, M., Leich, D.A., Russ, G.P., Bazan, J.M. and Borg, R.J. (1989) Direct determination of the half life of [187]Re. *Geochim. Cosmochim. Acta* **53**, 1597–1606.

Longerich, H.P., Fryer, B.J. and Strong, D.F. (1987a) Trace analysis of natural alloys by inductively coupled plasma mass spectrometry (ICP–MS): application to archeological native silver artifacts. *Spectrochim. Acta* **42B**, 101–109.

Longerich, H.P., Fryer, B.J. and Strong, D.F. (1987b) Determination of lead isotope ratios by inductively coupled plasma mass spectrometry (ICP–MS) *Spectrochim. Acta* **42B**, 39–48.

Longerich, H.P., Fryer, B.J., Strong, D.F. and Kantipuly, C.J. (1987c) Effects of operating conditions on the determination of the rare earth elements by inductively coupled plasma mass spectrometry (ICP–MS). *Spectrochim. Acta* **42B**, 75–92.

Luck, J.H. and Luders, V. (1989) Determination of trace metals in hydrothermal calcites by ICP-methods. *Mikrochim. Acta* **III**, 329–336.

L'vov, B.V. (1984) Twenty-five years of furnace atomic absorption spectroscopy. *Spectrochim. Acta* **39B**, 149–157.

Lyon, T.D.B., Fell, G.S., Hutton, R.C. and Eaton A.N. (1988a) Evaluation of inductively coupled plasma mass spectrometry (ICP–MS) for simultaneous multi-element trace analysis in clinical chemistry. *J. Anal. Atom. Spectrom.* **3**, 265–271.

Lyon, T.D.B., Fell, G.S., Hutton, R.C. and Eaton, A.N. (1988b) Elimination of chloride interference on the determination of selenium in serum by inductively coupled plasma mass spectrometry. *J. Anal. Atom. Spectrom.* **3**, 601–603.

Masuda, A., Hirata, T. and Shimizu, H. (1986) Determination of osmium isotope ratios in iron meteorites and iridosmines by ICP–MS. *Geochem. J.* **20**, 233–239.

Matthes, S.A. (1988) Guide-lines for developing microwave dissolution methods for geological and metallurgical samples. In *Introduction to Microwave Sample Preparation*, eds. Kingston, H.M. and Jassie, L.B., American Chemical Society, Washington, DC, pp. 33–51.

Matusiewicz, H. and Sturgeon, R.E. (1989) Present status of microwave sample dissolution and decomposition for elemental analysis. *Prog. Anal. Spectrosc.* **12**, 21–39.

Matusiewicz, H., Sturgeon, R.E. and Berman, S.S. (1989) Trace element analysis of biological material following pressure digestion with nitric acid-hydrogen peroxide and microwave heating. *J. Anal. Atom. Spectrom.* **4**, 323–327.

Matz, S.G., Elder, R.C. and Tepperman, K. (1989) Liquid chromatography with an inductively coupled plasma mass spectrometric detector for simultaneous determination of gold drug metabolites and related metals in human blood. *J. Anal. Atom. Spectrom.* **4**, 767–771.

McCurdy, E.J. (1990) The preparation of plant samples and their analysis by ICP–MS. In *Plasma Source Mass Spectrometry*, eds. Jarvis, K.E., Gray, A.L., Jarvis, I. and Williams, J.G., Special Publication of the Royal Society of Chemistry, Special publication no. 85, pp. 79–93.

McLaren, J.W. (1987) Applications: environmental. In *Inductively Coupled Plasma Emission Spectroscopy. Part II: Applications and Fundamentals*, ed. Boumans, P.W.J.M., Wiley, New York, pp. 48–64.

McLaren, J.W., Mykytiuk, A.P., Willie, S.N. and Berman, S.S. (1985) Determination of trace metals in seawater by inductively coupled plasma mass spectrometry with preconcentration on silica-immobilised 8-hydroxyquinoline. *Anal. Chem.* **57**, 2907–2911.

McLaren, J.W., Beauchemin, D. and Berman, S.S. (1987a) Determination of trace metals in marine sediments by inductively coupled plasma mass spectrometry. *J. Anal. Atom. Spectrom.* **2**, 227–281.

McLaren, J.W., Beauchemin, D. and Berman, S.S. (1987b) Application of isotope dilution inductively coupled plasma mass spectrometry to the analysis of marine sediments. *Anal. Chem.* **59**, 610–613.

McLaren, J.W., Beauchemin, D. and Berman, S.S. (1988) Analysis of the marine sediment reference material PACS-1 by inductively coupled plasma mass spectrometry. *Spectrochim. Acta* **43B**, 413–420.

McLaren, J.W., Siu, K.W.M., Lam, J.W., Willie, S.N., Maxwell, P.S., Palepu, A., Koether, M. and Berman, S.S. (1990) Applications of ICP–MS in marine analytical chemistry. *Fresenius Z. Anal. Chem.* **337**, 721–728.

McLeod, C.W., Date, A.R. and Cheung, Y.Y. (1986) Metal oxide ions in inductively coupled plasma-mass spectrometric analysis of nickel base alloys. *Spectrochim. Acta* **41B**, 169–174.

Meddings, B. and Ng, R. (1989) ICP–MS in the metallurgical laboratory. In *Applications of Inductively Coupled Plasma Mass Spectrometry*, eds. Date, A.R. and Gray, A.L., Blackie, Glasgow, pp. 220–241.

Meinhard, J.E. (1976) The concentric glass nebuliser. *ICP Inform. Newslett.* **2**, 163–165.

Miller, P.E. and Denton, M.B. (1986) The quadrupole mass filter: basic operating concepts. *J. Chem. Ed.* **63**, 617–622.

Millward, C.G. and Kluckner, P.D. (1989) Microwave digestion technique for the extraction of minerals from environmental marine sediments for analysis by inductively coupled plasma atomic emission spectrometry and atomic absorption spectrometry. *J. Anal. Atom. Spectrom.* **4**, 709–713.

Milne, T.A. and Greene, F.T. (1967) Mass spectrometric observations of argon clusters in nozzle beams, 1. General behaviour and equilibrium dimer concentrations. *J. Chem. Phys.* **7**, 4095–4101.

Mochizuki, T., Sakashita, A., Iwata, H., Kagaya, T., Shimamura, T. and Blair, P. (1988) Laser ablation for direct elemental analysis of solid samples by inductively coupled plasma mass spectrometry. *Anal. Sci.* **4**, 403–409.

Mochizuki, T., Sakashita, A., Iwata, H., Ishibashi, Y. and Gunji, N. (1989) Slurry nebulisation technique for direct determination of rare earth elements in silicate rocks by inductively coupled plasma mass spectrometry. *Anal. Sci.* **5**, 311–317.

Mochizuki, T., Sakashita, A. and Iwata, H. (1990) Laser ablation for direct elemental analysis of solid samples by ICP-atomic emission spectrometry and ICP-mass spectrometry. *NKK Tech. Rev.* **58**, 19–27.

Moenke-Blankenburg, L. (1989) *Laser Microanalysis*, Wiley, New York, 288 pp.

Moenke-Blankenburg, L., Gackle, M., Gunther, D. and Kammel, J. (1990) Processes of laser ablation and vapour transport to the ICP. In *Plasma Source Mass Spectrometry*, eds. Jarvis, K.E., Gray, A.L., Jarvis, I. and Williams, J.G., Royal Society of Chemistry, Special publication 85.

Mohamad, A., Creed, J.T., Davidson, T.M. and Caruso, J.A. (1989) Detection of halogenated compounds by capillary gas chromatography with helium plasma MS detection. *Appl. Spectrosc.* **43**, 1127–1131.

Moloughney, P.E. (1986) *Assay Methods used in CANMET for the Determination of Precious Metals*, Special Publication of the Mineral Sciences Laboratories Canada Centre for Mineral and Energy Technology (CANMET), **SP86-1E**, 33 pp.

Montaser, A. and Golightly, D.W. eds. (1987) *Inductively Coupled Plasmas in Analytical Atomic Spectrometry*, VCH Publishers, New York.

Montaser, A., Chan S. and D.W. Koppenaal, (1987) Inductively coupled helium plasma as an ion source for mass spectrometry. *Anal. Chem.* **59**, 1240–1242.

Moody, J.R. (1982) NBS clean laboratories for trace element analysis. *Anal. Chem.* **54**, 1358A–1376A.

Moore, G.L. (1989) Introduction to inductively coupled plasma emission spectrometry. In *Analytical Spectroscopy*, Library Vol. 3, Elsevier, Amsterdam.

Moore, J.A.F. McGuire, M.J. and Hart, P.A. (1990) The application of inductively coupled plasma mass spectrometry to the analysis of iron materials. In *Plasma Source Mass Spectrometry*, eds. Jarvis, K.E., Gray, A.L., Jarvis, I. and Williams, J.G., Royal Society of Chemistry, Special publication 85, pp. 163–169.

Morita, M., Ito, H., Uehiro, T. and Otsuka, K. (1989) High resolution MS with an ICP ion source. *Anal. Sci. (Jpn.)* **5**, 609–610.

Mukai, H., Ambe, Y. and Morita, M. (1990) Flow injection inductively coupled plasma mass spectrometry for the determination of platinum in airborne particulate matter. *J. Anal. Atom. Spectrom.* **5**, 75–80.

Mulligan, K.J., Davidson, T.M. and Caruso, J.A. (1990) Feasibility of the direct analysis of urine by inductively coupled argon plasma mass spectrometry for biological monitoring of exposure to metals. *J. Anal. Atom. Spectrom.* **5**, 301–306.

Munro, S., Ebdon, L. and Mcweeny, D.J. (1986) Application of inductively coupled plasma mass spectrometry (ICP–MS) for trace metal determination in foods. *J. Anal. Atom. Spectrom.* **1**, 211–219.

Murillo, M. and Mermet, J.M. (1989) Improvement in energy transfer with added-hydrogen in inductively coupled plasma atomic emission spectrometry. *Spectrochim. Acta* **44B**, 359–366.

Nakahara, T. (1983) Applications of hydride generation techniques in atomic absorption, atomic fluorescence, and plasma atomic emission spectroscopy, *Prog. Anal. Atom. Spectrosc.* **6**, 163–223.

Nakamura, N. (1974) Determination of REE, Ba, Fe, Mg, Na and K in carbonaceous and ordinary chondrites. *Geochim. Cosmochim. Acta* **38**, 575–775.

Nakashima, S., Sturgeon, R.E, Willie, S.N. and Berman, S.S. (1988a) Determination of trace elements in sea water by graphite furnace atomic absorption spectrometry after preconcentration by tetrahydroborate reductive precipitation. *Anal. Chim. Acta* **207**, 291–299.

Nakashima, S., Sturgeon, R.E., Willie, S.N. and Berman, S.S. (1988b) Determination of trace metals in seawater by graphite furnace atomic absorption spectrometry with preconcentration on silica-immobilized 8-hydroxyquinoline in a flow-system. *Fresenius Z. Anal. Chem.* **330**, 592–595.

Nakashima, S., Sturgeon, R.E., Willie, S.N. and Berman, S.S. (1988c) Acid digestion of marine samples for trace element analysis using microwave heating. *Analyst* **133**, 159–163.

Neas, E.D. and Collins, M.J. (1988) Microwave heating. Theoretical concepts and equipment design. In *Introduction to Microwave Sample Preparation. Theory and Practise*, eds. Kingston, H.H. and Jassie, L.B. American Chemical Society, Washington, DC, pp. 7–32.

Newman, R.A., Osborn, S. and Siddik, Z.H. (1989) Determination of tellurium in biological fluids by means of electrothermal vaporization-inductively coupled plasma-mass spectrometry (ETV–ICP–MS). *Clin. Chim. Acta* **179**, 191–196.

Ng, C.Y. (1983) Molecular beam photoionization studies of molecules and clusters. *Adv. Chem. Phys.* **52**, 263–362.

Ng, C.Y. (in press) State-selected and state-to-state ion-molecular reaction dynamics by photoionization and differential reactivity methods. *Adv. Chem. Phys.*, in press.

Ng, K.C. and Caruso, J.A. (1985) Electrothermal vaporization for sample introduction in atomic emission spectrometry. *Appl. Spectrosc.* **39**, 719–726.

Ng, K.C., Zerezghi, M. and Caruso, J.A. (1984) Direct powder injection of NBS coal fly ash in inductively coupled plasma atomic emission spectrometry with rapid scanning spectrometric detection. *Anal. Chem.* **56**, 417–421.

Nölter, T., Maisenbacher, P. and Puchelt, H. (1990) Microwave and digestion of geological and biological standard reference materials for trace element analysis by inductively coupled plasma-mass spectrometry. *Spectroscopy.* **5**(4), 49–53.

Ohls, K. and Sommer, D. (1987) Applications: metals and industrial materials. In *Inductively Coupled Plasma Emission Spectroscopy. Part II: Applications and Fundamentals*, ed. Boumans, P.W.J.M., Wiley, New York, pp. 1–26.

Olivares, J.A. and Houk, R.S. (1985a) Kinetic energy distributions of positive ions in an inductively coupled plasma mass spectrometer. *Appl. Spectrosc.* **39**, 1070–1077.

Olivares, J.A. and Houk, R.S. (1985b) Ion sampling for inductively coupled plasma mass spectrometry. *Anal. Chem.* **57**, 2674–2679.

Olivares, J.A. and Houk, R.S. (1986) Suppression of analyte signal by various concomitant salts in inductively coupled plasma mass spectrometry. *Anal. Chem.* **58**, 20–25.

Palmieri, M.D., Fritz, J.S., Thompson, J.J. and Houk, R.S. (1986) Separation of trace rare earths and other metals from uranium by liquid–liquid extraction with quantitation by inductively coupled plasma mass spectrometry. *Anal. Chim. Acta* **184**, 187–196.

Park, C.J. and Hall, G.E.M. (1987) Analysis of geological materials by inductively coupled plasma mass spectrometry with sample introduction by electrothermal vaporisation. Part 1. Determination of molybdenum and tungsten. *J. Anal. Atom. Spectrom.* **2**, 473–480.

Park, C.J. and Hall, G.E.M. (1988) Analysis of geological materials by inductively coupled plasma mass spectrometry with sample introduction by electrothermal vaporisation. Part 2. Determination of thallium. *J. Anal. Atom. Spectrom.* **3**, 355–361.

Park, C.J., Van Loon, J.C., Arrowsmith, P. and French, J.B. (1987a) Design and optimisation of an electrothermal vaporizer for use in plasma source mass spectrometry. *Can. J. Spectrosc.* **32**, 29–36.

Park, C.J., Van Loon, J.C., Arrowsmith, P. and French, J.B. (1987b) Sample analysis using plasma source mass spectrometry with electrothermal sample introduction. *Anal. Chem.* **59**, 2191–2196.

Patterson, K.Y., Veillon, C. and Kingston, H.M. (1988) Microwave digestion of biological samples. Selenium analysis by electrothermal atomic absorption spectrometry. In *Introduction to Microwave Sample Preparation*, eds. Kingston, H.M. and Jassie, L.B., American Chemical Society, Washington, DC, pp. 155–166.

Paulsen, P.J., Beary, E.S., Bushee, D.S. and Moody, J.R. (1988) Inductively coupled plasma mass spectrometric analysis of ultrapure acids. *Anal. Chem.* **60**, 971–975.

Pearce, J.A., Harris, N.B.W. and Tindle, A.G. (1984) Trace element discrimination diagrams of the tectonic interpretation of igneous rocks. *J. Petrol.* **25**, 956–983.

Pfannerstill, P.E., Creed, J.T., Davidson, T.M., Caruso, J.A. and Willeke, K. (1990) Introduction of powdered solid samples as aerosols into the inductively coupled plasma using a powder disperser. *J. Anal. Atom. Spectrom.* **5**, 285–291.

Pickford, C.J. (1981) Determination of As by emission spectrometry using an inductively coupled plasma source and the syringe hydride technique. *Analyst* **106**, 464–466.

Piper, D.Z. (1974) Rare-earth elements in the sedimentary cycle: a summary. *Chem. Geol.* **14**, 285–304.

Plantz, M.R., Fritz, J.S., Smith, F.G. and Houk, R.S. (1989) Separation of trace metal complexes for analysis of samples of high salt content by inductively coupled plasma mass spectrometry. *Anal. Chem.* **61**, 149–153.

Potts, P.J. (1987) *A Handbook of Silicate Rock Analysis*, Blackie, Glasgow, 622 pp.

Powell, M.J., Boomer, D.W. and McVicars, R.J. (1986) Introduction of gaseous hydrides into an inductively coupled plasma mass spectrometer. *Anal. Chem.* **58**, 2864–2867.

Price, W.J. and Whiteside, P.J. (1977) General method for analysis of siliceous materials by atomic absorption spectrophotometry and its application to macro- and micro-samples. *Analyst* **102**, 664–671.

Rantala, R.T.T. and Loring, D.H. (1989) Teflon bomb decomposition of silicate materials in a microwave oven. *Anal. Chim. Acta* **220**, 263–267.

Richardson, J.M., Dickin, A.P., McNutt, R.H., McAndrew, J.I. and Beneteau, S.B. (1989) Analysis of a rhenium-osmium solid solution spike by inductively coupled plasma mass spectrometry. *J. Anal. Atom. Spectrom.* **4**, 465–471.

Richardson, J.M., Dickin, A.P. and McNutt, R.H. (1990) Re–Os isotope ratio determinations by ICP–MS: a review of analytical techniques and geological applications. In *Plasma Source Mass Spectrometry*, eds. Jarvis, K.E., Gray, A.L., Jarvis, I. and Williams, J.G., Royal Society of Chemistry, Special Publication 85, pp. 120–144.

Ridout, P.S., Jones, H.R. and Williams, J.G. (1988) Determination of trace elements in a marine reference material of lobster hepatopancreas (TORT-1) using inductively coupled plasma mass spectrometry. *Analyst* **113**, 1383–1386.

Ripson, P.A.M. and de Galan, L. (1981) A sample introduction system for an inductively coupled plasma operating on an argon carrier gas flow of 0.1 l/min. *Spectrochim. Acta* **36B**, 71–76.

Robbins, W.B. and Caruso, J.A. (1979) Development of hydride generation methods for atomic spectroscopic analysis. *Anal. Chem.* **51**, 889A–899A.

Robert, R.V.D. (1987) The use of lithium tetraborate in the fire-assay procedure with nickel sulphide as the collector, Randburg Council for Mineral Technology Report **M324**, 9 pp.

Robert, R.V.D., Van Wyk, E. and Palmer, R. (1971) Concentration of noble metals by a fire assay technique using nickel sulphide as collector, Report of the National Institute of Metallurgy (South Africa) 1705.

Robert, R.V.D., van Wyk, E., Palmer, R. and Steele, T.W. (1971) Concentration of the noble metals by a fire assay technique using nickel sulphide as the collector, Johannesburg, National Institute of Metallurgy Report **1371**, 20 pp.

Roboz, J. (1968) *Introduction to Mass Spectrometry: Instrumentation and Techniques*, Wiley-Interscience, New York, 539 pp.

Ross, B.S., Chambers, D.M., Vickers, G.H., Yang, P. and Hieftje, G.M. (1990) Characterisation of a 9-mm torch for inductively coupled plasma mass spectrometry. *J. Anal. Atom. Spectrom.* **5**, 351–358.

Routh, M.W. and Tikkanen, M.W. (1987) Introduction of solids into plasmas. In *Inductively Coupled Plasmas in Analytical Atomic Spectrometry*, eds. Montaser, A. and Golightly, D.W., VCH Publishers, New York, pp. 431–486.

Russ, G.P. (1989) Isotope ratio measurements using ICP–MS. In *Applications of Inductively Coupled Plasma Mass Spectrometry*, eds. Date, A.R. and Gray, A.L., Blackie, Glasgow, pp. 90–114.

Russ, G.P. and Bazan, J.M. (1987) Isotopic ratio measurements with an inductively coupled plasma source mass spectrometer. *Spectrochim. Acta* **42B**, 49–62.

Russ, G.P., Bazan, J.M. and Date, A.R. (1987) Osmium isotopic ratio measurements by inductively coupled plasma mass spectrometry. *Anal. Chem.* **59**, 984–989.

Ruzicka, J. and Hansen, E.H. (1975) Flow injection analysis. Part 1. A new concept of fast continuous flow analysis. *Anal. Chim. Acta* **78**, 145–157.

Sansoni, B., Brunner, W., Wolff, G., Ruppert, H. and Dittrich, R. (1988) Comparative instrumental multielement analysis I: Comparison of ICP source mass spectrometry with ICP atomic emission spectrometry, ICP atomic fluorescence spectrometry and atomic absorption spectrometry for the analysis of natural waters from a granite region. *Fresenius Z. Anal. Chem.* **331**, 154–169.

Satzger, R.D. (1988) Evaluation of inductively coupled plasma mass spectrometry for the determination of trace elements in foods. *Anal. Chem.* **60** 2500–2504.

Schuette, S., Vereault, D., Ting, B.T.G. and Janghorbani, M. (1988) Accurate measurement of stable isotopes of magnesium in biological materials with inductively coupled plasma mass spectrometry. *Analyst* **113**, 1837–1842.

Scott, R.H., Fassel, V.A., Kniseley, R.N. and Nixon, D.E. (1974) Inductively coupled plasma-optical emission analytical spectrometry: a compact facility for trace analysis of solutions. *Anal. Chem.* **6**, 76–80.

Sen Gupta, J.G. (1989) Determination of trace and ultra-trace amounts of noble metals in geological and related materials by graphite-furnace atomic absorption spectrometry after separation by ion exchange or co-precipitation with tellurium. *Talanta* **36**, 651–656.

Sen Gupta, J.G. and Gregoire, D.C. (1989) Determination of ruthenium, palladium and iridium in 27 international reference silicate and iron formation rocks, ores and related materials by isotope dilution inductively coupled plasma mass spectrometry. *Geostandards Newslett.* **13**, 197–204.

Serfass, R.E., Thompson, J.J. and Houk, R.S. (1986) Isotope ratio determinations by inductively coupled plasma mass spectrometry for zinc bioavailability studies. *Anal. Chem. Acta* **188**, 73–84.

Serfass, R.E., Ziegler, E.E., Edwards, B.B. and Houk, R.S. (1989) Intrinsic and extrinsic stable isotopic zinc absorption by infants from formulas 1–3. *J. Nutr.* **119**, 1661–1669.

Shabani, M.B., Akagi, T., Shimizu, H. and Masuda A. (1990) Rapid and accurate determination of sub-parts per trillion lanthanides and yttrium in seawater by development of solvent extraction and back extraction using inductively coupled plasma mass spectrometry. *Anal. Chem.* **62**, 2709–271.

Shannon, R.D. (1976) Revised effective ionic radii and systematic studies of interatomic distances in halides and chalcogenides. *Acta Crystallogr.* **A32**, 751–767.

Sharp, B.L. (1988a) Pneumatic nebulisers and spray chambers for inductively coupled plasma spectrometry, A review, part 1. Nebulisers. *J. Anal. Atom. Spectrom.* **3**, 613–652.

Sharp, B.L. (1988b) Pneumatic nebulisers and spray chambers for inductively coupled plasma spectrometry, A review, part 2. Spray chambers. *J. Anal. Atom. Spectrom.* **3**, 939–963.

Shelkoph, G.M. and Milne, D.B (1988) Wet microwave digestion of diet and fecal samples for inductively coupled plasma analysis. *Anal. Chem.* **60**, 2060–2062.

Shen, W.L., Caruso, J.A., Fricke, F.L. and Satzger, R.D. (1990) Electrothermal vaporisation interface for sample introduction in inductively coupled plasma mass spectrometry. *J. Anal. Atom. Spectrom.* **5**, 451–455.

Sheppard, B.S., Shen, W.L., Caruso, J.A., Heitkemper, D.T. and Fricke, F.L. (1990) Elimination of argon chloride interference on arsenic speciation in inductively coupled plasma mass spectrometry using ion chromatography. *J. Anal. Atom. Spectrom.* **5**, 431–435.

Sheppard, B.S., Shen, W.-L. and Caruso, J.A. (1991) Investigation of matrix induced interferences in mixed gas He-Ar ICP–MS. *J. Am. Soc. Mass Spectrom.* **2**, in press.

Shibata, Y. and Morita, M. (1989) Speciation of arsenic by reversed-phase high performance liquid chromatography—inductively coupled plasma mass spectrometry. *Anal. Sci.* **5**, 107–109.

Sholkovitz, E.R. (1990) Rare earth elements in marine sediments and geochemical standards. *Chem. Geol.* **88**, 333–347.

Sing, R.L.A. and Salin, E.D. (1989) Introduction of liquid samples into the inductively coupled plasma by direct insertion on a wire loop. *Anal. Chem.* **61**, 163–169.

Smith, F.G. and Houk, R.S. (1991) Alleviation of polyatomic ion interferences for determination of chlorine isotope ratios by inductively coupled plasma mass spectrometry. *J. Am. Soc. Mass Spectrom.* **1**, 284–287.

Smith, R.G., Brooker, E.J., Douglas, D.J., Quan, E.S.K. and Rosenblatt, G. (1984) The typing of Aû and base metal occurrences by plasma mass spectrometry: initial results. *J. Geochem. Explor.* **21**, 385–393.

Spangenburg, K.R. (1984) *Vaccum Tubes*. McGraw-Hill, New York, 860 pp.

Spivack, A.J. and Edmund, J.M. (1986) Determination of boron isotope ratios by thermal ionisation mass spectrometry of the dicesium metaborate cation. *Anal. Chem.* **58**, 31–35.

Stearns, C.A., Kohl, F.J., Fryburg, G.C. and Miller, R.A. (1979) High pressure molecular beam mass spectrometric sampling of high temperature molecules. In *10th Materials Research Symp.*, ed., Hastie, J.W., National Bureau of Standards Special Publication 561, US Govt. Printing Office, Washington, DC, pp. 303–305.

Steele, T.W., Levin, J. and Copelowitz, T. (1975) Preparation and certification of a reference sample of a precious metal ore, Report of the National Institute of Metallurgy, South Africa **1696**, 50 pp.

Steger, H.F. (1983) CANMET Report No 83-3E, Canadian Government Publishing Centre, Quebec.

Stotesbury, S.J., Pickering, J.M. and Grifferty, M.A. (1989) Analysis of lithium and boron by inductively coupled plasma mass spectrometry. *J. Anal. Atom. Spectrom.* **4**, 457–460.

Strelow, F.W.E. and Jackson, P.F.S. (1974) Determination of trace and ultra-trace quantities of rare-earth elements by ion exchange chromatography-mass spectrography. *Anal. Chem.* **46**, 1481–1486.

Streusand, B.J., Allen, R.H., Coons, D.E. and Hutton, R.C. (1990) Reactive gas sample introduction system for an inductively coupled plasma mass spectrometer, United States patent, No. 4926021.

Strong, D.F. and Longerich, H.P (1985) The inductively coupled plasma mass spectrometer (ICP-MS). *Geosci. Can.* **12**, 72–75.

Sturgeon, R.E. and Berman, S.S. (1987) Sampling and storage of natural water for trace metals. *CRC Crit. Rev. Anal. Chem.* **18**, 209–244.

Suddendorf, R.F. and Boyer, K.W. (1978) Nebuliser for analysis for high salt content samples with inductively-coupled plasma emission spectrometry. *Anal. Chem.* **50**, 1769–1771.

Sugden, T.M. (1965) The direct mass spectrometry of ions in flames. In *Mass Spectrometry*, ed. Reed, R.I., Academic Press, New York, pp. 347–358.

Sulcek, Z. and Povondra, P. (1989) *Methods of Decomposition in Inorganic Analysis*, CRC Press, Boca Raton, FL, 325 pp.

Sun, X.F., Ting, B.T.G., Zeisel, S.H. and Janghorbani, M. (1987) Accurate measurement of stable isotopes of lithium by inductively coupled plasma mass spectrometry. *Analyst* **112**, 1223–1228.

Suyani, H., Heitkemper, D., Creed, J. and Caruso, J. (1989a) Inductively coupled plasma mass spectrometry as a detector for micellar liquid chromatography: speciation of alkyltin compounds. *Appl. Spectrosc.* **43**, 962–967.

Suyani, H., Creed, J., Davidson, T. and Caruso, J. (1989b) Inductively coupled plasma mass spectrometry and atomic emission spectrometry coupled to high-performance liquid chromatography for speciation and detection of organotin compounds. *J. Chromatogr. Sci.* **27**, 139–143.

Suzuki, M., Ohta, K. and Yamakita, T. (1981) Elimination of alkali chloride interference with thiourea in electrothermal atomic absorption spectrometry of Cu and Mn. *Anal. Chem.* **53**, 9–13.

Suzuki, S.-I., Tsuchihashi, H., Kunio, N., Matsushita, A. and Nagao, T. (1988) Analysis of impurities in methamphetamine by inductively coupled plasma mass spectrometry and ion chromatography. *J. Chromatogr.* **437**, 322–327.

Tan, S.H. and Horlick, G. (1986) Background spectral features in inductively coupled plasma mass spectrometry. *Appl. Spectrosc.* **40**, 445–460.

Tan, S.H. and Horlick, G. (1987) Matrix effect observations in inductively coupled plasma mass spectrometry. *J. Anal. Atom. Spectrom.* **2**, 745–763.

Taylor, H.E. (1989) Water resources. In *Applications of Inductively Coupled Plasma Mass Spectrometry*, eds. Date, A.R. and Gray, A.L., Blackie, Glasgow, pp. 71–89.

Thompson, J.J. and Houk, R.S. (1986) Inductively coupled plasma mass spectrometric detection for multielement flow injection analysis and elemental speciation by reversed-phase liquid chromatography. *Anal. Chem.* **58**, 2541–2548.

Thompson, J.J. and Houk, R.S. (1987) A study of internal standardisation in inductively coupled plasma mass spectrometry. *Appl. Spectrosc.* **1**, 801–806.

Thompson, M. and Walsh, J.N. (1989) *Handbook of Inductively Coupled Plasma Spectrometry*, Blackie, Glasgow, 316 pp.

Thompson, M., Pahlavanpour, B., Walton, S.J. and Kirkbright, G.F. (1978a) Simultaneous determination of trace concentrations of As, Sb, Bi, Se and Te in aqueous solution by introduction of the gaseous hydride into an inductively coupled plasma source for emission spectrometry. Part I: Preliminary studies. *Analyst* **103**, 568–579.

Thompson, M., Pahlavanpour, B., Walton, S.J, and Kirkbright, G.F. (1978b) Simultaneous determination of trace concentrations of As, Sb, Bi, Se and Te in aqueous solution by introduction of the gaseous hydride into an inductively coupled plasma source for emission spectrometry. Part II: Interference studies. *Analyst* **103**, 705–713.

Thompson, M., Goulter, M. and Sieper, F. (1981) Laser ablation for the introduction of solid samples into an inductively coupled plasma for atomic emission spectrometry. *Analyst* **106**, 32–39.

Thompson, M., Chenery, S. and Brett, L. (1989) Calibration studies in laser ablation microprobe—inductively coupled plasma atomic emission spectrometry. *J. Anal. Atom. Spectrom.* **4**, 11–16.

Thompson, M., Chenery, S. and Brett, L. (1990) Nature of particulate matter produced by laser ablation—implications for tandem analytical systems. *J. Anal. Atom. Spectrom.* **5**, 49–55.

Timothy, J.G. and Bybee, R.L. (1978) Performance characteristics of high-conductivity channel electron multipliers. *Rev. Sci. Instrum.* **49**, 1192–1196.

Ting, B.T.G. and Janghorbani, M. (1986) Inductively coupled plasma mass spectrometry applied to isotopic analysis of iron in human fecal matter. *Anal. Chem.* **58**, 1334–1340.

Ting, B.T.G. and Janghorbani, M. (1987) Application of ICP–MS to accurate isotopic analysis for human metabolic studies. *Spectrochim. Acta* **42B**, 21–27.

Ting, B.T.G., Mooers, C.S. and Janghorbani, M. (1989) Isotopic determination of selenium in biological materials with inductively coupled plasma mass spectrometry. *Analyst* **114**, 667–674.

Tingfa, F., Ming, Y. and Xuanhui, Y. (1990) ICP–MS: A study on the ultrasonic nebulisation with desolvation. *J. Anal. Atom. Spectrom.*, submitted for publication.

Tothill, P., Matheson, L.M., Smyth, J.F. and McKay, K. (1990) Inductively coupled plasma mass spectrometry for the determination of platinum in animal tissues and a comparison with atomic absorption spectrometry. *J. Anal. Atom. Spectrom.* **5**, 619–622.

Totland, M., Jarvis, I. and Jarvis, K. (1991) An assessment of dissolution techniques for the analysis of geological samples by plasma spectrometry. In *Plasma Spectrometry in the Earth Sciences.* eds. Jarvis, I. and Jarvis, K., Chemical Geology Special Issue (in press).

Tsukahara, R. and Kubota, M. (1990) Some characteristics of inductively coupled plasma-mass spectrometry with sample introduction by tungsten furnace electrothermal vaporization. *Spectrochim. Acta* **45B**, 779–787.

Tye, C., Gordon, J. and Webb, P. (1987) Analysis of uranium using laser abaltion ICP–MS. *Int. Lab.* **December**, 3–41.

Tyson, J.F. (1985) Flow injection analysis techniques for atomic-absorption spectrometry. A review. *Analyst* **110**, 419–429.

Uhrberg, R. (1982) Acid digestion bomb for biological samples. *Anal. Chem.* **54**, 1906–1908.

Van Delft, W. and Vos, G. (1988) Comparison of digestion procedures for the determination of mercury in soils by cold vapour atomic absorption spectrometry. *Anal. Chim. Acta* **209**, 147–156.

Van Eenbergen, A. and Bruninx, E. (1978) Losses of elements during sample decomposition in an acid-digestion bomb. *Anal. Chim. Acta* **98**, 405–406.

Van Loon, J.C. and Barefoot, R.R. (1989) *Analytical Methods for Geochemical Exploration*, Academic Press, San Diego, 344 pp.

Vanhoe, H., Vandecasteele, C., Versieck, J. and Dams, R. (1989) Determination of iron, cobalt, copper, zinc, rubidium, molybdenum and caesium in human serum by inductively coupled plasma mass spectrometry. *Anal. Chem.* **61**, 1851–1857.

Vanhoe, H., Vandecasteele, C., Versieck, J. and Dams, R. (1990) Evaluation of ICP–MS for the determination of trace elements and ultra-trace elements in human serum after simple dilution. In *Plasma Source Mass Spectrometry*, eds. Jarvis, K.E., Gray, A.L., Jarvis, I. and Williams, J.G., Royal Society of Chemistry, Special publication 85, pp. 66–78.

Van Wambeke, L. (1960) Geochemical prospecting and appraisal of niobium-bearing carbonatites by X-ray methods. *Econ. Geol.* **55**, 732–758.

Vaughan, M.A. and Horlick, G. (1986) Oxide, hydroxide and doubly charged species in inductively coupled plasma mass spectrometry. *Appl. Spectrosc.* **40**, 434–445.

Vaughan, M.A. and Horlick, G. (1987) A computerised reference manual for spectral data and interferences in ICP–MS. *Appl. Spectrosc.* **41**, 523–526.

Vaughan, M.A. and Horlick, G. (1989) Analysis of steels using inductively coupled plasma mass spectrometry. *J. Anal. Atom. Spectrom.* **4**, 45–50.

Vaughan, M.A. and Horlick, G. (1990a) Effect of sampler and skimmer orifice size on analyte and analyte oxide signals in ICP–MS. *Spectrochim. Acta* **45B**, 1289–1300.

Vaughan, M.A. and Horlick, G. (1990b) Ion trajectories through the input ion optics of an ICP mass spectrometer. *Spectrochim. Acta* **45B**, 1301–1312.

Vaughan, M.A., Horlick, G. and Tan, S.H. (1987) Effect of operating parameters on analyte signals in inductively coupled plasma mass spectrometry. *J. Anal. Atom. Spectrom.* **2**, 765–772.

Verbeek, A.A. and Brenner, I.B. (1989) Slurry nebulisation of geological materials into argon, argon-nitrogen and argon-oxygen inductively coupled plasmas. *J. Anal. Atom. Spectrom.* **4**, 23–26.

Vestal, M.L. and Ferguson, G.J. (1985) Thermospray liquid chromatograph/mass spectrometer interface with direct eletrical heating of the capillary. *Anal. Chem.* **57**, 2373–2378.

Vickers, G.H., Ross, B.S. and Hieftje, G.M. (1989) Reduction of mass-dependent interferences in ICP–MS by using flow injection analysis. *Appl. Spectrosc.* **43**, 1330–1333.

Viczian, M., Lasztity, A., Wang, X. and Barnes, R.M. (1990a) On-line isotope dilution and sample dilution by flow injection and inductively coupled plasma mass spectrometry. *J. Anal. Atom. Spectrom.* **5**, 125–133.

Viczian, M., Lasztity, A. and Barnes, R.M. (1990b) Identification of potential environmental sources of childhood lead poisoning by inductively coupled plasma mass spectrometry. Verification and case studies. *J. Anal. Atom. Spectrom.* **5**, 293–300.

Walsh, J.N. (1979) The simultaneous determination of the major, minor and trace constituents of silicate rocks using inductively coupled plasma spectrometry. *Spectrochim. Acta* **35B**, 107–111.

Walsh, J.N. and Howie, R.A. (1980) An evaluation of the performance of an inductively coupled plasma source spectrometer for the determination of the major and trace constituents of silicate rocks and minerals. *Mineral. Mag.* **43**, 967–974.

Walsh, J.N., Buckley, F. and Barker, J. (1981) The simultaneous determination of the rare-earth elements in rocks using inductively coupled plasma source spectrometry. *Chem. Geol.* **33**, 141–153.

Walton, S.J. (1989) Analysis of iron and steel. In *Handbook of Inductively Coupled Plasma Spectrometry*, 2nd edn., eds., Thompson, M. and Walsh, J.N., Blackie, Glasgow, 274–276.

Walton, S.J. and Goulter, J.E. (1985) Performance of a commercial maximum dissolved solids nebuliser for inductively coupled plasma spectrometry. *Analyst* **110**, 531–534.

Wang, J., Shen, W.-L., Sheppard, B.S., Evans, E.H., Caruso, J.A. and Fricke, F.L. (1990) Effect of ion lens tuning and flow injection on non-spectroscopic matrix interferences in inductively coupled plasma mass spectrometry. *J. Anal. Atom. Spectrom.* **5**, 445–449.

Wang, X., Viczian, M., Lasztity, A. and Barnes, R.M. (1988) Lead hydride generation for isotope analysis by inductively coupled plasma mass spectrometry. *J. Anal. Atom. Spectrom.* **3**, 821–827.

Watkins, P.J. and Nolan, J. (1990) Determination of rare-earth elements, scandium, yttrium and hafnium in 32 geochemical reference materials using inductively coupled plasma-atomic emission spectrometry. *Geostandards Newslett.* **14**, 11–20.

Watkins, P.J. and Nolan J. (1991) Determination of rare-earth elements, yttrium, scandium and hafnium using ICP–AES. In *Plasma Spectrometry in the Earth Sciences.* eds. Jarvis, I. and Jarvis, K., *Chemical Geology Special Issue* (in press).

Weast, R.C. (1987) *Handbook of Chemistry and Physics*, 68th edn., The Chemical Rubber Co., USA.

Weiss, D., Paukert, T. and Rubeska, I. (1990) Determination of rare earth elements and yttrium in rocks by inductively coupled plasma atomic emission spectrometry after separation by organic solvent extraction. *J. Anal. Atom. Spectrom.* **5**, 371–375.

Wendt, R.H. and Fassel, V.A. (1965). Induction-coupled plasma spectrometric excitation source. *Anal. Chem.* **37**, 920–922.

Whittaker, P.G., Lind, T., Williams, J.G. and Gray, A.L. (1989) Inductively coupled plasma mass spectrometric determination of the absorption of iron in normal women. *Analyst* **114**, 675–678.

Williams, J.G. (1989) Inductively coupled plasma mass spectrometry: analytical methodology and capability, unpublished PhD Thesis, University of Surrey.

Williams, J.G. and Gray, A.L. (1988) High dissolved solids and ICP-MS: are they compatible? *Anal. Proc.* **25**, 385-388.

Williams, J.G., Gray, A.L., Norman, P. and Ebdon, L. (1987) Feasibility of solid sample introduction by slurry nebulisation for inductively coupled plasma-mass spectrometry. *J. Anal. Atom. Spectrom.* **2**, 469-472.

Wilson, D.A., Vickers, G.H. and Hieftje, G.M. (1987a) Spectral and physical interferences in a new, flexible inductively coupled plasma mass spectrometry instrument. *J. Anal. Atom. Spectrom.* **2**, 365-368.

Wilson, D.A., Vickers, G.H. and Hieftje, G.M. (1987b) Use of the microwave-induced nitrogen discharge at atmospheric pressure as an ion source for elemental mass spectrometry. *Anal. Chem.* **59**, 1664-1670.

Wilson, D.A., Vickers, G.H. and Hieftje, G.M. (1987c) Ionization temperatures in the ICP determined by mass spectrometry. *Appl. Spectrosc.* **41**, 875-880.

Yoshinaga, J., Nakazawa, M., Suzuki, T. and Morita, M. (1989) Determination of trace elements in human liver and kidney by inductively coupled plasma mass spectrometry. *Anal. Sci.* **5**, 355-358.

Zhu, G. and Browner, R.F. (1987) Investigation of experimental parameters with a quadrupole ICP-MS. *Appl. Spectrosc.* **41**, 349-359.

Zhu, G. and Browner, R.F. (1988) Study of the influence of water vapor loading and interface pressure in ICP-MS. *J. Anal. Atom. Spectrom.* **3**, 781-789.

Additional references

American Chemical Society, Committee on Environmental Improvements (1980) Guidelines for data acquisition and data quality evaluation in environmental chemistry. *Anal. Chem.* **52**, 2242-2249.

Chambers, D.M. and Hieftje, G.M. (1991) Fundamental studies of the sampling process in an ICP mass spectrometer. II. Ion kinetic energy measurements. *Spectrochim. Acta* **46B**, 761-784.

Chambers, D.M., Poehlman, J., Yang, P. and Hieftje, G.M. (1991a) Fundamental studies of the sampling process in an ICP mass spectrometer. I. Langmuir probe measurements. *Spectrochim. Acta* **46B**, 741-760.

Chambers, D.M., Ross, B.S. and Hieftje, G.M. (1991b) Fundamental studies of the sampling process in an ICP mass spectrometer. III. Monitoring the ion beam. *Spectrochim. Acta* **46B**, 785-804.

Crock, J.G. and Lichte, F.E. (1982) Determination of rare-earth elements in geological materials by inductively coupled argon plasma-atomic emission spectrometry. *Anal. Chem.* **54**, 1329-1332.

Hart, S.R. and Kinloch, E.D. (1989) Osmium isotope systematics in Witwatersrand and Bushveld ore deposits. *Economic Geol.* **84**, 1651-165.

Heumann, K.G. (1985) Trace determination and isotopic analysis of the elements in life sciences by mass spectrometry. *Biomed. Mass Spectrom.* **12**, 477-488.

Jarvis, I. and Jarvis, K.E. (1991) Plasma spectrometry in the earth sciences: techniques, applications and future trends. In *Plasma Spectrometry in the Earth Sciences*, eds. Jarvis, I. and Jarvis, K.E., *Chemical Geology Special Issue* (in press).

Jarvis, I. and Jarvis, K.E. (1992) Inductively coupled plasma-atomic emission spectrometry in exploration geochemistry. In *Analytical Methods in Geochemical Exploration*, ed. Hall, G.E.M., *J. Geochem. Explor. Special Issue* (in press).

Jarvis, K.E. and Jarvis, I. (1988) Determination of the rare-earth elements and yttrium in 37 international silicate reference materials by inductively coupled plasma-atomic emission spectrometry. *Geostand. Newslett.* **12**, 1-12.

Niu, H.S., Hu, K. and Houk, R.S. (1991) Langmuir probe measurements of electron temperature and electron density behind the skimmer of an ICP mass spectrometer. *Spectrochim. Acta* **46B**, 805-817.

Potts, P.J., Thorpe, O.W. and Watson, J.S. (1981) Determination of the rare-earth element abundances in 29 international rock standards by instrumental neutron activation analysis. *Chem. Geol.* **34**, 331-352.

Robinson, P., Higgins, N.C. and Jenner, G.A. (1986) Determination of rare earth elements, yttrium and scandium in rocks by an ion exchange–X-ray fluorescence technique. *Chem. Geol.* **55**, 121–137.

Ross, B.S. and Hieftje, G.M. (1991) Selection of solvent load and first stage pressure to reduce interference effects in ICP–MS. *J. Am. Soc. Mass Spectrom.* **2**, (in press).

Smith, F.G., Wiederin, D.R. and Houk, R.S. (1991) Ar–Xe plasma for alleviating polyatomic ion interferences in ICP–MS. *Anal. Chem.* **63**, 1458–1462.

Index

abundance sensitivity   41, 313, 314
accuracy   154, 181, 235, 287, 288, 299
acid
  boric, use of   194
  fuming nitric   174
  high purity   259
  hydrochloric   130, 176
  hydrofluoric   176, 177
  nitric   130, 174
  perchloric   177, 178
  phosphoric   179
  sulphuric   178
acids, mineral   174
aerosol   58
  desolvation   25
alkali
  fluxes   197
  fusion   185, 196, 221
alloys   179, 191, 253, 296
  high carbon   191
  Ni-based   254
analogue mode   157
analysis
  least squares regression   162
  qualitative   160
  quantitative   162
analyte suppression   252
analyte vapour   85
analytical procedure   227
animal tissue   261
anion exchange   114
  column   114
  pairing   114
anti-freeze   74
apatite   234
aqua regia   179
arc nebulisation   309
argon dimmer   165
argon-nitrogen plasma   284
argon oxide   130
ashing
  dry   182, 183
  wet   182, 183
atomic emission spectrometry   1

backstreaming   56

barrel shock   26
biological
  applications   260
  materials   204, 247
  monitoring   97
biomethylation   118
blank   323
blank spectra   285
blanks   158, 304, 335
  gas   304
  procedural   158
blood   334
blood serum   322
bomb dissolution   195
bond strength   136
boundary layer   22

calcium-rich samples   127
calibration   153, 167, 299
  curves   291
  external   162, 236, 276
  for laser ablation   296
  mass scale   153
  of solids   280
  semi-quantitative   160
  strategies   276
carbide   84
  formation of   94
Ca-rich matrices   246
catalysts   250, 284
cation-exchange
  chromatography for REE   212, 215
  procedure   214
  separation   211
cation pairing   114
central channel   15, 18, 19
  temperature   12
ceramics   296
certified reference materials   281
chemical separation   271
chromatography
  gas   112
  ion exchange   116
  ion-pair   116
  liquid   112
  separation   263
  size exclusion   116

cleanroom facilities   173
commercial systems   6
  ETV   87
  optimisation of   89
contamination   158, 173, 175, 192, 198, 216, 217, 239, 267, 276, 279
co-precipitation   210, 272
correction
  external drift   163
  raw data procedures   162
count rate
  background   4
  dark   50
counting statistics   323
coupling, RF   13
crucibles
  carbon   197
  platinum-gold   197
  vitreous carbon   197
cyropump   7, 57
cupferron separation   221–224, 241

data
  acquisition   44
  collection   153
  collection for ETV–ICP–MS   91
dead time   312, 313, 335
detection limits   180, 287, 288, 301
detectors
  Coniphot   51
  Daly   51
  other   51
determination limits   243
detoxification   118
digestion
  bombs   192, 193, 194
  closed-vessel   181, 192
  microwave   182, 202, 203–207, 250
  open   187
  open-vessel   181, 182, 188, 200
  procedures   181
  techniques, partial   190

dilution factors 180
direct sample insertion 124, 269, 270, 308
direct water analysis 269
discrete sample volume 119
dispersing agents 283
drift monitor 227
drinking waters 265, 269
drugs, anti-tumour 96
dusts 189
dwell time 238, 325
dynamic range 8, 50, 51

effect of laser energy 303
electron multiplier 42, 48
  Channeltron 48
  discrete dynode 51
  lifetime of 49
electrothermal
  evaporation 8
  vaporisation 82
  vaporiser designs 85
elements
  incompatible 240
  platinum group 176
  rare earth 210, 229
  volatile 192, 196, 209, 256, 279, 288
environmental
  applications 247
  materials 211
  studies 331
ETV–ICP–MS
  application of 93
  biological applications of 96
  calibration for 92
  data collection for 91
  environmental applications of 97
  geological applications of 94
ETV system 87
  optimisation 89
ETV tube
  ceramic 97

Faraday cup 51
Fe alloys 178
FI–ICP–MS, applications of 122
filament temperature 89
fire assay 210, 216, 217, 219, 235, 328
  method for 217
  nickel sulphide 216, 219, 220, 236
flow injection 119, 189, 249, 250, 271, 272, 274
  dispersion in 121
  transient peak in 120
  valves 120
fluorides, volatile 177
foods 260
free electron path 12
Freon 23, 84, 94, 97
freshwater 265, 266, 268, 273

fringing field 43
furnace programme 90

galena 331
gas dynamics 28
gas flow
  central 11
  nebuliser 11
  outer 10
gas peaks 129, 130
gas phase injection 270
gases
  mixed 146
  molecular 146
  properties and flow of 51
gel standards 281, 296, 297
generator
  crystal controlled 13
  free running 13
  RF 13
geological applications 228
graphite platform 95
gravitational deposition 302
grinding techniques 281

halogens 200, 202, 248, 249, 288–290, 337
heating rate 84
HeICP 147
HeMIP 147
high pressure asher 195, 196
high resolution 126
  ICP–MS 147
high transport efficiency 82
HPLC–ICP–MS 113
hydride generation 98, 250, 270
  applications of 103
  batch 99
  continuous 99
  interference effects 102
  system optimisation 100
hydrocarbons 256
hydrogen 77
hydroxide formation 143

ICP–MS
  high resolution 35, 41, 46, 147
  high resolution spectra 47
  history of 5
industrial applications 253
instrument
  design 146, 147
  optimisation 146, 151, 230
  performance 3
instrumental
  drift 200
  operation conditions 231
integration 155
  constant mass width 155, 156
  valley 156, 157
interfact 78
interference 13, 225, 229
  barium oxide 214

corrections 262, 274, 335
  automatic 134
correction procedures 142
drift 262
in laser ablation 299
non-spectroscopic 125, 148
polyatomic ion 147
spectroscopic 125, 145
internal standardisation 164, 170
internal standards 152, 227, 232, 241, 243, 252, 299
  choice of 164
  natural 164
ion
  detection 48
  detector 129
  distributions 18, 20, 28
  energies 30
  exchange 210
  exchange separation 214, 221, 231, 235
  extraction 19, 21, 22, 23, 149, 159
  focusing 31, 32
  lenses 31, 32, 34, 320
  molecule reactions 129
  optics 31, 151
  source 2, 39
  trajectories 38
  transmission 35
  velocity 32
ionisation
  degree of 14, 15, 16, 17, 160
  first energy 167
  second energy 143, 144
ions
  doubly charged 143, 145, 251
  polyatomic 29, 129, 132, 146, 147, 169, 175, 176, 178, 179, 185, 189, 197, 198, 206, 246, 247, 254, 257, 321
iron isotope ratios 96
  suppression 18
  temperature 29
iron ores 194
isobaric overlap 125, 140
  correction for 126
isobutyl methyl ketone extraction 244
isotope
  dilution 113, 167–169, 170, 171, 219, 237, 244, 245, 248, 277, 311
  preferred 225, 226
  radiogenic 310
  ratio
    boron 316
    bromine, chlorine 337
    copper 324
    iron 320
    lead 331
    lithium 315
    magnesium, potassium, chromium, selenium 336

osmium 329
rhenium 328
rhenium and osmium 327
uranium 334
zinc 324
ratios 304, 310
reference 169
spike 168, 244
stable 310, 315, 322

laboratory equipment for sample preparation 173
Langmuir probes 24
lanthanide 229
laser
  ablation 8, 280, 290, 291, 305
  ablation cell 292–294
  definition 291
  modes of operation of 291
  normal mode 292
  operation 294
  Q-switching mode 291, 292
lead additives in petrol 331
lead isotope ratios 126
limestone 127, 148, 194, 211, 246
limit of quantitation 180
linearity of response 157
liquid binders 296
lithium and boron matrices 252
lithium metaborate fusions 199–201, 211, 241
lithium tetraborate 196
load coil 23, 24, 30
  centre tapped 25
  grounding 6

Mach disc 26, 28
magnetic sector mass anlyser, resolution of 9
manganese nodules 143, 194
mass
  bias 315, 319, 320, 330, 331, 333, 335
  discrimination 44, 170, 318
  fractionation 310, 315
  spectra 41
mass spectrometers
  higher resolution 45
  magnetic sector 45
matrix effects 125, 165, 243, 245, 246, 304, 318, 326
  non-spectroscopic 165, 255
matrix elements 1, 151
  interference 8, 37
  separating 113
matrix matched standards 182, 249, 255
matrix matching 152, 162, 192, 303, 326
matrix modifier 260
medical applications 263
memory effects 70, 151, 316, 317

mercury speciation 249
metallurgical reference materials 191
metals 191, 209, 254
  noble 217–220
  precious 216, 219
micro-heater merging chamber 239
micro-volume 82
minifusions 211
monazite 234
monoxide bond strength 134, 142
multi-channel analyser 8
multi-element
  analysis 1
  determinations 225
multiplier gain 50

natural waters 247
  reference materials 266
  sampling procedures for 268
  storage of 267
nebuliser 58
  Babington type 65
  blockage 62
  concentric 59
  cross flow 64
  De Galan 286
  flow rate 132–135
  Frit type 66
  gas flow 11, 137
  high dissolved solids 184
  Meinhard glass concentric 59
  pneumatic 14
  recirculating 61
  ultrasonic 67
  V-groove 65
Ni-based alloys 126, 140, 141, 306, 307
normal mode 295, 302, 304
nuclear applications 251
nuclear industry 97

on-line separation 274
operating conditions 319
operating parameters for laser ablation 302
organic
  analysis 256, 257
  materials 193
  solvent 256
Os-tetroxide 238, 329
oxide formation 135, 140
oxide ions in laser ablation 300
oxide levels 230
oxidising agents 175, 178
oxygen addition 256, 257

particle size
  distribution 282, 283
  reduction 282
peak
  hopping 44, 154, 155, 238
  hopping mode 295, 315

shape 41
tailing 155
petrogenetic discriminators 221
phase separator 100
phosphorites 194
photon stop 7, 33, 47
plant reference materials 186
plasma 10
  alternative sources 147
  argon-nitrogen plasma 284
  DC type 3, 4
  dry 124, 323
  equilibrium 136
  populations 15
  potential 6, 23, 30, 34
  support gas 11
  torches 58
plasmas
  helium microwave 57
  other 21
platinum group metals 235, 236, 327
plutonium 335
pneumatic nebulisation 269
polyatomic ions 29, 129
powdered solids, introduction of 308
precision 124, 148, 154, 171, 181, 225, 235, 283, 287, 299, 312, 325, 328
  intersample 287
  intrasample 287
pre-concentration 271, 273, 274
  methods 209, 210
  methods for waters 275
  procedures 189
pressed powder pellets 294, 296
procedural blanks 231, 269
pulse counting 50
  mode 157
pump
  diffusion 53–56
  mechanical 53, 54
  peristaltic 60
  rotary 55
  turbomolecular 54–56
pyrolitic graphite furnace 95
pyrolitically-coated graphite 84

Q-switching 294
Q-switching mode 291, 295, 302, 303
quadrupole 42
  configuration 37
  rest mass 332
  RF only 42, 43
  rods 38
  stability region 40, 43
quantitation limit 233, 234

reactive gases 110
reagents, pure 173
recombination processes 140
red blood cells 316

REE
  detection limits   233
  determination   290, 300
  method for   212
reference materials   162
reference standard   333
refractrory
  minerals   172
  oxide formation   229
  oxides   134, 251
  phases   209
  zircons   211
regulatory detection limits   266
relative sensitivity factor   298,
  299
reproducibility   154
resolution   40, 42, 46, 248, 258,
  313, 314, 332
response curve   160, 161
rheumatoid arthritis   118

Saha equation   15
sample
  biological   206, 207, 211, 248
  botanical   206, 211
  calcium-rich   245
  dissolution   172
  environmental   189–191, 202,
    206, 209
  geological   188, 206, 234,
    241, 285, 321
  history   14
  inhomogeneity   280
  iron-rich   247
  organic   204, 205
  plant   306
sample preparation   125
  for laser ablation   295
  geological samples   185
  laboratory equipment for   173
  plant and animal tissue   182
sampling cone   7, 23, 58, 78
  cleaning   79
  construction of   78
  corrosion of   179
  failure   79
  interface   21
sapphire injectors   177
scan speed   47, 155
scanning   44, 154, 155
  mode   332
seawater   189, 239, 240, 265,
  271, 272, 274, 320
secondary discharge   23, 24
sediments   189, 209, 243, 244,
  247, 290, 307, 308
sensitivity counting statistics
  312
separation methods   209, 210

serum   316
sewage sludges   189
signal
  drift   241
  enhancement   125, 238
  fluctuation   165
  stability   158
  suppression   125, 150, 189,
    196
silane   110
silica phytoliths   184, 206
skimmer cone   7, 36, 58, 78, 80
slurry nebulisation   280, 281,
  286, 287, 289
  applications of   284
slurry preparation method   285
sodium carbonate   196
soils   189, 209, 284
solids
  high dissolved   148, 158, 159,
    194, 287
  total dissolved   140, 148, 194,
    196
solution introduction   58
solvent extraction   210
space charge   36
  effects   35–37
spark source mass spectrometry
  (SSMS)   2
speciation   112
  studies   263
species
  dioxide   138
  trioxide   138
spectra, high resolution   47
spray chambers   58, 68
  corrosion resistant   252
  double pass   69, 73
  PTFE   177
  Scott double pass   283
  single pass   73
  temperature regulated   134
  thermally stabilised   71
stable isotope spike   218
standard additions   167, 244,
  248, 255, 277
standard reference materials
  181
steels   178, 179, 296, 307
sulphide minerals   187
supersonic expansion   34, 151
supersonic jet   22, 26, 27, 30, 35
suppression
  analyte   152
  effects   183
synthetic standard   232
system configuration for laser
  ablation   292, 293
system optimisation   133

tantalum filament   95
Teflon PTFE   173
tellurium co-precipitation   219,
  238
Tesla coil   11
tissue
  animal   203
  plant   203
torch   10
  alignment   76
  Scott Fassel   10
torches   75
  demountable   75
  for hazardous gases   76
  low flow   76
  mixed gas and sheath gas   77
toxicological properties   118
transient signal   323
transport efficiency   83

ultrasonic transducer   58
uranium matrices   251
urine   316
  analysis   97

vacuum measurement   53
  system   52
vaporisation cell materials
  graphite   84
  rhenium   84
  tantalum   84
  tungsten   84
vaporisation
  cells   83
  electrothermal   189, 237, 243,
    269, 270, 322, 335
vapour generation
  and phase introduction   98
  mercury   98
  osmium tetroxide   98, 105
vapour generator
  discrete batch osmium
    tetroxide   107
  memory effects of osmium
    109

wash-out   70
wastewater   265
  industrial   266
water sampling   267
water vapour loading   133

xenon   128, 129

zircon-bearing rocks   187
zone
  initial radiation   14, 15, 29
  normal analytical   14
  of silence   26